Cell Biology and Genetics

Cell Biology and Genetics

Edited by **Gloria Doran**

SYRAWOOD
PUBLISHING HOUSE

New York

Published by Syrawood Publishing House,
750 Third Avenue, 9th Floor,
New York, NY 10017, USA
www.syrawoodpublishinghouse.com

Cell Biology and Genetics
Edited by Gloria Doran

International Standard Book Number: 978-1-68286-079-3 (Hardback)

Printed in the United States of America.

Contents

Preface

Genetics is the branch of biology which deals with the variation and inheritance patterns in organisms. It is an expanding field of research contributing to new discoveries in cellular biology and genomic studies. The book focuses on cellular differentiation, cellular metabolism and DNA markers. It contains latest researches and case-studies on molecular genetics, gene regulation and DNA sequencing. It delves into the historical, theoretical and experimental approaches to understand cell biology and genetics, and brings forth new avenues for further discussion. This book will prove to be an invaluable source of reference for students and researchers interested in this field.

This book has been the outcome of endless efforts put in by authors and researchers on various issues and topics within the field. The book is a comprehensive collection of significant researches that are addressed in a variety of chapters. It will surely enhance the knowledge of the field among readers across the globe.

It gives us an immense pleasure to thank our researchers and authors for their efforts to submit their piece of writing before the deadlines. Finally in the end, I would like to thank my family and colleagues who have been a great source of inspiration and support.

Editor

Factors determining the load of *Staphylococci* species from raw bovine milk in Khartoum State, Khartoum North, Sudan

Adil M. A. Suliman[1] and Tawfig El tigani Mohamed[2]

[1]Faculty of Veterinary Science, University of Bahr Elgazal, Department of Preventive Medicine and Veterinary Public Health, Sudan.
[2]Faculty of Veterinary Medicine, Khartoum University, Khartoum North, Sudan.

In this study, 644 raw milk samples were collected from various milk sources of Khartoum state during May 2003 till April 2004. Using the API kits, both coagulase positive and negative *Staphylococci* (CPS and CNS) were identified. Most of the CPS isolates were *Staphylococcus aureus*, while the majority of CNS were *Staphylococcus epidermidis*. 23.8% of the samples were found to be positive to both CPS and CNS while 33.7% of the samples were negative to both CPS and CNS. Forty one percent of CPS isolates was found to be *S. aureus* and 39% of CNS were *S. epidermidis*. The coagulase positive *Staphylococci* (CPS) were determined in 45% of the tested samples while coagulase negative *Staphylococci* (CNS) were found in 44.7% of them, of which 16.5% exceeded the recommended bacterial safety limits for CPS. The highest of the samples that exceeded this limit was found in vendor milk samples (19.4%) while the market milk was 18.4%.

Key words: Staphylococci, bovine, raw, milk, Sudan.

INTRODUCTION

According to their reaction with Coagulase test, *Staphylococcus* spp are divided into coagulase positive (CPS) which includes the pathogenic *Staphylococcus aureus* (S. aureus) and coagulase negative (CNS) which includes the pathogenic and non pathogenic species of the genus *Staphylococci* (Devriese, 1990).

The infected mammary gland is the primary and most important reservoir of *S. aureus* (Jerry, 2003). Davidson (1961) isolated *S. aureus* from the teat skin, udder skin, nose lips, sacral regions and belly of cows. *S. aureus* was also isolated from teat skin of healthy cattle without being shedding in milk (Davidson, 1961).

API system was evaluated as a means of identifying the species of bovine strain of several groups of micro-organisms isolated from milk samples (Bruce et al., 1983) with similar values reported in human isolates (Gemmel and Dowson, 1982).

Bran et al. (1978) found the API STAPH identification system to be more accurate in identifying *S. aureus* (93.3%) than non *S. aureus*. Gaery et al. (1989) found the overall accuracy of the API STAPH identification system as 80.9%, when 188 strains of *Staphylococci* were tested, the accuracy among CNS was 86%. Wesley and Jana (1982) reported a high degree of congruence > 90% between the API system and the conventional methods for most species.

Importance of *Staphylococci*

Food or water contaminated with infectious organisms or toxins secreted by such organisms act as a source of food poisoning (Haeghebaert et al., 2003). Bacteria are the leading cause of food born diseases; they cause diseases through the pathogenesis of the organisms in host cells. The pathogenesis of the bacteria causing food born poisoning depends on their capacity to produce toxins after ingestion or before (Haeghebaert et al., 2003).

*Corresponding author. E-mail: adilsal4@yahoo.com.

Table 1. Number and percentage of milk samples from three regions of Khartoum State during summer and winter according to the source.

	Summer				Winter			
	Kh.	Kh.N.	Omd.	Total	Kh.	Kh.N.	Omd.	Total
Ind.	5	12	17	34	33	19	13	65
	07.9%	22.2%	27.9%	19.1%	24.8%	09.3%	10.1%	13.9%
Bulk	13	15	19	47	35	109	53	197
	20.6%	23.8%	31.1%	26.4%	26.3%	53.4%	41.1%	42.3%
Vendor	20	11	15	46	42	51	16	109
	31.8%	20.4%	24.6%	25.8%	31.6%	25.0%	12.4%	23.4%
Market	25	16	10	51	23	25	47	95
	39.7%	29.6%	16.4%	28.7%	17.3%	12.3%	36.4%	10.4%
Total	63	54	61	178	133	204	129	466

Kh = Khartoum; Kh.N.= Khartoum North; Omd. = Omdurman; Ind. = Individual.

Some *S. aureus* strains can produce staphylococcal enterotoxins and are the causative agents of Staphylococcal food poisoning (Yves et al., 2003). Although their contamination can be readily avoided by heating of milk, nevertheless, it remains a major cause of food-born diseases because it can contaminate milk after milking and during processing. About 10% of *Staphylococci* causing mastitis in cows produce enterotoxins which are heat stable and may cause food toxicity in man (I.D.F, 1980). *S. aureus* is the most important pathogen of CPS due to its combination of toxin mediated virulence, invasiveness and antibiotics resistance (Yves et al., 2003).

Among coagulase negative species, Jay (1986) and Bautista et al. (1988) reported that *Staphylococcus Cohnii, Staphylococcus xylosus* and *Staphylococcus haemolyticus* can produce also one or several staphylococcal enterotoxins. Coagulase positive *Staphylococcus Intermedius* have been shown to produce enterotoxins. Kambaty et al. (1994) reported *S. Intermedius* as the only non-*S. aureus* species that has been clearly involved in Staphylococcal food poisoning outbreaks. Ibrahim (1973) showed that an out break of Staphylococcal enterotoxins poisoning in Khartoum North was due to milk purchased from milk vendors.

In Sudan, the traditional ways of milking and distribution of milk are still in use. The majority of the dairy farms do not use milking machines. No system of milk cooling is applied. Traditional ways using donkeys or cars (vendors) are used to distribute milk to consumers which can take more than five hours after milking.

This study was conducted to achieve the following objectives: 1) Enumerate the *Staphylococci* of raw milk produced and sold in the three regions of Khartoum State. 2) Study the effect of seasons on the number of *Staphylococci* in milk. 3) Isolate some of the most important CPS and CNS that can affect human health and milk quality. 4) Propose some quality limits of acceptable raw milk which can be adopted by the

Ministry of Agriculture and Animal Recourses of Khartoum state.

MATERIALS AND METHODS

Collection of milk samples

The study was conducted in Khartoum State, using stratified random sampling for the dairy farms. The state was divided into three regions; Khartoum, Khartoum North and Omdorman. Random samples were collected from each region (Table 1).

The samples were collected in sterile Glass bottles either directly from the udder in cases of individual cows or from the milk bulk tanks or milk containers from milk markets and milk vendors. Samples were then kept in an ice box and transported directly to the laboratory at Shambat which is usually ranging between half an hour to one hour in the icebox.

Enumeration and isolation of *Staphylococci*

The test steps to enumerate *Staphylococci* were performed according to (ISO/DIS, 1997). Five decimal dilutions were prepared by adding 10 ml of the milk sample to 90 ml of peptone water in the first glass bottle resulting in 10^1 dilution, the process was continued till a dilution of 10^5 is obtained. For each dilution to be plated, aseptically, 1 ml suspension was spread on 3 plates of Baird Parker Agar medium (0.4, 0.3 and 0.3 ml) of agar plate usi and then incubated for 48 h at 35°C. Plates containing 20 - 200 colonies with appearance of coagulase positive *Staphylococci* were selected. Five to seven colonies from each plate were selected and inoculated into 5 ml brain heart infusion broth and then incubated for 24 h at 37°C. 0.1 ml of brain heart infusion broth was added to a test tube containing 0.3 ml of rabbit plasma. The tubes were incubated for 4 - 6 h at 37°C.

The Baird Parker Agar medium was prepared according to the manufacture instruction (bioMérieux sa). Bacto EY tellurite enrichment (egg yolk emulsion and 3 ml of 3.5% aqueous solution of potassium tellurite) were added and mixed well (avoiding bubbles). The isolates were classified as CPS or CNS using coagulase test.

API STAPH kits

It was prepared and performed according to the manufacturer

Table 2. Seasonal CPS count in the three regions of Khartoum state.

	Khartoum			Khartoum North			Omdurman		
Range	Winter	Summer	Total	Winter	Summer	Total	Winter	Summer	Total
Negative	44	75	119	30	105	135	37	65	102
(%)	69.8	56.4	60.7	55.6	51.5	52.3	60.7	50.4	53.7
< 100	01	05	06	05	11	16	03	12	15
(%)	01.6	3.8	3.1	09.3	05.4	6.2	04.9	9.3	7.9
100 - >500	04	19	23	05	19	24	04	11	15
(%)	06.3	14.3	11.7	09.3	09.3	9.3	06.6	8.5	7.9
500 - <10^3	08	16	24	04	23	16	08	16	24
(%)	12.7	12.0	12.2	07.4	11.3	06.2	13.1	12.4	12.6
10^3 - 10^5	06	12	18	10	39	49	09	25	34
(%)	09.6	9.0	9.2	18.5	19.1	19.0	14.7	19.4	17.9
Total	063	133	196	054	204	258	061	129	190

At 0.05 level of significance there was no significant) difference between Khartoum North and Omdurman. The difference was significant between Khartoum and Khartoum North, Khartoum and Omdurman. The count was significantly (P < 0.05) higher during summer than winter.

□ Negative ■ <100 □ 100 - 500 □ 500 -10^3 ■ 10^3–10^5

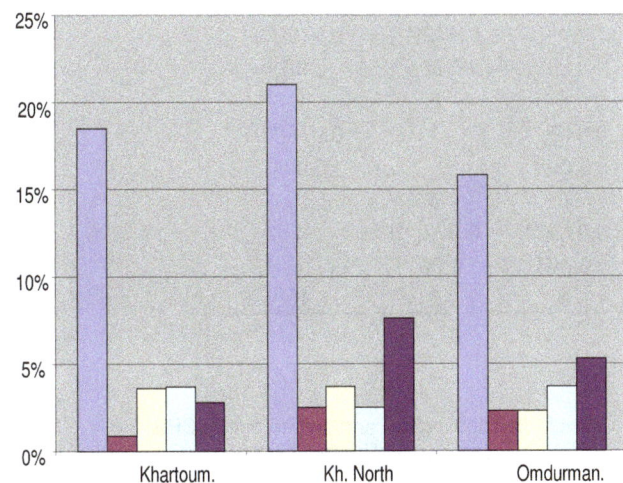

Figure 1. The % of CPS counts in the three regions of the state.

manual API STAPH REF 20 500 (identification system for *Staphylococci*, micro cocci and related genera bioMérieux sa).

The API strip system combining about 20 miniaturized biochemical tests many of which are derived from the conventional methods (Wesley and Jana, 1982). API system is particularly advantageous in providing preformed strips containing the test substrates and made available necessary reagents and also in most cases reaction could be interpreted after incubation at 37°C for 24 h (Bran et al., 1978).

Statistical analysis

Microsoft excel 2003 and the spss for windows version 11 were used for data analysis. t-test was selected to calculate the significance levels. Descriptive statistics were used.

RESULTS

Enumeration of CPS

The percentage of negative samples to CPS test in Khartoum State in the two seasons was 55.3%, of which 17.2% was in winter and 38.1% in summer. The percentages of samples with CPS count between 10^3 - 10^5 cell/ml were 15.7, 3.9 and 11.8% in Khartoum, Khartoum North and Omdorman, respectively (Table 2 and Figure 1). Statistically at 0.05 level, there was no significant difference between the three regions. However, the difference between the two seasons was statistically significant (Table 2).

In Khartoum state, the percentage of samples of negative counts during the two seasons were 65.7, 57.8, 46.9 and 51.3%, respectively in individual, bulk, vendor and market milk. In winter, the percentages of such counts were 73.5, 63.8, 58.6 and 56.9%, respectively and in summer, the percentages were 61.5, 56.3, 50.5 and 48.4% in individual, bulk, vendor and market milk, respectively (Table 3). A count between 10^3 - 10^5 cell/ml in Khartoum state was found to be 17.2, 19.3, 19.4 and 18.5% in individual, bulk, vendor and market milk, respectively during the two seasons, while in winter, the percentages of the same counts were 14.7, 17.0, 17.4 and 17.6%, in summer these counts were 18.5, 19.8, 20.2 and 18.9%, in individual, bulk, vendor and market milk, respectively (Table 3). Statistically, there was a significant correlation (at 0.01 levels) between individual, farm bulk milk, vendor and market milk counts (Table 4 and Figure 2).

The CPS and CNS

23.8% of the samples were positive to both CPS and

Table 3. Seasonal Total CPS count in the State, according to the source.

	Summer					Winter				
Range	**Ind.**	**Bulk**	**Vendor**	**Market**	**Total**	**Ind.**	**Bulk**	**Vendor**	**Market**	**Total**
Negative	025	030	027	029	111	040	111	048	046	245
(%)	73.5	63.8	58.6	56.9	62.4	61.5	56.3	50.5	48.4	52.6
< 100	01	02	03	03	09	03	11	08	06	28
(%)	2.9	4.3	6.5	5.9	5.1	4.6	5.6	7.3	6.3	6.0
100 - < 500	01	03	04	05	13	04	14	18	13	49
(%)	2.9	6.5	10.9	7.8	7.3	6.2	7.1	6.5	13.7	10.5
$500 - <10^3$	02	04	05	04	15	06	22	13	12	53
(%)	5.9	8.5	10.9	7.8	8.4	9.2	11.2	11.9	12.6	11.4
$10^3 - 10^5$	05	08	08	09	30	12	39	22	18	91
(%)	14.7	17.0	17.4	17.6	16.9	18.5	19.8	20.2	18.9	19.5
Total	034	047	046	051	178	065	197	109	095	466

Table 4. Correlations of CPS between the individual, Bulk, Vendor and market milk.

		Individual	**Bulk**	**Vendor**	**Market**
Individual	Pearson correlation	1	0.962(**)	0.960(**)	0.995(**)
	Sig. (2-tailed)	.	0.002	0.002	0.000
	N	6	6	6	6
Bulk	Pearson correlation	0.962(**)	1	0.942(**)	0.972(**)
	Sig. (2-tailed)	0.002	.	0.005	0.001
	N	6	6	6	6
Vendor	Pearson correlation	0.960(**)	0.942(**)	1	0.973(**)
	Sig. (2-tailed)	0.002	0.005	.	0.001
	N	6	6	6	6
Market	Pearson correlation	0.995(**)	0.972(**)	0.973(**)	1
	Sig. (2-tailed)	0.000	0.001	0.001	.
	N	6	6	6	6

** Correlation is significant at the 0.01 level (2-tailed).

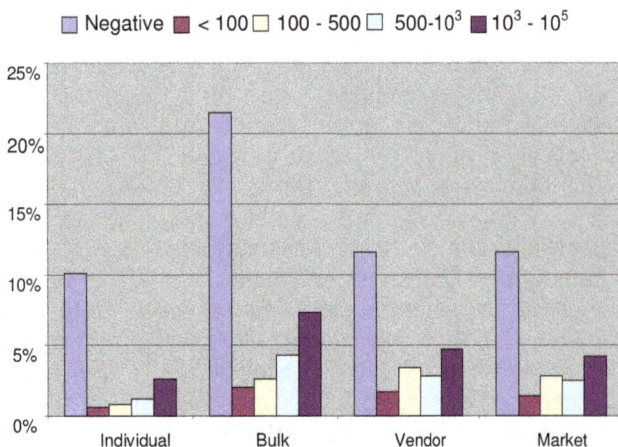

Figure 2. The % of the CPS counts at different sources.

CNS during the two seasons, 33.7% of the samples were negative to both CPS and CNS during the two seasons.

Table 5. The Number and percentage of CPS and CNS milk samples in Khartoum state.

		CPS		
CNS		**Positive**	**Negative**	**Total**
	Positive	153	139	292
	(%)	23.8	21.5	45.3
	negative	135	217	352
	(%)	21.0	33.7	54.7
	Total	288	356	644
	(%)	44.7	55.3	

45.3% of the whole samples were positive to CNS and 44.7% were positive to CPS (Table 5).

Identification of the CPS and CNS isolates

Using the API STAPH system, 275 isolates of the total

Table 6. The percentages of CPS and CNS isolated from the vendor and market milk from Khartoum state.

Species	Kh..	(%)	Kh.N	(%)	Omd	(%)	Total	(%)
S. aureus	26	9.50	55	020	32	11.6	113	41.10
S. intermedius	01	0.40	01	00.4	02	00.7	004	01.50
S. epidermidis	29	10.5	36	13.1	44	016	109	39.60
S. lentus	02	0.70	03	01.1	01	00.4	006	02.20
S. capitis	01	0.40	03	01.1	00	00.0	004	01.50
S.Simulans	00	1.50	02	00.7	02	00.7	004	01.50
S. caprae	00	0.00	00	00.0	02	00.7	002	00.70
S. xylosus	02	0.70	01	00.4	02	0.70	005	01.80
S. chromogens	02	0.70	03	01.1	00	00.0	005	01.80
S. sciuri	03	1.10	00	00.0	00	0.00	003	01.10
S. hyicus	05	1.80	02	00.7	02	0.70	009	03.30
S. hominis	01	0.40	02	00.7	00	0.00	003	01.10
Micrococcus spp	04	1.50	02	00.7	02	0.70	008	02.90

samples were identified of which 41.1% were *S. aureus* whereas 39.6% of the CNS isolates were *S. epidermidis* (Table 6). Other isolates includes *S. intermedius, S. lentus, S. capitis, S. xylosus, S. hyicus, S. caprae, S. sciuri, S. simulan, S. hominis* and *Micrococcus* spp.

DISCUSSION

The high prevalence of *Staphylococci* could be due to the fact that the herdsmen milk wet udders, using same cloth for washing the udders, the cloth was washed in the same water bucket and milkers usually clean their hands between milking by passing the hand through their cloths. Milking is usually performed inside the same shade for there were no separate milking areas. All these unhygienic conditions contribute a lot to the bacterial count before it reaches the milking buckets. The chain may last for five hours till milk reaches the consumer.

Due to all these factors which may affect the count and elevate it at any level, it was expected that milk would have a moderate to high bacterial count.

CPS were primarily mastitis pathogens, their presence in the milk is favored by udder infection, the hygienic measure before, during and after milking and the environmental conditions (Anderson, 1982).

During this study in Khartoum state, percentage of the samples which were positive to CPS were higher in summer (47.4%) compared to 37.6% in winter. Khartoum north was the region with higher percentage of positive samples in winter (44.4%) and Omdurman has the highest percentage during summer (49.6%) while Khartoum region has the lowest percentage during the two seasons (30.2%) in winter and (43.6%) in the summer. The difference between the CPS positive sample in Khartoum in one side and Khartoum north and Omdurman on the other side was significant at 0.05, but

the difference between Khartoum north and Omdurman was insignificant. This may be partially due to the better hygienic practices in most of the Khartoum region farms, since most of the farmers were educated retired government employees. Also, they can use the mastitis medicine properly and can educate their farmers. The percentage of CPS positive samples in vendor milk was the highest (53.1%), then market milk (48.8%), then the bulk tank farm milk, which was 42.2%, since the number of farms from which milk is collected varies between vendor and market milk, while the milk in the farms bulk tank is almost collected from the apparently mastitis free cows, but addition of milk of the neighboring farms to farm bulk milk is practiced in some farms. Nada (2000) found these percentages to be 20% in vendor milk and 25% in farm bulk milk. These differences may be due to the differences in sample size between the two studies.

The bacterial safety limits for the CPS was recommended by Bruce (2003) not to exceed 10^3. In this study, percentage of samples which exceeded this limit was 16.5% in the state, Khartoum north has the highest percentage followed by Omdurman then Khartoum region. The vendor milk which exceeds this limit was the highest (19.4%) then market milk (18.4%), Bulk (18%) and then the individual which was 17.1%. These higher percentages of counts reflect the poor hygienic measure at milking time and during distribution.

Statistically, there is significant correlation at 0.01 significance level between the Individual, farm bulk tank, market and vendor milk; this is again due to subclinical mastitis, milking of unhygienic udder, poor sanitation of milkers distributors and milk containers.

In conclusion, it was found that most of the raw milk sold in Khartoum State is of low hygienic quality, so, the Ministry of Agriculture and Animal Resources of Khartoum State should enforce all the regulations needed for producing and purchasing raw milk with acceptable hygienic quality. Some of these regulations should include: 1) Small producing units should be grouped together to facilitate veterinary services and direct supervision. 2) Khartoum State should establish collection centers to receive and cool the milk immediately after production. 3) The practice of selling raw milk through the pick-up cars and carriages without least hygienic measures should be stopped. Milk must be treated before reaching the consumers. 4) The presence of *Staphylococci* must be more investigated to determine their impact on public health and their pathogenicity.

REFERENCES

Anderson JC (1982). Progressive pathology of *staphylococcus mastitis* with vote in control immunization and therapy. Vet. Rec. 110(16): 372- 376.

Bautista L, Gaya P, Medina M, Nunezm (1988). A quantitative assay of enterotoxin production by sheep milk Staphylococci. Appl. Environ. Microbial. 54: 566-569.

Bran Y, Flewette J, Orey F (1978). Micro method for biochemical identification of coagulase negative Staphylococci. J. Clini. Microbiol. (1982). 16.

Bruce EL, Robert JH, Katherine A (1983). Identification of Staphylococcus species of Bovine origin with the API staph. Ident. System. J. Clin. Micro. 17: 984-986.

Bruce H (2003). Report to the technical consultative committee (TCC) - on recommended product safety limits for the NZ Dairy industry 59: 82-92.

Davidson A (1961). Observation of the pathogenic *Staphylococci* in dairy herd during a period of six years. Res. Vet. Sci. 2: 22.

Devriese LA, Hommez J, Leevers H, Pot B, Vandamme P,Haesbrock (1999). Identification of Esculin hydrolysis *Streptococci*, Lactococci, Aerococci and *Enterococci* from subclinical intramammary infection in dairy cows. Vet. Microbiol. 94.

Devriese L (1999). Staph in healthy and disease animals, J. Appl. Bact. Symp. Supply pp. 71-81.

Geary C, Stevens M, Sneath PH, Mitchell CJ (1989). Construction of database to identify Staphylococcus species. J. clinical pathology 42: 289-294.

Gemmel CG, Dowson JE (1982). Identification of coagulase Negative *Staphylococci* with the API- staph. System. J. Clin. Microbiol. (1982). 16 /5/874 (abstract).

Haeghaert S, Le Querrce, Gallay F, Bouvet A, Gomez PM, Valio V (2002). Les Toxi infection alimentaires collectives enfrance en 1999 et 2000. Bull. Epidemiol. Hebdo 23. In Yves LL et al. (2004), (Abstract).

Ibrahim EA (1973). A note on some characteristics of the raw fluid milk available in the three towns. Sudan. J. Vet. Sci. Anim. Husbandary 14(1): 36-41.

IDF (1980). Bovine mastitis symposium. IDF bull. 84 International Dairy Federation Brussels- Belgium- chapter I (1980).

ISO/DIS (1997). Milk and milk products. Enumeration of Coliforms. Part2-MPN technique. International organization for standardization. 5541-2-1997 Geneva.

Jay JM (1986). Modern food microbiology, 3 ed. New York. Van Nostrand Rein hold (Abstract).

Jerry RR (2003). The Epidemiology of *Staphylococcus aureus* on dairy farms. Virginia technical Report, Blacksburg Virginia 9: 1217-1220.

Kambaty FM, Bennet RW, Shah DB (1994). Application of Purified Gel Electrophoresis to the epidemiological characterization of Staphylococcus intermedius implicated in a food related outbreak. Epidemiol. Infection 113: 75-80.

Kelous WE, Wolfshon JF (1982). Identification of *Staphylococci* species with the API- ident. System. J. Clin. Microbiol. 52: 45-52.

Nada AA (2000). Studies on the Sanitary Quality of Raw Fluid Market Milk in Khartoum State . MVSc thesis. University of Khartoum pp. 77 -78.

Olarae G, Caharilaose C, Charilaou T (1983). Detection of Staph with conventional method used for identification of negative *Staphylococci*. J. Clin. Micro. (31): 2683-2688.

Wesley EK, Jana FW (1982). Identification of staphylococcus species with API staph. Ident system. J. Clin. Microbiol., (1982).

Yves LL, Florence B, Michel G (2003). *Staphylococcus aureus* and food poisoning. Genet. Mol. Res. 2(1): 63-76.

2

Analysis of the domestic animal reservoir at a micro-geographical scale, the Fontem sleeping sickness focus (South-West Cameroon)

G. R. Njitchouang[1,5], F. Njiokou[1]*, H. C. Nana Djeunga[1], P. Moundipa Fewou[5], T. Asonganyi[3], G. Cuny[4] and G. Simo[2]

[1]General Biology Laboratory, Department of Animal Biology and Physiology, Faculty of Science, University of Yaoundé 1, P. O. Box 812, Yaoundé, Cameroon.
[2]Department of Biochemistry, Faculty of Science, University of Dschang, P. O. Box 67, Dschang, Cameroon.
[3]Faculty of Medicine and Biomedical Sciences, University of Yaoundé 1, Yaoundé, Cameroon.
[4]Laboratoire de Recherche et de Coordination sur les Trypanosomoses IRD, UMR 177, CIRAD, TA 207/G Campus International de Baillarguet, 34398 Montpellier Cedex 5, France.
[5]Department of Biochemistry, University of Yaoundé 1, P. O. Box 812, Yaoundé, Cameroon.

To better understand the epidemiology of sleeping sickness in two Human African Trypanosomiasis (HAT) sub foci (central and northern sub foci) of the Fontem focus where diversity in the prevalence of *Trypanosoma bucei gambiense* was reported in domestic animals and man, 397 domestic animals were sampled in eight villages. Parasitological tests revealed trypanosomes in 86 (21.60%) animals. The CATT test was positive in 254 (64%) animals with the lowest value in dogs. The PCR test revealed *T. b. gambiense* in 11.55% of pigs, 3.45% of goats and 15.38% of sheep. The *T. b. gambiense* infection rates were not significantly different between the two sub foci. However, *T. b. gambiense* was found in animals from all villages of the Northern sub focus while only animals from Menji and Nsoko (Central sub focus) revealed this infection. The detection of *T. b. gambiense* in animals of the central sub focus was in line with results of medical surveys where HAT patients were detected in the same villages. The absence of patients in the northern sub focus despite the circulation of *T. b. gambiense* in animals from all villages of this sub focus since several years is surprising and needed more investigations.

Key words: Domestic animals, *T. b. gambiense,* sleeping sickness, animal reservoir.

INTRODUCTION

Under the combined effects of socio-political, economic, environmental and genetic factors, there has been a recrudescence of sleeping sickness in many historic foci in the Central African region (Cattand, 1994; Grébaut et al., 2001; Kaba et al., 2006; Brun et al., 2009). About 70,000 new cases are reported recently, with three quarters coming from the Democratic Republic of Congo (DRC) and Angola (Seed, 2001; WHO, 2007). Many hypotheses including the genetic diversity of trypanosomes and tsetse flies as well as the presence of an animal reservoir were suggested to explain the reactivation and the maintenance of the disease in the various foci. Genetic studies have shown that *Trypanosoma brucei* s.l. has a flexible mode of reproduction, and that genetic exchanges between *T. brucei* isolates can occur in the *Glossina* vector.

These exchanges can be at the origin of new *T. brucei* genotypes which introduce a genetic diversity that lead sometimes to the emergence of different epidemiological

---

*Corresponding author. E-mail: njiokouf@yahoo.com.

Abbreviations: PCR, Polymerase chain reaction; **HAT,** Human African Trypanosomiasis.

profiles of HAT (Tait et al., 1984; Paindavoine et al., 1989; Gibson, 2001; MacLeod et al., 2001; Njiokou et al., 2004).

In West and Central Africa, investigations on the animal reservoir of the Gambian sleeping sickness has shown that *T. brucei gambiense* infects a variety of domestic and wild animals (Molyneux, 1973; Mehlitz et al., 1982; Herder et al., 2002; Njiokou et al., 2006). The role of some animals in the epidemiology of HAT has been shown experimentally since cyclical transmission of *T. b. gambiense* in pig and cattle for example does not affect its virulence and its pathogenicity for humans (Van Hoof et al., 1942; Moloo et al., 1986). In spite of such experimental finding, definitive evidence that these animals play an important role in the maintenance or the resurgence of HAT still not yet well elucidated. Such evidence would help to define efficient control strategies by integrating the control of animal reservoirs. Indeed, domestic animals have a crucial importance as they live close to humans, and enjoy close social relationship with them (Laveissière et al., 2000).

In Cameroon, investigations on the animal reservoir using a combination of parasitological, immunological and PCR based methods have shown that domestic and wild animals harbour *T. b. gambiense* DNA and antibodies against *T. b. gambiense* Litat 1.3 antigen in almost all HAT foci (Njiokou et al., 2006, 2010). At this macro-geographical scale, animal infection rates by *T. b. gambiense* differed significantly among the foci, reflecting the level of the transmission of HAT in these localities. The Fontem sleeping sickness focus of the South West region of Cameroon seems to have a particular status because medical surveys performed by the national sleeping sickness control team detect HAT patients only in the Central sub focus, whereas preliminary studies on animal reservoirs showed that pigs were infected by *T. b. gambiense* in both Central and Northern sub foci (Nkinin et al., 2002; Simo et al., 2006). Furthermore, the Fontem focus is a remote zone, isolated from the other Cameroonian HAT foci (Doumé, Campo and Bipindi) currently investigated.

To better understand the epidemiology of sleeping sickness in different HAT sub foci of the Fontem focus, we undertook to study the domestic animal reservoir status at a micro-geographical level, notably the villages of the Central and the Northern sleeping sickness sub foci of the Fontem focus, in order to precise the implication of those animals in the transmission cycle of HAT in this focus.

MATERIALS AND METHODS

Study zone

This study was carried out in Fontem (5°40'12 N, 9°55'33 E), a sleeping sickness focus in the South West region of Cameroon. Known since 1949, this focus is subdivided in three sub foci: the North, the Centre and the South sub foci (Figure 1). The Fontem focus was previously amongst the most active foci in Cameroon (Asonganyi and Ade, 1994).

It remains active with very few cases (8 patients out of > 16 000 persons examined; OCEAC, MINSANTE, unpublished data) detected only in the Centre sub focus. Preliminaries studies on animal reservoir revealed *T. b. gambiense* in pigs from the centre and North sub foci (Nkinin et al., 2002; Simo et al., 2006). The main population activities in this focus are agriculture, palm oil extraction, animal husbandry and poultry farming at a small scale. The Fontem focus is characterized by a tropical humid climate with varied topography of hills and valleys through which several high speed rivers flow (Asonganyi et al., 1990).

Collection of samples

Sampling of the domestic animals was done in July 2006 and June 2007 in eight villages: Besali, Bechati, Folepi and Agong in the Northern sub focus and Nsoko, Fossung, Menji and Azi in the central sub focus (Figure 1). Geographic coordinates of each of these villages were recorded using a Global Positioning System (Table 1). The objective of the study was explained to the local authorities and villagers of the study zone. After obtaining their approval, villagers were asked to catch and/or keep their domestic animals. In each village, all domestic animals that had spent at least 3 months in the study zone were selected. From each animal, about 5 ml of blood was collected into EDTA coated tubes. Bleeding was performed from the jugular vein in goats, pigs and sheep and from the saphena vein in dogs. Blood samples were then processed and analysed using direct and indirect methods to detect trypanosomes (0MS, 1986).

Direct methods of detection of trypanosomes

Parasitological tests: A drop of each blood sample was used to make a thick blood film (TBF), which was further, examined for the presence of trypanosomes under light microscope at 1,000X magnification.

The detection of trypanosomes using Capillary Tube Centrifugation (CTC) was performed according to the method described by Woo (1970). Briefly, about 70 µl of each blood sample was taken up in a heparinised capillary tube and its extremities were sealed with potty. After centrifugation at 11,000 rpm for 5 min, trypanosomes were detected at the red blood cell/platelet boundary at 100X magnification.

Polymerase chain reaction (PCR)

DNA was extracted from blood samples using (DNeasy Tissue Kit) (QIAGEN). Initially, 1 ml of blood was mixed with 1 ml of sterile water and vortexed before centrifuging at 14,000 rpm for 10 min. The supernatant was discarded and the pellet containing parasites was re-suspended in 200 µl PBS. Then, the DNA was extracted following manufacturer's instructions. The DNA extract was stored at -20°C until use.

The detection of *T. brucei* s.l. was carried using the TBR1/TBR2 primers (Masiga et al., 1992). PCR reactions were performed in 20 µl (final volume) of mixture containing 10 mM Tris-HCl (pH9), 50 mM KCl, 1.5 mM MgCl$_2$, 15 pmol of each primer, 200 µM of each dNTP, 0.4 Units of Taq DNA polymerase and 5 µl of DNA solution (or 5 µl of water for negative controls). Amplification was done according to the protocol described by Masiga et al. (1992). The amplification program contains an initial denaturation step at 94°C for 3 min 30 s, followed by 40 amplification cycles composed of a

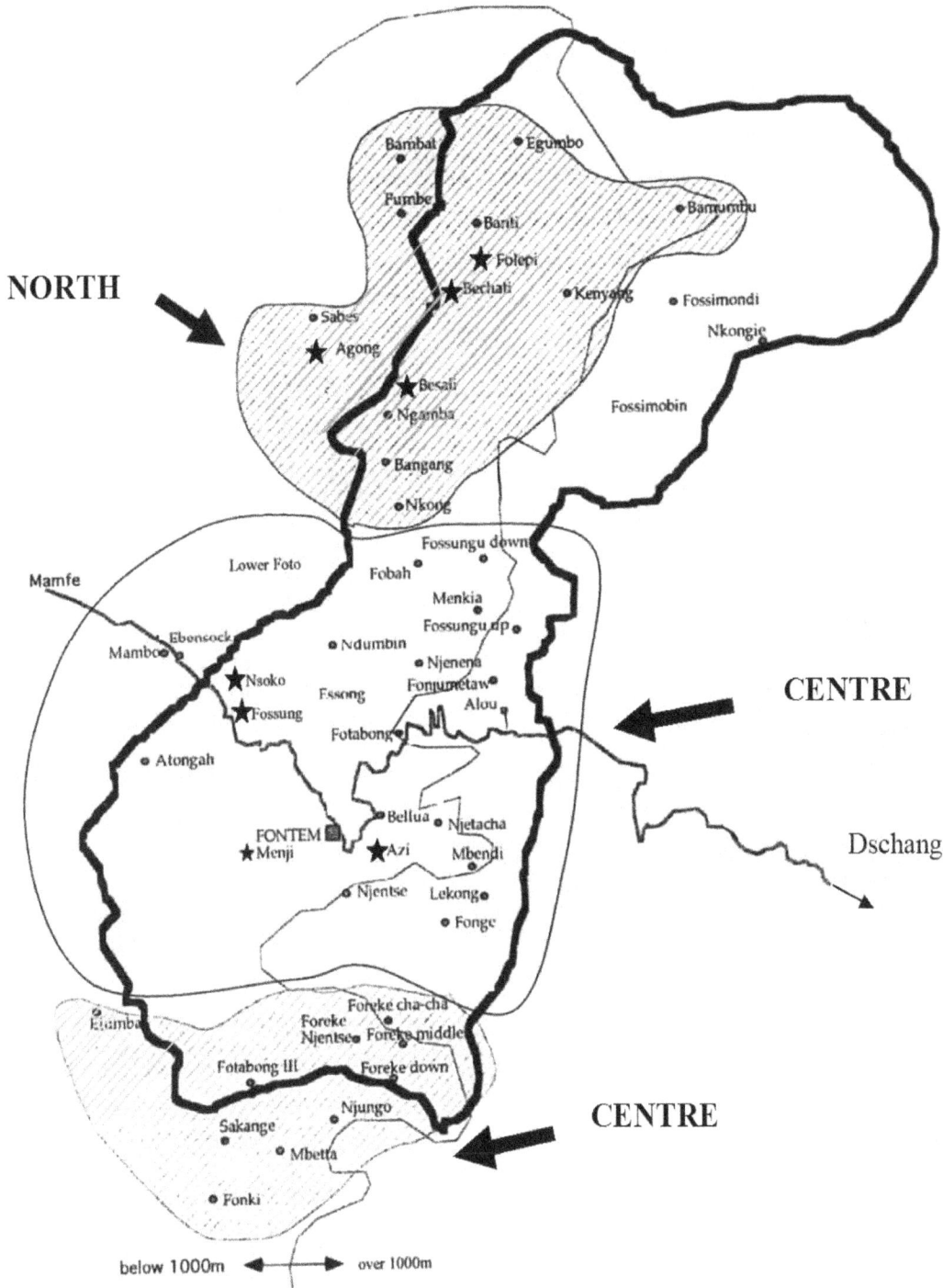

Figure 1. Map of Fontem focus (Simo et al., 2006).

(★): Villages where animals were sampled.

denaturation step at 94 °C for 30 s, an annealing step at 58 °C for 30 s and an extension step at 72 °C for 1 min. A final extension was performed at 72 °C for 5 min. The PCR products were resolved on 2% agarose gel containing ethidium bromide and visualised under UV light.

All samples positive for *T. brucei* s.l. were submitted to a second PCR as described by Herder et al. (2002). During this second amplification, primers TRBPA1/TRBPA2 (Herder et al., 2002) which amplify a DNA sequence of 149 bp characteristic of *T. b. gambiense* group 1 were used. The PCR conditions were identical to those described above, except that the annealing step was performed at 62 °C. The PCR products were resolved on 10%

Table 1. Geographic coordinates of the villages where animals were sampled.

Sub foci	Villages	Geographic coordinates	
		Latitude	Longitude
North	Besali	05°38'044"N	009°54'790"E
	Bechati	05°40'060"N	009°55'187"E
	Folepi	05°40'074"N	009°56'041"E
	Agong	05°39'295"N	009°53'750"E
Centre	Menji	05°29'257"N	009°51'194"E
	Nsoko	05°31'670"N	009°49'671"E
	Fossung	05°30'878"N	009°50'024"E
	Azi	05°28'798"N	009°52'816"E

polyacrylamide gel. Electrophoresis was done at 90 V for 17 h in TBE X1. The gels were stained in ethidium bromide, washed in water and the bands were identified under UV light.

Indirect methods of detection of trypanosomes

Card agglutination test for trypanosomiasis (CATT): The immunological test was performed on plasma using CATT 1.3 test (Card agglutination test for trypanosomiasis) as described by Magnus et al. (1978). Briefly, the blood was centrifuged at 3000 rpm for 5 min, and 50 μl of each plasma was mixed with a drop of CATT 1.3 reagent (lyophilised bloodstream *T. b. gambiense* forms with LiTat 1.3 antigen expressed on their surface and suspended in phosphate buffer) and then shaken at 60 rpm for 5 min on an orbital agitator. Samples presenting agglutination were concluded as positive (Magnus et al., 1978). The remaining blood sample was stored at 4°C for molecular ana.

Statistical analysis

Statistical analysis was done using the programme XLSTAT-PRO version 2009.3. The χ^2 test was used to compare the antibody rates, the *T. brucei* s.l. and *T. b. gambiense* infection rates as well as the unidentified trypanosome infection rates. The comparisons were done according to the village, the sub foci and to the animal species.

The concordance between the different tests was assessed by comparing, for a given test, the percentage of positive samples to that of negative ones, according to the percentage of positive samples obtained with the other test, using the χ^2 test. The threshold for significance was set at 5%.

RESULTS

During the field trips performed in July 2006 and June 2007, 397 animals including 225 pigs, 87 goats, 65 sheep and 20 dogs were sampled in the Fontem focus: 184 were from Central sub focus and 213 from Northern sub focus. Detailed results concerning the number of animals, the species and the villages where these animals were sampled are reported in Tables 2 and 3.

Parasitology

Out of the 397 animals sampled, Thick Blood Film and Capillary Tube Centrifugation tests revealed trypanosomes in 86 (21.66%) of them. The four animal species sampled here were found with trypanosome infections, with significant different infection rates (Table 2). Animals from all the studied villages harboured trypanosomes with significant different infection rates.

In addition, in the Central sub focus, animals from Menji were significantly more infected compared to other villages. In the contrary, the trypanosome infection rates were not statistically different between the animals sampled in:

(1) The villages of the Northern sub focus (Table 3),
(2) The Northern (19.25%) and the Central (24.46%) sub foci (Table 4).

Serology

The CATT 1.3 test was positive in 254 (64%) animals. All the animal species harboured antibodies against the LiTat 1.3 antigen. The antibody rates were statistically different between the four animals species sampled in the Northern sub focus and when considering the entire focus (Tables 2). In addition, no significant difference was observed between the antibody rates in animals from the two HAT sub foci (Table 4). Identical results were obtained when comparing the antibody rates between villages of each sub focus and between pigs, goats and sheep in the entire focus (Table 3).

PCR

The PCR test identified 140 (35.26%) animals harbouring *T. brucei* s.l . This trypanosome species was found in the four animal species as well as in all villages of the

Table 2. Number and percentage of positive animals for different tests, by animal species.

Animal species	Zones	Number examined	TBF/CTC (%)	CATT (%)	TBR (%)	TBG (%)
Pigs	North	91	24 (26.37)	53 (58.24)	50 (54.95)	14 (15.38)
	Centre	134	35 (26.12)	94 (70.15)	42 (31.34)	12 (08.95)
	Total 1	225	59 (26.22)	147 (65.33)	92 (40.88)	26 (11.55)
Goats	North	46	08 (17.39)	30 (65.21)	20 (43.48)	0.0 (0.00)
	Centre	41	09 (21.95)	28 (68.29)	05 (12.20)	03 (07.32)
	Total 2	87	17 (19.54)	58 (66.67)	25 (28.73)	0.3 (03.45)
Sheep	North	58	06 (10.34)	40 (68.96)	17 (29.31)	10 (17.24)
	Centre	07	01 (14.28)	03 (42.85)	02 (28.57)	0.0 (0.00)
	Total 3	65	07 (10.77)	43 (66.15)	19 (29.23)	10 (15.38)
Dogs	North	18	03 (16.67)	05 (27.78)	04 (22.22)	0.0 (0.00)
	Centre	02	0.0 (0.00)	01 (50.00)	0.0 (0.00)	0.0 (0.00)
	total 4	20	03 (15.00)	06 (30.00)	04 (20.00)	0.0 (0.00)
Total	North	213	41 (19.24)	128 (60.09)	91 (42.72)	24 (11.26)
	χ^2		6.11	10.37	12.92	11.78
	P		0.10	0.016	0.005	0.008
	Centre	184	45 (24.45)	126 (68.47)	49 (26.63)	15 (8.15)
	χ^2		1.37	2.61	6.63	0.95
	P		0.71	0.45	0.08	0.81
	All	397	86 (21.66)	254 (63.97)	140 (35.26)	39 (9.82)
	χ^2		8.05	10.60	7.82	9.20
	P		0.045	0.014	0.05	0.027

TBF/CTC: Thick blood film and capillary tube centrifugation, TBR: *Trypanosoma brucei* s.l., TBG: *Trypanosoma brucei gambiense*, CATT: Card agglutination test for trypanosomiasis.

Fontem HAT focus. The infection rates by this trypanosome species differs significantly between animal species and villages of the northern sub focus (Tables 2 and 3). Similar results were observed in the entire focus (Table 4).

T. b. gambiense was detected in 39 (9.82%) animals: 26 (11.55%) in pigs, 10 (15.38%) in sheep and 3 (03.45%) in goats (table 2). In the two sub foci, no *T. b. gambiense* infection was found in dogs. The infection rates differed significantly between animal species, pigs being the most infected. In the Northern sub focus, only pigs and sheep were found positive, and the infection rates differed significantly between these animal species. In the Central sub focus, pigs and goats carried *T. b. gambiense* infections, but the infection rates did not differ significantly (Table 2). Analysis by village revealed *T. b. gambiense* infections in animals of all villages of the northern sub focus.

No significant difference was observed between the *T. b. gambiense* infection rates in the villages of this sub focus. In the Central sub focus, only animals from Menji and Nsoko were found with *T. b. gambiense* infections with significantly differences according to the villages

(Table 3). No significant difference was observed between the *T. b. gambiense* infection rates of the two sub foci (Table 4).

Concordance test

Out of the 86 animals that carried trypanosome infections, 19 were positive for the two parasitological tests (Thick Blood Film and CTC). Moreover, 53 and 14 trypanosome infections were revealed only by CTC and Thick Blood Film, respectively. The Chi square test between CTC and TBF was significant ($\chi^2 = 32.11$; $P < 0.0001$) in favour of negative samples, showing that results of CTC and TBF are not linked.

A total of 24 animals were positive for both CATT test and *T. b. gambiense*-PCR, whereas 230 were only positive for the CATT test and 15, only positive for PCR. The Chi square test between CATT and PCR was significant ($\chi^2 = 334.14$; $P < 0.0001$) in favour of negative samples; suggesting that positive CATT tests are not linked absolutely to the presence of *T. b. gambiense* in animals.

Table 3. Number and percentage of positive animals for different tests, by study area.

Study sites	No animals	TBF/CTC (%)	CATT (%)	TBR (%)	TBG (%)
North					
Besali	67	11 (16.42)	38 (56.72)	19 (28.36)	11 (16.42)
Bechati	90	17 (18.89)	57 (63.33)	45 (50.00)	10 (11.11)
Folepi	37	10 (27.02)	22 (59.46)	17 (45.95)	01 (02.70)
Agong	19	03 (15.79)	11 (57.89)	10 (52.63)	02 (10.52)
Sub-total	213	41 (19.25)	128 (60.09)	91 (42.72)	24 (11.27)
$\div^2$		1.93	0.75	8.51	4.50
P		0.58	0.86	0.036	0.21
Centre					
Menji	86	32 (37.21)	66 (76.74)	28 (32.56)	12 (13.95)
Nsoko	38	06 (15.79)	21 (55.26)	07 (18.42)	03 (07.89)
Fossung	21	03 (14.28)	12 (57.14)	06 (28.57)	0.0 (0.00)
Azi	39	04 (10.25)	27 (69.23)	08 (20.51)	0.0 (0.00)
Sub-total	184	45 (24.46)	126 (68.48)	49 (26.63)	15 (08.15)
$\div^2$		14.54	7.05	3.64	9.19
P		0.002	0.07	0.30	0.027
Total		86 (21.66)	254 (64.00)	140 (35.26)	39 (09.82)
$\div^2$	397	19.20	10.40	23.44	13.93
P		0.008	0.16	0.001	0.052

No: Number of, TBF/CTC: Thick blood film and capillary tube centrifugation, TBR: *Trypanosoma brucei* s.l., TBG: *Trypanosoma brucei gambiense,* CATT: Card agglutination test for trypanosomiasis.

Table 4. Number and percentage of positive animals for different tests, between the two sub-foci.

Sub focus	No animals examined	TBF/CTC (%)	CATT (%)	TBR (%)	TBG (%)
North	213	41 (19.24)	128 (60.09)	91 (42.72)	24 (11.26)
Centre	184	45 (24.45)	126 (68.47)	49 (26.63)	15 (08.15)
$\div^2$		1.57	3.01	11.19	1.082
P		0.20	0.083	0.001	0.29

No: Number of, TBF/CTC: Thick blood film and capillary tube centrifugation, TBR: *Trypanosoma brucei* s.l., TBG: *Trypanosoma brucei gambiense,* CATT: Card agglutination test for trypanosomiasis.

DISCUSSION

The parasitological tests showed that 21.66% of animals carry trypanosomes. This result is in line with those reported previously in domestic animals of the Fontem focus (Nkinin et al., 2002; Simo et al., 2006) as well as in animals of the other HAT foci of West and Central Africa (Scott et al., 1983; Noireau et al., 1986; Asonganyi et al., 1990). No significant difference was found between the trypanosome infection rates in domestic animals sampled in the two HAT sub foci as well as in villages of the northern sub focus whereas there was a significant difference between villages of the Central sub focus and when comparing villages of the entire focus. These results suggest that, although the relative homogeneity in the ecological and bioclimatic conditions (same climate,

vegetation, topography and unique vector), epidemiological considerations such as human activities are slightly different between villages. CTC identified many positive animals compared to thick blood film confirming its high sensitivity. A high discordance was observed between the two tests, suggesting their complementarity.

The high proportion of animals found positive for the CATT test confirms the result of Nkinin et al. (2002) in the Fontem focus and those of Njiokou et al. (2010) in other HAT foci of Cameroon. This indicates the presence of antibodies directed against the LiTat 1.3 antigens of *T. b. gambiense* in these animals. However, the high antibody rates obtained here does not corroborate the low prevalence (11.89%) of *T. b. gambiense* revealed by PCR. Indeed, the analysis of CATT and PCR-*gambiense* results revealed a high discordance between these two tests; suggesting that many CATT tests are positive in animals that do not carry *T. b. gambiense* infection. It is probably the case of animals harbouring other trypanosomes of *T. brucei* complex species or *T. congolense* as reported by Noireau et al. (1986). Animals positive to PCR-*gambiense* and negative for the CATT test are probably infected by *T. b. gambiense* isolates which do not express the LiTat 1.3 antigenic variant as already reported in the Fontem focus (Dukes et al., 1992; Asonganyi; Ade, 1994; Kanmogne et al., 1996). It is also possible that some of these animals were recently infected (less than 10 days) and the *T. b. gambiense* antibodies were not produced at time of sampling. Similar results were obtained during experimental infections of pigs by *T. b. gambiense* (Penchenier et al., 2005). Moreover, serological analysis showed that the antibody rate is not significantly important in pigs than in sheep and in goats. However, analysis of blood meals of *Glossina palpalis palpalis* in the same area revealed that 55% were from pigs, 23% from human and very little from other domestic animals (Njitchouang, pers. Comm.). These results can be explained by the fact that pigs are reared within a maximum of 12 months while other domestic animals are maintained for many years in the villages and accumulate infections (Njiokou et al., 2010).

The percentage of animals infected by *T. brucei* s.l. (35.66%) and *T. b. gambiense* (9.82%) are comparable to those reported by Simo et al. (2006) in the Fontem focus, and superior to the values obtained by Njiokou et al. (2010) in the four HAT foci of southern Cameroon. These results confirm a high circulation of trypanosomes of *T. brucei* s.l. complex and that of *T. b. gambiense* in domestic animals in Fontem, despite the relative low prevalence of sleeping sickness in humans compared to other HAT foci like Bipindi and Campo (Penchenier et al., 1999; Grébaut et al., 2001). The presence of *T. brucei* s.l. and *T. b. gambiense* respectively in 4 and 3 animal species examined confirms a direct affinity between these hosts and trypanosomes of the *T. brucei* s.l. complex, and suggesting their direct implication in the epidemiology of sleeping sickness (Nkinin et al ., 2002 ; Simo et al., 2006 ; Njiokou et al., 2010).

Looking at *T. b. gambiense* infections in animals between the two HAT sub foci, no significant difference was observed. This result is surprising given the results of medical surveys that revealed HAT patients in the central sub focus and no patient in the northern sub focus. The identification of *T. b. gambiense* in animals in two villages (Menji and Nsoko) of the central sub focus are in line with results obtained during medical surveys where HAT patients were detected in these two villages during the last decade (National Trypanosomiasis control program unpublished data). This indicates an active transmission of HAT in both man and animal in these villages; confirming several disease transmission cycles including man, pig and goats in the central sub focus. In the northern sub focus where *T. b. gambiense* was identified in animal of all villages, no HAT patient was detected during these last decades. However, *T. b. gambiense* infections were identified in animals from this sub focus since several years (Nkinin et al., 2002); indicating a circulation of this trypanosome sub species and also a predominant animal transmission cycle in all villages of the northern sub focus. Furthermore, the presence of human blood meals in tsetse flies captured in this sub focus indicates a contact between tsetse flies and man. This contact is illustrated by the high positivity of inhabitants of this sub focus to immunological test like CATT 1.3 (Magnus et al., 1978) and Latex *T. b. gambiense* (OCEAC, unpublished data). These results show clearly that human of the Northern sub focus are in contact with *T. b. gambiense*. However, the absence of HAT patients in this sub focus is difficult to explain given the fact that transmission conditions found in this sub focus are similar to those of the central sub focus. Moreover, the genetic characterization of *T. b. gambiense* isolated in pigs and human of these sub foci showed considerable genetic homogeneity (Nkinin et al., 2002; Njiokou et al., 2004). One interesting aspect that needs to be addressed in order to improve knowledge on the epidemiology of HAT in these sub foci requires investigations on human susceptibility to *T. b. gambiense* infection. In West Africa, diversity in the clinical evolutions associated to human genetic factors has been recently suspected (Jamonneau et al., 2004; Garcia et al., 2006).

ACKNOWLEDGEMENTS

This study was supported by IRD through JEAI trypanosomiasis and UMR 177 funds. We would like to thank the Trypanosomiasis Team of the Yaoundé 1 University for their contribution towards the realisation of this work and the population of Fontem for their collaboration.

REFERENCES

Asonganyi T, Ade SS (1994). Sleeping sickness in Cameroon. J. Camerounais de Méd., 3(2): 30-37.

Asonganyi T, Suh S, Tetuh MD (1990). Prevalence of domestic animal trypanosomiasis in Fontem sleeping sickness focus, Cameroon. Revue Elevage et Médecine Vétérinaire des pays Tropicaux, 43(1): 69-74.

Brun R, Blum J, Chappuis F, Burri C (2009). Human African Trypanosomiasis. The Lancet, in press

Cattand P (1994). Trypanosomiase Humaine Africaine : Situation épidémiologique actuelle, une recrudescence alarmante de la maladie. Bulletin de la Société de Pathologie Exotique, 87: 307-310.

Dukes P, Gibson WC, Gashumba KM, Bromidge TJ, Kaukas A, Asonganyi T, Magnus E (1992). Absence of the LiTat 1.3 (CATT antigen) gene in *Trypanosoma brucei gambiense* stocks from Cameroun. Acta Tropica, 51: 123-134.

Garcia A, Courtin D, Solano P, Koffi M, Jamonneau V (2006). Human African Trypanosomiasis: connecting parasite and host genetics. Trends Parasitol., 22(9): 405-409.

Gibson W (2001). Sex and evolution of trypanosomes. Int. J. Parasitol., 31: 643-647.

Grébaut P, Bodo JM, Assona A, Foumane Ngane V, Njiokou F, Ollivier

G, Soula P, Laveissière C (2001). Recherche des facteurs de risqué de la trypanosomose humaine africaine dans le foyer de Bipindi au Cameroun. Médecine Tropicale, 61: 377-383.

Herder S, Simo G, Nkinin S, Njiokou F (2002). Identification of trypanosomes in wild animals from south Cameroon using PCR. Parasite, 9: 345-349.

Jamonneau V, Ravel S, Koffi M, Truc P, Laveissère C, Herder S, Grébaut P, Cuny G, Solano P (2004). Characterization of *Trypanosoma brucei* s.l. infecting asymptomatic sleeping-sickness patients in Côte d'Ivoire: a new genetic group? Ann. Trop. Med. Parasitol., 98(4): 329-337.

Kaba D, Dje NN, Courtin F, Oke F, Koffi M, Garcia A (2006). L'impact de la guerre sur l'évolution de la THA dans le Centre-Ouest de la Côte d'Ivoire. Trop. Med. Int. Health, 11(2): 136-143.

Kanmogne GD, Stevens JR, Asonganyi T, Gibson WC (1996). Characterisation of *Trypanosoma brucei gambiens* isolates using restriction fragment length polymorphism in 5 variant surface glycoprotein genes. Acta Tropica, 61: 239-254.

Laveissière C, Grébaut P, Herder S, Penchenier L (2000). Les glossines vectrices de la Trypanosomiase Humaine Africaine. Institut de Recherches pour le Développement, OCEAC Yaoundé Cameroun, 240 pp.

MacLeod A, Andy T, Michael RT (2001). The population genetics of *Trypanosoma brucei* and the origin of human infectivity. Phil. Trans. R. Soc. Lond. B., 356: 1035-1041.

Magnus E, Vervoort T, Van Merveinne N (1978). A Card Agglutination Test with stained Trypanosome (CATT) for serological diagnosis of *Trypanosoma brucei gambiense* trypanosomiasis. Annales de la Société Belge de Médécine Tropicale, 58: 169-179.

Masiga DK, Smyth AJ, Hayes P, Bromidge TJ, Gibson WC (1992). Sensitive detection of trypanosomes in tsetse flies by DNA amplification. Int. J. Parasitol., 21(7): 909-918.

Mehlitz D, Zillmann D, Scott CM, Godfrey DG (1982). Epidemiological studies on animal reservoir of gambiense sleeping sickness; part III: Characterization of *Trypanozoon* stocks by isoenzyme and sensitivity to human serum. Tropenmedizin und Parasitol., 33: 113-118.

Moloo SK, Asonganyi T, Jenni L (1986). Cyclical development of *Trypanosoma brucei gambiense* from cattle and goats in Glossina. Acta Tropica, 43: 407-408.

Molyneux DH (1973). Animal reservoirs and gambian trypanosomiasis. Annales de la Société Belge de Médecine Tropicale, 53(6): 605-618.

Njiokou F, Laveissière C, Simo G, Nkinin S, Grébaut P, Cuny G, Herder S (2006). Wild fauna as a probable animal reservoir for *Trypanosoma brucei gambiense* in Cameroon. Inf. Gene. Evol. 6: 147-153.

Njiokou F, Nimpaye H, Simo G, Njitchouang GR, Asonganyi T, Cuny G Herder S (2010). Domestic animals as potential reservoir hosts of *Trypanosoma brucei gambiense* in sleeping sickness foci in Cameroon. Parasite, in press.

Njiokou F, Nkinin SW, Grébaut P, Penchenier L, Barnabé C, Tibayrenc M, Herder S (2004). An isoenzyme survey of *Trypanosoma brucei* s.l. from the Central African subregion: population structure, taxonomic and epidemiological considerations. Parasitology, 128: 645-653.

Nkinin SW, Njiokou F, Penchenier L, Grébaut P, Simo G, Herder S (2002). Characterization of *Trypanosoma brucei* s.l. subspecies by isoenzymes in domestic pigs from the Fontem sleeping sickness focus of Cameroon. Acta Tropica, 81: 225-232.

Noireau F, Gouteux JP, Frezil JL (1986). Sensibilité du test d'agglutination sur carte (TESTTRYP[R] CATT) dans les infections porcines à *Trypanosoma (Nannomonas) congolense* République Populaire du Congo. Annales de la Société Belge de Médecine Tropicale, 66: 63-68.

OMS (1986). Trypanosomiase Humaine Africaine. Rapport annuel de la division de lutte contre les maladies tropicales. Genève, Suisse 32 pp.

Paindavoine P, Zampetti-Bosseler F, Coquelet H, Pays E, Steinert M (1989). Different allele frequencies in *Trypanosoma brucei brucei* and *Trypanosoma brucei gambiense* populations. Molecular and Biochem. Parasitol., 32(1): 61-71.

Penchenier L, Alhadji D, Bahébégué S, Simo G, Laveissière C, Cuny G (2005). Spontaneous cure of domestic pigs experimentally infected by *Trypanosoma brucei ganbiense*. Implications for the control of sleeping sickness. Veterinary Parasitol., 133(1): 7-11.

Penchenier L, Grébaut P, Ebo'o Eyenga V, Bodo JM, Njiokou F, Binzouli JJ, Simaorro P, Soula G, Herder S, Laveissière C (1999). Le foyer de Trypanosomose humaine de Campo (Cameroun). Historique et situation de l'endémie en 1998. Bulletin de la Société de Pathologie Exotique, 92(3): 185-190.

Scott CM, Frezil JL, Toudic A, Godfrey DG (1983). The sheep as a potential reservoir of human trypanosomiasis in the Republic of Congo. Trans. Royal Society Trop. Med. Hygiene, 77: 397 - 401.

Seed JR (2001). African trypanosomiasis research: 100 years of progress, but questions and problems still remain. Int. J. Parasitol., 31: 434-442.

Simo G, Asonganyi T, Nkinin SW, Njiokou F, Herder S (2006). High prevalence of *Trypanosoma brucei gambiense* group 1 in pigs from Fontem sleeping sickness focus in Cameroon. Veterinary Parasitol., 139: 57-66.

Tait A, Eldirdiri A, Babiker Le Ray D (1984). Enzyme variation in *Trypanosoma brucei* spp. I evidence for the sub-speciation of *Trypanosoma brucei gambiense*. Parasitology, 89: 311-316.

Van Hoof L, Henrard C, Peel E (1942). Recherche sur le comportement du *Trypanosoma gambiense* chez les porcs. Recueil des Travaux de Science et de Médecine au Congo Belge, 1: 53-68.

WHO (2007). Report of the first meeting of WHO strategic and technical advisory group on neglected tropical diseases. 22 pp.

3

Condition and population size of *Macaca fascicularis* (long-tailed macaque)

Karimullah* and Shahrul Anuar

School of Biological Sciences, University Sains Malaysia, 11800 Penang, Malaysia.

This study discusses the population size of *Macaca fascicularis* in Penang Botanical Gardens, Malaysia. The scan sampling method was used to observe the groups of *M. fascicularis* in the gardens. The study was carried out from February 2007 to June 2007. Chi-square test was used to find the correlation of individual appears on the specific area of the gardens. The total number of observations that were carried out during this study was 1134. Among these observation the adult females were observed as 22% (P = 0.15), adult males 17% (P < 0.05), juveniles 56% (P < 0.05) and infants 5% (P = 0.34). This study revealed that the population of long tailed macaques is decreasing in Penang Botanical Gardens. In arranging to get better human contact with macaque and at the similar time to keep up a developed macaque inhabitants in Botanical Gardens Penang, there is an imperative requirement for wildlife department to enhance their safety, food availability and predator's threats.

Key words: *Macaca fascicularis*, scan sampling, botanical gardens Penang, wild life.

INTRODUCTION

Macaca fascicularis (long tailed macaques) is one of the foremost biologically well-known and plentiful anthropoid species of primate in the world (Wheatley, 1999). The long tailed macaque inhabits from west to east, as of Myanmar toward the Philippines, and north to south as of Northern Thailand to the southern islands of Indonesia (Fooden, 1995). They are living near the sea sides at the height of 1524 to 2000 m (Medway, 1970) and in the edges of primary and secondary forests (Lekagul and McNeely, 1977). They use to live and adopt in the areas where other primates are not living (Angst, 1975). These types of monkeys have great similarities with human behaviors in their natural condition. This has occur on Mauritius (Sussman and Tattersall, 1986), Hong Kong (Southwick and Manry, 1987; Wong and Ni, 2000), Tinjil Island, Java (Kyes, 1993), Jayapura area, Ngeuar Island, Republic of Palau (Wheatley et al., 2002), West Papua, (Kemp and Burnett, 2003) and Kabaena Island, Sulawesi (Froehlich et al., 2003). *M. fascicularis* has been established obviously in a broad range of habitats with naturally, built-in, mangrove, coastal, swamp and riverine forest (Wolfheim, 1983). They adapt well to changed their

environments (Wheatley, 1999) and frequently develop atmosphere exaggerated by human arrangement and farming (Fuentes et al., 2005). In Malaysia, they are dominant near sea sides' that is the sea sides of Penang, Langkawi, Tioman, Singapore, tall bamboo, and beaches (Medway, 1983). *M. fascicularis* is mostly isolated, reserved, highly awaken and emotional in comparison to lion-tailed macaques (*Macaca silenus*) and rhesus macaques (*Macaca mulatta*) (Clarke et al., 1988). These are because of diet, habitat and social factors, which are based on group development (Clarke et al., 1994).

M. fascicularis is grey-black and brown-gold in color. Its lower part of the body is brown-pale, their face, hand's palms and feet are hairless and coloured like red meat. Its tail is hairy and longer than *Macaca nemestrina* (pig-tailed) (Medway, 1983). In old animals (*M. fascicularis*) the tail may be shortened by accident (Adams et al., 1985). Head length is different among *M. fascicularis* from different locations (Fooden and Albrecht, 1993).

Macaque's populations have declined (Corlett, 1992). The current inhabitants of macaques are extremely noticeable and they oftenly facing in habitat lost in the region of Penang Botanical Gardens. In this area, the predators, loss of natural environment and less availability of food occur over declining of species. Other aspects, such as human acts of violations and behaviors have great manipulation on long tailed macaques (e.g.

*Corresponding author. E-mail: karimullah76@yahoo.com.

giving food, etc) can be increasing influence on clashes toward macaques. In view of long-tailed macaques' conflicts with predators, it is vital to assess in dispose to find the main cause of these clashes. As described by Achmad et al. (2009) that the population of long tailed macaques are decreasing in Southeast Asia, because of habitat variation and loss.

In discussion with above, relevant studies, it is also not clear that the population of long tailed macaques in Penang Island has increased since this population was counted. In order to know the current position of macaques in Penang Botanical Gardens, a survey was conducted on the population size of long tailed macaques for suitable judgment. This research study provides important and detail information about the long tailed macaques in Penang Botanical Gardens, and the factors that influence population.

MATERIALS AND METHODS

Study area

Penang Botanical Gardens was selected as study area. Botanical gardens are very famous, it gives out plant resources and it is the main center for the protection of wild species (Heywood, 1991). It is situated eight kilometers off the north-western coast of Peninsular Malaysia. George Town, the capital of Penang. This study was conducted from February 2007 until June 2007. Permission for this research was confirmed by the appropriate institutional animal care committee (Wild Life Department Pulau Penang, Malaysia). For moral cure of these primates, the Management of Botanical Gardens Penang, Malaysia was allowed the observer for the observation of long tailed macaques.

Study subject

The study was carried out only on M. fascicularis spp. because this specie has lot of behaviours and activities, as compared to other primates, that is:

(1) Easy availability, as this specie oftenly used to come down to the Gardens in search of food.
(2) There was more probability of accuracy to count the group's members.
(3) Simply familiar with humans (the gardens have unique attraction for tourists and M. fascicularis is abundant in number).
(4) M. fascicularis was selected as study subject because they are so similar to humans.

Field method and population study

The method of observation in the botanical gardens used in this study was quite slow with gentle walking and waiting in different corners of the forest (Cooper, 2000). Maximum time was spent waiting outside the jungle for the troop to appear. The dense jungle was difficult to access and noisy because of thick bushes, and not as satisfactory for finding monkeys like in open spaces such as those that walks on open grounds or tall trees standing alone. Another procedure used was to stay and wait for a long time near the fruiting trees and location where food was available for them. This method usually requires more time for observer to walk around in the whole gardens. The observer took all observations, though, occasionally information was provided by other sources either gardeners or visitors regarding the presence of troops elsewhere in the gardens. This study took five months from 3^{r}d of February to 24th of June 2007, and most of the fieldwork was done in the morning from 09:00 to 12:00 h and in the evening from 14:00 to 17:00 h. After 17.00 h the troops would be heading their way back towards the forest. All possible ways were carried out to investigate, identify and count the monkeys. Counting the troops was usually difficult because of shy responses from monkeys and thick foliage of forest. The sighting of two or three monkeys indicated that there is a group hiding. The troops used to enter and leave the forest by many ways as it is next to the gardens. Most of the troops used to rush for food in the rubbish near the palm garden.

Population size of M. fascicularis in botanical gardens Penang

This paper estimated the population of M. fascicularis in Botanical Gardens Penang. The population size of M. fascicularis in the gardens was projected by counting all troops and individuals examined throughout the study. The observer verified the size of all five groups. The total population size in botanical gardens was projected from troops and individual count up of identified groups from the study as well as reported numbers from the various areas in the gardens. The population from out-side the boundaries of gardens were not considered and no survey was conducted in this study. This gardens consist of all adjoining forest maintain areas, such as bushes, forest and visitors' parks surrounded by the boundaries of gardens.

Statistical analysis

The data was first uploaded to the computer through excel's spreadsheet. Percentages were calculated among different ages/sexes with the use of pivot tables in excel. Pearson correlation used to analyze the association between the number of macaque troops and the different area of the gardens. Chi-square test was used to find the correlation of individual appears on the specific areas of the gardens. The significant value for the findings must be less than 0.05 ($p < 0.05$). All analyses were made by the statistical package for the social sciences version 16.0 (SPSS) (Sri, 1997).

RESULTS

Population size and distribution

This study counted the population size in the form of groups in the Penang Botanical Gardens. Five habituated groups that always came down and have close interaction with visitors and four nonhabituated groups; they were running away from observer. Based on counts and reports of this study, it was estimated that the total population for Penang Botanical Gardens was 164 individuals in 9 groups. Present study calculated the size of population in the different area of the gardens. The mean of population derived as 18 individuals per group. The population exhibit high attentiveness within two unambiguous areas such as Main Gate and Orchid Garden. In these areas the troops enlarged considerably with increasing nearness to food and visitors. The statistical analyses illustrated the value of population in the form of Pearson correlation for juveniles, adult males, adult females and infants as:

Pearson correlation: r (juveniles) = -0.187, P = 0.047 (df = 20).
Pearson correlation: r (males) = -0.206, P = 0.045 (df = 10).
Pearson correlation: r (females) = -0.264, P = 0.152 (df = 10).
Pearson correlation: r (infants) = -0.228, P = 0.343 (df = 4).

The total number of *M. fascicularis* in botanical gardens was 164 as shown in Table 3. The long tailed macaques were classified in four categories, such as infants, juveniles, adult males and adult females. The percentage of infants was 5%, juveniles 49%, adult males 21% and adult females 25% (Table 3).

The number of observation of categories was divided in relation to each month. The total number of observations counted in February was 145; March was 290, April proved 245, May provided 204 and June 250, respectively. The total number of infants was 54, juveniles 635, adult females 252, adult males 193 and the total number of categories was 1134 (Table 1).

Group description of population

The study was conducted from February until June 2007. A total of 218 groups were observed. In these groups 19 individuals, 78 small groups and 121 large groups were thoroughly observed and studied. The members of large groups were more than the members of other groups. These groups consisted of fourteen to thirty members (Table 2).

Total numbers of *M. fascicularis* in botanical gardens The composition of the groups of *M. fascicularis* was observed in different areas of the botanical gardens, Table 3.

Number of individuals

The locations of adult males, adult females, juveniles and infants were evaluated in different areas of the gardens (Table 4).

Comparison by sex

M. fascicularis was calculated on the basis of sex (males and females). Total males and females populations of long-tailed macaques in five months of period were observed. During data collection only the adult individuals were observed because of their sexual characteristic that distinguished them from other members of the group (Table 5).

DISCUSSION

The present study found a projected total macaque population of 164 individuals in Penang Botanical Gardens. The inhabitants of *M. fascicularis* at the present come into view to be mainly plentiful in areas closeness to human surroundings and gardens, competence in the conflicting habitats. Macaques associate with human surrounding, so the observation is more clear compare to the jungle. Although, the results are reliable with recognized distributions of *M. fascicularis* and thus it is suspect that observing good organization only can account for the difference in distribution as examined. *M. fascicularis* exhibit solitary riverine behavior (Wheatley, 1980; Van Schaik et al., 1996), they are mostly sharing these behaviors in beach side near the jungle in 100 m of rivers (Crockett and Wilson, 1980; Bismark, 1991) and have a tendency to take place at fewer size in interior jungles (McConkey and Chivers, 2004). It was found that long tailed macaques have an inclination for forest fringe environment, and therefore, the atmosphere provided to them in Penang Botanical Gardens is almost similar to their natural habitat. *M. fascicularis* has higher members in the group with great proportion of male and female ratio. Moreover, they are frequently inhabitants near the human settlement where food is easily available and therefore, they are paying attention to human behaviors

Table 1. Total Observed *M. fascicularis* in botanical gardens.

Age/sex	February	March	April	May	June	Percentage	Total
Infants	8	12	12	9	13	5	54
Juveniles	77	157	151	119	131	56	635
Adult females	36	72	48	44	52	22	252
Adult males	24	49	34	32	54	17	193
Total	145	290	245	204	250	100	1134

Table 2. Total observed groups of *M. fascicularis* in botanical gardens.

No.	Month	Observed groups			Total
		Individual	Small group	Large group	
1	February	3	11	14	28
2	March	7	21	29	57
3	April	3	17	25	45
4	May	2	15	30	47
5	June	4	14	23	41
Total		19	78	121	218

Table 3. Groups sizes and compositions of long-tailed macaques in botanical gardens.

Location	Adult females	Adult males	Juveniles	Infants	Total group size
Main gate	3	2	7	1	13
Plants nursery	6	5	9	0	20
Rubbish side	2	2	4	0	8
Picnic garden	7	5	13	2	27
Orchid garden	8	7	15	3	33
Herbal Garden	2	3	8	0	13
Sun Rockery	1	1	3	0	5
Dam (P.B.A)	5	4	9	1	19
Japanese garden	7	5	12	2	26
Total	41	34	80	9	164
Means	4.5	3.7	8.8	1	18
Percent composition (%)	25	21	49	5	100

(Sha et al., 2009; Fuentes et al., 2008).

Although with this effort, food is giving by the visitors still happens usually in the gardens. However, the growth of population of this specie is not appearing to be the main issue in the raise of human-macaque clash in Penang Botanical Gardens. According to this study it is originated that there is no obvious confirmation of a large inhabitants amplify from earlier population estimated. The information is not provided yet according to the cause of not increasing the population of long tailed macaques in Penang Botanical Gardens, but it is probable that population is not increased by inappropriate conservation and unavailability of natural environment, also by the mortality of natural causes. The size of long tailed macaques found of 18 individual per area of the gardens and its margin is measured very less in contrast to density estimated in other published data as described in previous study (Southwick and Cadigan, 1972), that the number of individuals in a group of *M. fascicularis* in Penang, Kuala Lumpur, Cape Rachado and Singapore varied from 7 to 44 individual and the average group of *M. faticularis* consisted of 24 individuals.

Many people described that they are having fun and enjoying with the long tailed macaques in the surrounding environment and consider essential to preserve these macaques (Sha et al., 2009). Addition to these, another feature that amplified consideration require to be paid to *M. fascicularis* as they are reducing in some areas of their range due to the animal trade and environmental alteration (Eudey, 2008). Moreover, it is essential to

Table 4. Number of adult males, adult females, juveniles and infants in various locations of botanical gardens.

Sex	Main gate	Orchid garden	Japanese garden	Rubbish side	Plant nursery	Dam (P.B.A)	Sun rockery	Herbal garden	Picnic garden	Total	P-value*
Male (%)	61 (31.61)	14 (7.25)	7 (3.63)	23 (11.92)	33 (17.1)	16 (8.29)	23 (11.9)	9 (4.7)	7 (3.6)	193	0.045
Female (%)	87 (34.5)	16 (6.35)	17 (6.75)	30 (11.9)	33 (13.09)	12 (4.8)	24 (9.52)	21 (8.3)	12 (4.8)	252	0.152
Juvenile (%)	204 (32.12)	35 (5.51)	33 (5.2)	82 (12.9)	72 (11.33)	45 (7.1)	79 (12.44)	52 (8.2)	33 (5.2)	635	0.047
Infant (%)	21 (38.88)	3 (5.55)	4 (7.4)	5 (9.26)	6 (11.11)	4 (7.4)	7 (13.0)	2 (3.7)	2 (3.7)	54	0.343

*Chi-square analysis.

Table 5. Sex proportion of M. fascicularis in botanical gardens.

No.	Month	Sex compositions				Total	Percent	Ratio
		Females	Percent	Males	Percent			
1	February	36	08.09	24	05.39	60	13.48	3:2
2	March	72	16.18	49	11.01	121	27.19	1.47:1
3	April	48	10.79	34	07.64	82	18.42	1.41:1
4	May	44	09.88	32	07.19	76	17.08	1.37:1
5	June	52	11.68	54	12.13	106	23.82	1:1.03
	Total	252	56.62	193	43.37	445	100	1.3:1

provide them housing in recreational areas, for education purpose. Careful consideration also wants to be paid to the expansion of behavioral management and preservation contact programs to control human performance in periphery areas since it is recognized that human settlement areas is sources of food for the macaques and these macaques are more attractiveness to human and human made environments. Improved development with amplified systematic investigate on the activities and environmental science of Penang Botanical Gardens, M. fasciclaris will give better and faster to a perfect supervision model for an uphold macaque inhabitants to the landscape of human settlement.

Population size in Penang Botanical Gardens assessed in this study. The observations prove that like the other primates the long-tailed macaques also prefer living in large groups to receive protection and show riverine refuging behaviour since they have learned to associate with humans or food. These animals preferred to live in area where food is available in abundance. In this study, the observer found a projected total macaque population of 164 individuals inside the Penang Botanical Gardens. Furthermore, it was shown that the ratio of the females was 0.3 times higher than the males, which agrees with the findings for most of the primates. Present study explains that the population of long tailed macaques is decreasing in Penang Botanical Gardens. In dispose to get better macaque-human contact and at the similar keeping up a maintain macaque population in Botanical Gardens Penang, it is an essential need for the department of wildlife to enhance their safety, food availability and predator's threats. It is recommended that any future planning or improvement within Penang Botanical Gardens should take into consideration the existing wildlife population in the area.

ACKNOWLEDGEMENTS

The authors acknowledge the immense contribution of Associate Professor Dr. Shahrul Anuar their supervisor at School of Biological Sciences, University Sains Malaysia, Management of Wild Life Penang, Mr. Lim Boon Tiong Superintendent of the Botanical Gardens Penang for their facilitation and cooperation and special thanks to Mr.

Ganish School of Biological Sciences for providing guidance and assistance in 'Gardens' during data collection, and at the end we would like to thank the School of Biological Sciences, University Sains Malaysia.

REFERENCES

Adams MR, Adams JR, Kaplan TB, Clarkson DR, Koritnik (1985). Ovariectomy, social status and atherosclerosis in cynomolgus monkeys. Arteriosclerosis, 5: 192-200.

Achmad Y, David JC, Jito S, Deborah JM, Jeremy TH (2009). The population distribution of pig-tailed macaque (*Macaca nemestrina*) and long tailed-tailed macaque (*Macaca fascicularis*) in West Central Sumatra, Indonesia. Asian Prim. J., 1(2): 2-11.

Angst W (1975). Basic data and concepts on the social organization of *Macaca fascicularis*. Primate Behav. Dev. Field Lab. Res., 4: 325-388.

Bismark M (1991). Analisis populasi monyet ekor panjang (Macaca fascicularis) pada beberapa tipe habitat hutan. Bull. Penelitian Hutan, 532: 1-9.

Corlett RT (1992). The ecological transformation of Singapore (1819-1990). J. Biogeogr., 19: 411-420.

Clarke AS, Mason WA, Moberg GP (1988). Differential behavioral and adrenocortical responses to stress among three macaque species. Am. J. Primatol., 14: 37-52.

Clarke AS, Mason WA, Mendoza SP (1994). Heart rate patterns under stress in three species of macaques. Am. J. Primatol., 33: 133-148.

Cooper G (2000). The Behaviour and Ecology of the Buton Macaque: LIPI project report. Operation Wallacea website (www.opwall.com/2000_research_section14.htm).

Crockett CM, Wilson WL (1980). The Ecological Separation of *Macaca nemenstrina* and *Macaca fascicularis* in Sumatra. In: The Macaques: Studies in Ecology, Behaviour and Evolution, Lindburg, D.G. (Ed.). Van Nostrand Reinhold, New York, pp. 182-214.

Eudey AA (2008). The crab-eating macaque (*Macaca fascicularis*): Widespread and rapidly declining. Primate Conserv., 23: 129-132.

Froehlich J, Schillaci M, Jones-Engel L, Froehlich D, Pullen B (2003). A Sulawesi beachhead by longtail monkeys (*Macaca fascicularis*) on Kabaena Island, Indonesia. Anthropologie, 41: 17-24.

Fooden J, Albrecht GH (1993). Latitudinal and insular variation of skull size in crab-eating macaques (Primates, Cercopithecidae: Macaca fascicularis). Am. J. Phys. Anthropol., 92: 521-538.

Fooden J (1995). Systematic review of Southeast Asian longtail macaques, *Macaca fascicularis* (Raffles 1821). Fieldiana Zool., 81: 1-206.

Fuentes A, Southern M, Suaryana KG (2005). Monkey Forests and Human Landscapes: Is Extensive Sympatry Sustainable for Homo Sapiens and *Macaca fascicularis* on Bali? Am. Soc. Primatol.Pub., San Diego, pp. 168-195.

Fuentes A, Kalchik S, Gettler L, Kwiatt A, Konecki M, Jones-Engel L (2008). Characterizing human–macaque interactions in Singapore. Am. J. Primatol., 70: 879-883.

Heywood VH (1991). Developing a Strategy for Germplasm Conservation in Botanic Gardens. In Tropical Botanic Gardens: Their Role in Conservation and Development. Academic Press Limited, London, pp. 11-23.

Kemp NJ, Burnett JB (2003). Final report: A biodiversity risk assessment and recommendations for risk management of long-tailed Macaques (*Macaca fascicularis*) in New Guinea. Indo-Pacific Conservation Alliance, Washington, DC. http:/www.indopacific.org/papuamacaques.pdf.

Kyes RC (1993). Surve of the long-tailed macaques introduced onto Tinjil Island, Indonesia. Am. J. Primatol., 31: 77-83.

Lekagul B, McNeely JA (1977). Mammals of Thailand. 1st Edn., Kurusapha Ladprao Press, Bangkok, Thailand.

McConkey KR, Chivers (2004). Low mammal and hornbill abundance in the forests of Barito Ulu, Central Kalimantan, Indonesia. Oryx Int. J. Conserv., 38: 439-447.

Medway L (1983). The Wild Mammals of Malaya (Peninsular Malaysia) and Singapore. 2nd Edn., Oxford University Press, Oxford, p. 131.

Medway L (1970). The Monkeys of Sundaland: Ecology and Systematic of the Cercopithecids of a Humid Equatorial Environment. Academic Press, New York, pp. 513-554.

Sha JCM, Gumert MD, Lee BPYH, Fuentes A, Rajathurai S, Chan S, Jones-Engel L (2009). Status of the long-tailed macaque *Macaca fascicularis* in Singapore and implications for management. Biodivers. Conserv., 18: 2909-2926.

Sussman RW, Tattersall I (1986). Distribution, abundance and putative ecological strategy of *Macaca fascicularis* on the island of Mauritius, south western Indian Ocean. Folia Primatologica, 46: 28-43.

Southwick CH, Manry D (1987). Habitat and population changes for the Kowloon macaques. Primate Conserv., 8: 48-49.

Southwick CH, Cadigan FC (1972). Population studies of Malaysian primates. Primates, 13: 1-18.

Sri S, Serge U, Elisabeth AW, Jan HMS, Van ARAM, Hooff (1997). Food Competition Between Wild Orangutans in Large Fig Trees. Int. J. Primatol., 18(6): 909-927.

Van Schaik CP, van Amerongen A, van Noordwijk MA (1996). Riverine Refuging by Wild Sumatran Long-Tailed Macaques. In: Evolution and Ecology of Macaque Societies, Fa, JA, and D.G. Lindburg (Eds.). Cambridge Univ. Press, New York. ISBN: 0521416809 9780521416801, pp. 160-181.

Wheatley BP (1980). Feeding and Ranging of East Bornean. In: The Macaques: Studies in Ecology, Behavior and Evolution, Lindburg, D. (Ed.). Van Nostrand Reinhold Co., New York, pp. 215-246.

Wheatley BP (1999). The Sacred Monkeys of Bali. Waveland Press, Long Grove, IL, US. ISBN 1577660595, p. 189.

Wheatley B, Stephenson R, Kurashina H, Marsh-Kautz K (2002). A Cultural Primatological Study of *Macaca fascicularis* on Ngeaur Island, Republic of Palau. In: Primates Face-To-Face: Conservation Implications of Human and Nonhuman Primate Interconnections, Fuentes, A. and L. Wolfe (Eds.). Cambridge University Press, Cambridge, pp. 240-253.

Wong CL, Ni IH (2000). Population dynamics of the feral macaques in the Kowloon Hills of Hong Kong. Am. J. Primatol., 50: 53-66.

Wolfheim J (1983). Primates of the World: Distribution, Abundance, Conservation. Seattle and London: ORYX. Int. J. Conserv., 18: 252-253.

4

Molecular genetic analysis of male alternative strategy and reproductive success in the polygynous mating bat *Cynopterus sphinx*

T. Karuppudurai and K. Sripathi*

Department of Animal Behaviour and Physiology, Centre for Excellence in Genomic Sciences, School of Biological Sciences, Madurai Kamaraj University, Madurai 625 02, India.

Cynopterus sphinx is known to use polygynous mating system based on availability of resources, called resource defense polygyny. It is the primary mating strategy adopted by *C. sphinx*. In addition to such harem groups, a number of single adult males roost solitarily, nearer to the harems. Identifying the reasons behind the solitary roosting behaviour of such adult males is essential to further understand further the details of mating strategy in *C. sphinx*. In this context incomplete monopolization of harem females by harem males and nonharem male's access to harem females is to be observed. The role of nonharem males as probable fathers has not been tested. In the present study, PCR based RAPD markers were used to assess the paternity of harem males and nearby nonharem males to the young born in the harems. A total of 30 arbitrary primers were used to assign the parentage of offsprings. Samples from a total of 651 individuals (41 harem males, 295 females, 267 suckling pups and 48 solitary males) from 41 harems (dry season 14 harems and wet season 27 harems) of *C. sphinx* were tested for their RAPD-PCR patterns. The molecular results suggest that the nonharem males also gain access to harem females and sire more offspring in July-August breeding season (wet) than March-April breeding season (dry). These results suggest that nonharem males are reproductively active and enjoy some reproductive success.

Key words: *Cynopterus sphinx*, nonharem male, alternative strategy, mating system, paternity assessment, RAPD markers.

INTRODUCTION

Studies on mating strategies have been one of the core aspects of behavioural ecology (Alcock, 2001). Understanding the evolutionary causes and consequences of social organization in a species requires an in depth knowledge of the mating system. In mammals, reproductive behaviour of females can be determined by observation of parturition and maternal care, which is a good indicator of motherhood. Therefore, a female's reproductive success is often determined by behavioural observations. In contrast, a male's reproductive success is much more difficult to determine. It has been reported that even detailed observations on male mating success may resulted in inaccurate estimates of reproductive success (Pemberton et al., 1992) and complex social

systems with many competing males challenge the quantification of mating behaviour through observations. This is true for small-bodied, nocturnal and highly mobile animals like bats, wherein observations of behavior are rather difficult. Bats are of particular interest in sociobiology because of their peculiar life history. They form the second largest mammalian order, representing about a quarter of all mammals (Nowak, 1994). Interestingly, most species are social despite enormous ecological differences among them (Bradbury, 1977; Kunz, 1982).

In bats, most known mating associations are composed of a single male and several females (Kleiman, 1977; McCracken and Wilkinson, 2000). Such groups are usually called harems, although female composition is often unstable or only temporarily stable (Storz et al., 2000b; Dechmann et al., 2005). This has been observed in other polygynous mammals such as ungulates

*Corresponding author. E-mail: tkdurai1@rediffmail.com.

(Ruckstuhl and Neuhaus, 2000). In some cases monopolization of paternity by the dominant males are incomplete due to alternative strategies performed by satellite males to gain copulation. These alternative strategies include coalitions, forced copulations, or sperm competition (Clutton-Brock et al., 1979). Such parameters are frequently difficult to identify in wild populations by direct observation alone as mating system may be difficult to observe because of the nocturnal activity and the high mobility exhibited in this taxon.

As a result, the use of genetic techniques to accurately determine kinship in wild populations is increasingly common. One taxonomic group, which particularly benefits from the use of such techniques is the bats (Rossiter et al., 2000). Wide array of molecular markers are available to study the genetic variation within and among populations, to establish the phylogenetic relationship, identification of a taxon, genetic mapping and paternity assessment (Avise, 1994). The most common DNA fingerprinting strategy currently used for genetic analyses of natural populations is PCR based Random Amplified Polymorphic DNA (RAPD) analysis (Williams et al., 1990). RAPD is a DNA polymorphism assay based on the amplification of random DNA segments with single primers of arbitrary nucleotide sequence (Williams et al., 1990; Welsh and McClelland, 1990). It is widely used in the conservation, population and evolutionary biology because of its swiftness of results, cost-effectiveness and reproducibility (Williams et al., 1990; Hadrys et al., 1992).

Additionally, RAPDs are more cost-effective and less labour-intensive. The technique does not require a prior knowledge of DNA sequence information or the use of radioisotopes and generates DNA markers from much lesser tissue than needed for microsatellites. The results are directly visualized from the gels by screening the entire genome (Williams et al., 1993).

Recent studies on several bat species have employed molecular genetic methods to analyze the reproductive success within natural populations (Petri et al., 1997; Burland et al., 2001; Heckel et al., 1999, 2003; Ortega et al., 2003; Dechmann et al., 2005). However, knowledge about the mating systems in bats is far from understanding. The Indian short-nosed fruit bat, *Cynopterus sphinx*, belongs to the Old World fruit bats (Megachiroptera: Pteropodidae). It is a common plant-visiting bat, found throughout the Indo-Malayan region (Storz and Kunz, 1999). It weighs about 45 – 70 g. This bat roosts in the foliages either as solitarily or in 1 small group consisting of about 2 - 30 individuals (Balasingh et al., 1995; Bhat and Kunz, 1995; Storz et al., 2000a; Gopukumar et al., 2005; Karuppudurai et al., 2006, 2008). *C. sphinx* is a polygynous mating bat with polyestrous reproductive cycle having two well-defined and highly synchronous parturition periods per year (Krishna and Dominic, 1983).

C. sphinx is known to exhibit polygynous mating system (that is, prolonged association of one male with more than one female) based on resource availability and such behavior is popularly known as resource defence polygyny (Storz et al., 2000b). In *C. sphinx*, adult males are categorized into two groups, harem males and non-harem males. Harem males construct and defend tents (resource). Only those males who are in possession of a tent recruit females and gain mating access. This organization of bats is called harem. During breeding seasons these harem male bats defend critical resources to attract females, thereby facilitating a harem-polygynous mating system.

However, recent studies have shown that breeding population also consists of non-harem males and most of the time they occupy roosts that are adjacent to the harems (Storz et al., 2000b; Gopukumar et al., 2005; Karuppudurai et al., 2006, 2008). However, the role of non-harem males as probable fathers has not been examined in detail. Therefore, in the present study a PCR based RAPD strategy was used to study the paternity of harem males and nearby nonharem males to the young born in the harems.

MATERIALS AND METHODS

Study area

Fieldwork was conducted in Madurai (lat: 9° 58′ N; long: 78° 10′ E) and Palayamkottai (lat: 8° 44′ S; long: 77° 42′ E), Tamil Nadu, South India from January 2003 to December 2004 over a span of 2 years (4 breeding seasons). *C. sphinx* is known to construct tents from the leaves of several tree species found in the habitat. Within the study area, *Polyalthia longifolia* (mast tree) and *Borassus flabellifer* (palm tree) trees served as potential foliage-roosting sites for *C. sphinx* (Gopukumar et al., 1999; Balasingh et al., 1993, 1995). The breeding population of *C. sphinx* is subdivided into diurnal roosting colonies called "harems" consisting of a single male and one or more females and often one or more satellite males in adjacent trees in the study area.

Sample collection

Bats were collected from the foliage tents of *P. longifolia* (mast tree) and *B. flabellifer* (palm tree) using a hoop net with an extensible aluminium pole. The entire tree was enveloped with a 6 x 9 m nylon mist net (Avinet-Dryden, New York, USA) to prevent bats from escaping. In this study, 41 complete harem groups were captured and 48 adjacent solitary males were trapped in their diurnal roosts and pups and adults were individually marked. Bats were sampled over a period of four weeks immediately following each of four annual parturition periods: March – April 2003 and 2004 (dry season) and July – August 2003 and 2004 (wet season).

Table 1. List of primers and their sequences used in the present study.

S/No	Primer code	Primer sequence 5'-3'
1	A01	CAGGCCCTTC
2	A02	TGCCGAGCTG
3	A03	AGTCAGCCAA
4	A04	AATCGGGCTG
5	A05	AGGGGTCTTG
6	A06	GGTCCCTGAC
7	A07	GAAACGGGTG
8	A08	GTGACGTAGG
9	A09	GGGTAACGCC
10	A10	GTGATCGCAG
11	SK1	GTGTCTCAGG
12	SK2	GTGGGCTGAC
13	SK3	GTCCATGCCA
14	SK4	ACATCGCCCA
15	SK5	GTGGTCCGCA
16	SK6	TCCCGCCTCA
17	SK7	AACGCGTCGG
18	SK8	AAGGGCGAGT
19	SK9	GGAAGCCAAC
20	SK10	GGCTTGGCCT
21	OPA1	GTTTCGCTCC
22	OPA2	AGTCAGCCAC
23	OPA3	CATCCCCCTG
24	OPA4	AATCGGGCTG
25	OPA5	TGCGCCCTTC
26	OPA6	TGCTCTGCCC
27	OPA7	GAAACGGGTG
28	OPA8	GTGACGTATG
29	OPA9	GGTGACGCAG
30	OPA10	GTGATCGCAG

The breakdown of collection was as follows: 2003 dry season, 6 harems (76 adult females and 72 pups); 2004 dry season, 8 harems (79 adult females and 70 pups) and 2003 wet season, 15 harems (79 adult females and 71 pups); 2004 wet season, 12 harems (61 adult females and 54 pups). Additionally, all males that defended territories within the study area were sampled over a two year period (2003 - 2004) that spanned the dates of conception of sampled pups.

Bats were sampled when nearly all females had given birth but pups had not weaned as yet. Most pups were two to three weeks old at the time of sampling and all were matched with known mothers. Blood and or wing membrane biopsy samples of harem males, solitary males, harem females and pups of *C. sphinx* were collected during the breeding seasons (March/April and July/August). A medical punch was used for the excision of tissue (4 mm^2) and care was taken to place it in an

area between the blood vessels to avoid injury (wing membranes healed within 3 - 4 weeks). After each sampling, the punched hole and the punch were disinfected with 70% ethanol. No negative effects of this treatment on the health of the bats were observed. It should also be noted that the bats frequently have natural injuries of this type in their wing membranes. The collected blood samples were immediately mixed with anticoagulant ACD, transferred to microcentrifuge tubes and sealed with parafilm. The blood and tissue samples were stored in ice, transported to the lab and stored at - 20°C until DNA extraction (Worthington Wilmer and Barratt, 1996; Karuppudurai et al., 2007). Bats were held in net cages and released at their roosts during the evening of the same day they were captured.

Genomic DNA isolation and primer screening

Genomic DNA was isolated from wing-membrane biopsy samples using standard proteinase K digestion and phenol: chloroform extraction method (Sambrook et al., 1989). The quality and quantity of extracted DNA were checked using 0.7% agarose gel electrophoresis and spectrophotometric measurement at A260 and A280 nm (Hitachi U-2000, Tokyo, Japan). Finally the DNA pellets were stored at 4°C until further analysis. Alternately, DNA extraction was also performed with the DNeasy Tissue kit (Qiagen), following the manufacturer's instructions.

In the present study, RAPD-PCR was performed by using three series of primers (Table 1) namely A (A-01-A-10), SK (SK1-SK10) and OPA (OPA1-OPA10) each comprising ten primers (Microsynth, Switzerland). PCR conditions were optimized by varying concentrations of template DNA, primer, MgCl2 and Taq DNA polymerase.

Initial screening was done with all 30 primers using DNA from four (2 colonies from wet season and 2 colonies from dry season) colonies. PCR-RAPD analysis was repeated at least three times and the primers producing prominent reproducible bands were used for the analysis of 41 colonies. "Colony" or "Harem" is a group of individuals that roosted together regularly, which consisted of single adult male, adult females (3 - 21 individuals) and their pups.

Polymerase chain reaction (PCR)

PCR was carried out in 20 μl reaction containing 100 ng of template DNA, 2 μl of 10X PCR buffer (100 mM Tris HCl, 500 mM KCl, 0.8% Nonidet P40) with 1.5 mM MgCl2, 2 μl of 2 mM dNTP mixture, 5 μl of 2 μM primer, 1U of Taq DNA polymerase and 10 μl of H2O. All DNA amplifications were performed using an Applied Biosystems GeneAmp 2700 PCR system, with following cycling conditions including initial denaturation at 94°C for 5 min, followed by 35 cycles of denaturation at 94°C for

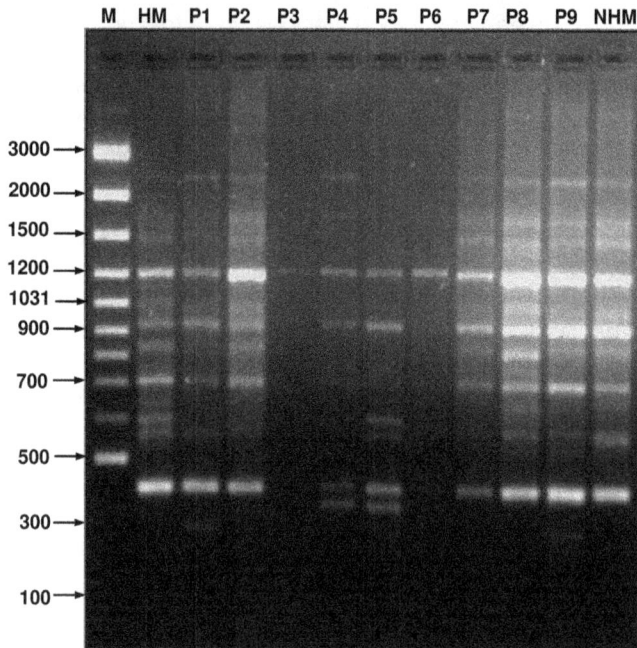

Figure 1. RAPD profile of *C. sphinx* (colony no: 1) obtained with primer A-05. Lane M, marker (100 bp DNA ladder); lane HM, harem male; lanes P1-P9, pups; lane NHM, nonharem male.

40 s, annealing at 30 - 36°C for 2 min (annealing temperature varying with the primers), extension at 72°C for 3 min and final extension at 72°C for 10 min. PCR products were electrophoresed on 2% agarose gel in 1x Tris-acetate-EDTA (TAE) buffer, stained with ethidium bromide (0.5 µg/ml) observed and photographed using gel documentation system (Biorad, USA, model 2000, Quantity One Software).

Data analysis

The RAPD data were analysed using NTSYS-pc version 2.0 (Numerical Taxonomy and Multivariate Analysis System) computer package (Rohlf, 1998). A genetic similarity (GS) between fathers and pups was computed based on Jaccard's coefficient of similarity as follows.

$$GS\ (ij) = a\ /\ (a + b + c)$$

Where:

GS (ij) is the measure of genetic similarity between individuals i and j.
a is the number of polymorphic bands that are shared by i and j.
b is the number of bands present in i and absent in j.
c is the number of bands present in j and absent in i.

Each RAPD fragment was treated as a unit character and was scored as 1 (present) or 0 (absent). The 1/0 matrix

was prepared for all fragments scored and the data were used to generate Jaccard's similarity coefficients for RAPD bands (Jaccard, 1908). The Jaccard's coefficients were used to construct a dendrogram using the unweighted pair group method with arithmetic averages (UPGMA) or Weighted Average Linkage with the following formula.

$$dki = (np/n) \times dpi + (nq/n) \times dqi$$

where:

p, q = Indices indicating two clusters that are to be joined into a single cluster.
k = Index of the cluster formed by joining clusters p and q.
i = Index of any remaining clusters other than clusters p, q, or k.
np = Number of samples in the pth cluster.
nq = Number of clusters in the qth cluster.
n = Number of clusters in the kth cluster formed by joining the pth and qth cluster (n = np + nq).
dpq = Distance between cluster p and cluster q.

RESULTS

Over the course of two-year survey, 41 complete harem groups and 48 adjacent solitary males were trapped in their diurnal roosts. A total of 651 individuals (41 harem males, 295 females, 267 suckling pups and 48 solitary males) were sampled for the genetic analysis. Mother /pup pairs were sampled when pups were still attached to the teats of their mothers so that their relatedness was unambiguous. We tried to capture all females and their pups in a harem. In some capture attempts one or more females escaped and those pups were not included in our analysis.

In this study, only the harem males, pups and nonharem males were subjected to the paternity analysis. During the dry season 14 harem males, 142 offsprings and 18 nonharem males were captured and analyzed to assign the paternity of harem and nonharem males. The pairwise Jaccard's coefficients of genetic similarity matrix were generated for all the harem and nonharem males and offsprings. Based on the pairwise Jaccard's coefficients genetic similarity matrix, 132 of 142 offsprings were sired by harem males (average 94%) and the nonharem males sired only 10 offsprings (average 6%). A higher proportion of pups were sired by harem males in the March - April (dry) breeding season. Representative RAPD patterns generated by dry season colony 1 and their RAPD gel picture, pairwise Jaccard's coefficients genetic similarity matrix and dendrogram (UPGMA) are shown in Figure 1, Table 2 and Figure 2 respectively.

For example, the colony number 1 comprised a harem male, 9 females and 9 offsprings during the capture, among the 9 offsprings the harem male sired 7 offsprings

Table 2. Similarity matrix for Jaccard's coefficients for colony no: 1.

	HM	P1	P2	P3	P4	P5	P6	P7	P8	P9	NHM
HM	100.0	49.8	72.6	59.2	73.5	46.5	35.9	67.6	49.1	50.4	46.2
P1	49.8	100.0	58.2	31.8	49.4	74.3	63.0	48.5	31.7	41.5	38.4
P2	72.6	58.2	100.0	66.8	81.0	45.8	45.2	81.5	33.6	39.8	39.3
P3	59.2	31.8	66.8	100.0	60.4	25.6	29.7	64.1	31.9	22.3	35.0
P4	73.5	49.4	81.0	60.4	100.0	52.6	40.1	75.1	40.8	38.3	41.8
P5	46.5	74.3	45.8	25.6	52.6	100.0	52.0	45.1	38.3	40.3	40.1
P6	35.9	63.0	45.2	29.7	40.1	52.0	100.0	34.4	26.4	25.9	28.7
P7	67.6	48.5	81.5	64.1	75.1	45.1	34.4	100.0	33.4	40.9	36.0
P8	49.1	31.7	33.6	31.9	40.8	38.3	26.4	33.4	100.0	44.0	69.9
P9	50.4	41.5	39.8	22.3	38.3	40.3	25.9	40.9	44.0	100.0	47.1
NHM	46.2	38.4	39.3	35.0	41.8	40.1	28.7	36.0	69.9	47.1	100.0

Table 3. Similarity matrix for Jaccard's coefficients for colony no: 15.

	HM	P1	P2	P3	P4	NHM
HM	100.0	68.2	47.3	66.6	42.6	40.5
P1	68.2	100.0	51.8	89.1	49.5	42.1
P2	47.3	51.8	100.0	48.0	36.3	30.0
P3	66.3	89.1	48.0	100.0	44.9	36.6
P4	42.6	49.5	36.3	44.9	100.0	91.5
NHM	40.5	42.1	30.0	36.6	91.5	100.0

and the adjacent nonharem male sired only 2 offsprings (Figures 1, 2 and Table 2). The same way, we have analysed all the 14 harems captured during the dry season (March - April).

In the wet season 27 harem males, 125 offsprings and 30 nonharem males were captured and analyzed to assign the reproductive success of harem and nonharem males. Of the 125 offsprings the harem males sired only 52 offsprings (average 42%) and the nonharem males sired the rest 73 offsprings (average 58%). In the July - August (wet) breeding season the nonharem males sired a higher proportion of pups compared to harem males. Representative RAPD patterns generated by wet season colony 15 and their RAPD gel picture, pairwise Jaccard's coefficients genetic similarity matrix and dendrogram (UPGMA) are shown in Figure 3, Table 3 and Figure 4 respectively. The same way, we have analysed all the 27 harems captured during the wet season (July - August). The cumulative offsprings sired by harem and nonharem males from the 27 colonies are presented in the Table 4.

Over the course of 2 years (four breeding seasons) the nonharem males sire more offspring (58%) in July - August breeding season (wet) than March - April breeding season (dry) (6%) and the harem males sire more offspring (94%) in March - April breeding season (dry) than July - August breeding season (wet) (42%) (Table 4).

DISCUSSION

In the present study, the paternity assignments based on 30 RAPD random primers, revealed an unequal distribution of reproduction between harem and nonharem males. The results indicated that the C. sphinx study population is characterized by an extremely high within-season variance in male mating success, as expected from the harem-forming mode of social structure (Storz et al., 2000a, b). The monopolization of paternity by the harem males is incomplete due to alternative strategies used by satellite males to gain access to harem females and obtain some reproductive success.

During the dry season in our study area the average harem size was slightly higher compared to wet season because the dispersion of female C. sphinx is highly clumped due to limited roosting sites and the harem male sires 94% offsprings conceived during this period. However, during the wet season more roost sites are available and the harem size decreased because the females are widely dispersed as a result the harem males sire only 42% of offsprings, while nonharem males sire the rest of 58%.

Similar results have been reported in this species by Storz et al. (2000b, 2001). Apart from the mating success of nonharem males, low paternity for harem males can also occur as a result of female choice. Heckel et al. (1999)

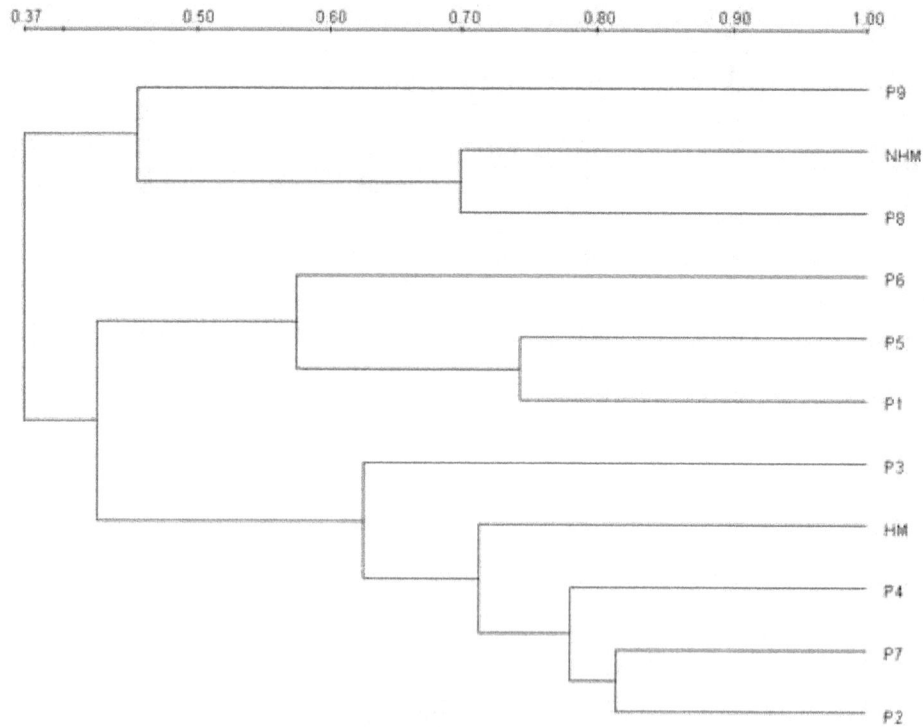

Figure 2. Dendrogram of genetic relationships among harem male, nonharem male and pups identified by RAPD analysis using UPGMA for colony no: 1.

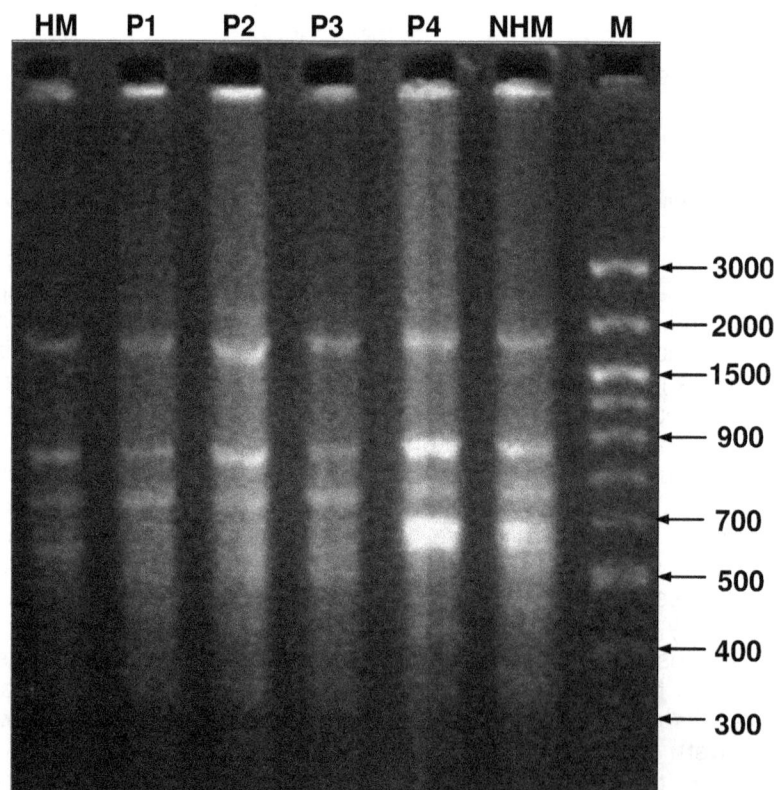

Figure 3. RAPD profile of *C. sphinx* (colony no: 15) obtained with primer SK7. Lane HM, harem male; lanes P1-P4, pups; lane NHM, nonharem male; lane M, marker (100 bp DNA ladder).

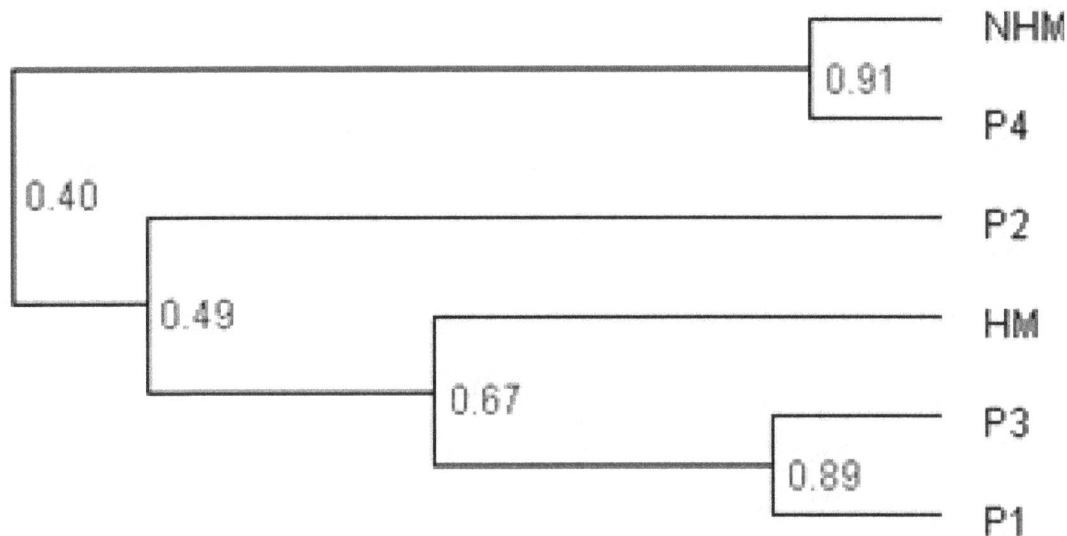

Figure 4. Dendrogram of genetic relationships among harem male, nonharem male and pups identified by RAPD analysis using UPGMA for colony no: 15.

reported the importance of female choice especially in highly mobile animals with harem system. It appears that female *Saccopteryx bilineata* actively select their roosting location and are highly mobile; some females shift roosting territories during the course of a day and some disperse to other colonies. Our recent radio-telemetry studies lend support to the observation of Heckel et al. (1999). We observed three postpartum estrus females (*C. sphinx*) visit a nonharem male exclusively during the night hours and engage in mating.

Balasingh et al. (1995) reported fluctuations in the harem size on a day-to-day basis, indicating that females periodically shifted their tents. Similarly, among the polygynous bats Artibeus jamaicensis (Ortega and Arita, 1999; Ortega et al., 2003), Phyllostomus hastatus (McCracken and Bradbury, 1977), Desmodus rotundus (Wilkinson, 1985) and *S. bilineata* (Heckel et al., 1999; Heckel and von Helversen, 2002), incomplete monopolization of females by harem males has been observed. The incomplete control of harem males over harem females increases the chances for nonharem males to fertilize some of the females.

The mating system most commonly described in bat species has been polygyny. However, a recent comprehensive review of bat mating systems has recognized the wide diversity of mating behaviour and has emphasized the importance of alternative strategy by males and multiple mating by females (McCracken and Wilkinson, 2000). This is supported by several genetic analyses which have shown that paternity is biased in polygynous mating systems. In harem groups of *S. bilineata*, it was demonstrated that 71% of offsprings born into a harem are not sired by the resident harem male, but are instead fathered by a number of different males, either adjacent harem males or peripheral males. However, harem males

do gain greater overall reproductive success than peripheral males, as they achieve fertilization success both in their own and in other harems (Heckel et al., 1999; Heckel and von Helversen, 2002, 2003).

The dominant males in other polygynous species studied achieved slightly higher reproductive success than S. bilineata, although complete monopolization of females is rare. In the spear-nosed bat P. hastatus, the harem male fathers 60 - 90% of offsprings (McCracken and Bradbury, 1977, 1981), while the harem male in D. rotundus fathers approximately 45% of young (Wilkinson, 1985). In the later species, many different males, including those from other colonies contribute to the 55% offsprings. Similarly, the estimated paternity for dominant males of A. jamaicensis ranged from 33 - 90% followed by satellite (22%) and subordinate males (9%). Overall, most adult males belonging to a harem remained as dominant in the same group at least for two reproductive seasons (Ortega and Arita, 1999, 2000; Ortega et al., 2003). Comparable parentage studies on polygynous temperate-zone species have till date been restricted to some species such as the greater horseshoe bat *Rhinolophus ferrumequinum*. A study of one maternity colony demonstrated that no male fathered more than 12.5% of young in any cohort, despite the highly polygynous mating behaviour reported at mating sites in this species.

However, this finding may be at least partially explained by females from the same maternity colony visiting different mating sites (Rossiter et al., 2000). In those species for which less polygynous mating behaviour is predicted, genetic studies have generally confirmed low levels of male reproductive skew. The high number of different paternal alleles in the cohorts of young mouse-eared bats Myotis myotis and low proportion of paternal

Table 4. Summary of molecular data and number and percentage of pups sired by harem and nonharem males over a span of two years (four breeding seasons) between January 2003 and December 2004.

Breeding season	Total no. harem males	Total no. of nonharem males	Total no. females	Total no. pups	Total no. of pups sired by		Total % of pups sired by	
					Harem males	Nonharem males	Harem males	Nonharem males
Dry (March - April)	14	18	155	142	132	10	94	6
Wet (July - August)	27	30	140	125	52	73	42	58
Total	41	48	295	267	184	83	68	32

half siblings identified among brown long-eared bats *Plecotus auritus* suggest that many males contribute to the gene pool in these species. Interestingly, the genetic data in both species also indicate that males do not generally achieve reproductive success within their own colony, suggesting that males and females from different maternity colonies mix during the mating season (Petri et al., 1997; Burland et al., 2001).

Taken together, the present results suggest that the nonharem males gain access to females and sire more offspring in July - August breeding season (wet) than March - April breeding season (dry). These results suggest that nonharem males are reproductively active, gain access to harem females and enjoy some reproductive success. Further investigations are needed to understand reproduction of nonharem males. The relatively high reproductive success of some nonharem males may indicate that solitary behaviour can be an acceptable alternative to territoriality. These solitary males had no costs for roost defense but sired number of juveniles. Reproduction by nonharem males is possible because harem males provide no paternal care. The social dominance of harem males exhibited by the persistent maintenance of a tent might indicate male quality and could therefore explain why most females reproduced with harem males. Investigations that follow the behaviour and reproductive success of individual males over their lifetime could clarify whether some nonharem males could potentially compensate lower reproductive success per year with longer persistence in the harem.

ACKNOWLEDGEMENTS

This work was supported by grants from University Grants Commission (UGC), New Delhi and Centre for Excellence in Genomic Sciences, School of Biological Sciences, Madurai Kamaraj University, Madurai to K. S. We thank Profs. G. Shanmugam and G. S. Selvam, Cancer Biology Division, Department of Biochemistry for providing laboratory facilities and also we thank Prof. G. Marimuthu, Department of Animal Behaviour and Physiology and Dr. N. Gopukumar, University Grants Commission, South-Western Regional Office, Bangalore, India for their valuable suggestions. We are grateful to the assistance rendered by R. Dhanabalan, C. Sekar, Dr. T. Karuppanapandian and Dr. P. T. Nathan. This work is cleared by the Institutional Ethical and Bio-safety Committee of Madurai Kamaraj University, Madurai, India.

REFERENCES

Alcock J (2001). *Animal Behaviour*, Sinauer Associates, Sunderland, MA, USA. p. 560.

Avise JC (1994). Molecular markers, natural history and evolution. Chapman and Hall Ltd, New York p. 511.

Balasingh J, Isaac SS, Subbaraj R (1993). Tent-roosting by the frugivorous bat, *Cynopterus sphinx* (Vahl 1797) in southern India. Curr. Sci. 65: 418.

Balasingh J, Koilraj A, Kunz TH (1995). Tent construction by the short-nosed fruit bat *Cynopterus sphinx* (Chiroptera: Pteropodidae) in southern India. Ethology 100: 210-229.

Bhat HR, Kunz TH (1995). Altered flower/fruit clusters of the kitul palm used as roosts by the short-nosed fruit bat, *Cynopterus sphinx* (Chiroptera: Pteropodidae). J. Zool. Lond. 235: 597-604.

Bradbury JW (1977). Social organization and communication. In: Wimsatt WA (ed) Biology of Bats III: Academic Press, New York pp. 1-73.

Burland TM, Worthington Wilmer J (2001). Seeing in the dark: molecular approaches to the study of bat populations. Biol. Rev. 76: 389-409.

Burland TM, Barratt EM, Nichols RA, Racey PA (2001). Mating patterns, relatedness and the basis of natal philopatry in the brown long-eared bat, *Plecotus auritus*. Mol. Ecol. 10: 1309-1321.

Clutton-Brock TH, Rose KE, Guinness FE (1997). Density-related changes in sexual selection in red deer. Proc. R. Soc. Lond. B264: 1509-1516.

Dechmann DKN, Kalko EKV, Konig B, Kerth G (2005). Mating system of a Neotropical roost-making bat: the white-throated, round-eared bat, *Lophostoma silvicolum* (Chiroptera: Phyllostomidae). Behav. Ecol. Sociobiol. 58: 316-325.

Gopukumar N, Elangovan V, Sripathi K, Marimuthu G, Subbaraj R (1999). Foraging behaviour of the Indian nosed fruit bat *Cynopterus sphinx*. Z. Saugetierkunde 64: 187-191.

Gopukumar N, Karuppudurai T, Nathan PT, Sripathi K, Arivarignan G, Balasingh J (2005). Solitary adult males in a polygynous mating bat (*Cynopterus sphinx*): a forced option or a strategy? J. Mammal. 86: 281-286.

Hadrys M, Balick M, Schierwater B (1992). Applications of random amplified polymorphic DNA (RAPD) in molecular ecology. Mol. Ecol. 1: 55-63.

Heckel G, Voigt GG, Mayer F, von Helversen O (1999). Extra harem paternity in the white-lined bat *Saccopteryx bilineata* (Emballonuridae). Behaviour 136: 1173-1185.

Heckel G, von Helversen O (2002). Male tactics and reproductive success in the harem polygynous bat *Saccopteryx bilineata*. Behav. Ecol. 13: 750-756.

Heckel G, von Helversen O (2003). Genetic mating system and the significance of harem associations in the bat *Saccopteryx bilineata*. Mol. Ecol. 12: 219-227.

Jaccard P (1908). Nouvelles recherches sur la distribution florale. Bull. Soc. Vaud. Sci. Nat. 44: 223-270.

Karuppudurai T, Gopukumar N, Sripathi K (2006). Solitary or non-territorial adult males in bats are "Making the best of a bad job"? Bat Net 7: 30-33.

Karuppudurai T, Sripathi K, Gopukumar N, Elangovan V, Marimuthu G (2007). Genetic diversity within and among populations of the Indian short-nosed fruit bat *Cynopterus sphinx* assessed through RAPD analysis. Curr. Sci. 93: 942-950.

Karuppudurai T, Sripathi K, Gopukumar N, Elangovan V, Arivarignan G (2008). Transition of nonharem male to harem male status in the short-nosed fruit bat *Cynopterus sphinx*. Mamm. Biol. 73: 138-146.

Kleiman DG (1977) Monogamy in mammals. Q. Rev. Biol. 52: 39-69.

Krishna A, Dominic CJ (1983). Reproduction in the female short-nosed fruit bat *Cynopterus sphinx* (Vahl). Period. Biol. 85: 23-30.

Kunz TH (1982) Roosting ecology of bats. In: Kunz TH (ed) Ecology of Bats: Plenum Press, New York pp. 1-50.

McCracken GF, Bradbury JW (1977). Paternity and heterogeneity in the polygynous bat, Phyllostomus hastatus. Sci. 198: 303-306.

McCracken GF, Bradbury JW (1981). Social organization and kinship in the polygynous bat Phyllostomus hastatus. Behav. Ecol. Sociobiol. 8: 11-34.

McCracken GF, Wilkinson GS (2000). Bat mating systems. In: Krutzsch PH, Creighton EG (eds) Reproductive Biology of Bats: Academic Press, New York pp. 321-362.

Nowak RM (1994). Walker's bats of the world, Johns Hopkins University Press, Baltimore, Maryland p. 287.

Ortega J, Arita HT (1999). Structure and social dynamics of harem groups in Artibeus jamaicensis (Chiroptera: Phyllostomidae). J. Mammal. 80: 1173-1185.

Ortega J, Arita HT (2000). Defence of female by dominant males of Artibeus jamaicensis (Chiroptera: Phyllostomidae). Ethology 106: 395-407.

Ortega J, Maldonado JE, Wilkinson GS, Arita HT, Fleischer RC (2003). Male dominance, paternity and relatedness in the Jamaican fruit-eating bat (Artibeus jamaicensis). Mol. Ecol. 12: 2409-2415.

Pemberton JM, Albon SD, Guiness FE, Clutton-Brock TH, Dover GA (1992). Behavioral estimates of male mating success tested by DNA fingerprinting in a polygynous mammal. Behav. Ecol. 3: 66-75.

Petri B, Paabo S, Von Haeseler A, Tautz D (1997). Paternity assessment and population subdivision in a natural population of the larger mouse-eared bat Myotis myotis. Mol. Ecol. 6: 235-242.

Rohlf FJ (1998). NTSYS/PC Numerical Taxonomy and Multivariate Analysis System, version 2.0, Exeter Publications, Setauket, New York.

Rossiter SJ, Jones G, Ransome RD, Barratt EM (2000). Parentage, reproductive success and breeding behaviour in the greater horseshoe bat (*Rhinolophus ferrumequinum*). Proc. R. Soc. Lond. B267: 545-551.

Ruckstuhl KE, Neuhaus P (2000). Causes of sexual segregation in ungulates: a new approach. Behaviour 137: 361-377.

Sambrook J, Fritsch EF, Maniatis T (1989). Molecular Cloning: a laboratory manual, 2nd edition, Cold Spring Harbor Laboratory Press, Cold Spring Harbor, New York, USA p. 1659.

Storz JF, Balasingh J, Nathan PT, Emmanuel K, Kunz TH (2000a). Dispersion and site-fidelity in a tent-roosting population of the short-nosed fruit bat (*Cynopterus sphinx*) in southern India. J. Trop. Ecol. 16: 117-131.

Storz JF, Bhat HR, Kunz TH (2000b). Social structure of a polygynous tent-making bat *Cynopterus sphinx* (Megachiroptera). J. Zool. Lond. 251: 151-165.

Storz JF, Bhat HR, Kunz TH (2001). Genetic consequences of polygyny and social structure in an Indian fruit bat, *Cynopterus sphinx*. II. Variance in male mating success and effective population size. Evolution 55: 1224-1232.

Welsh J, McClelland M (1990). Fingerprinting genomes using PCR with arbitrary primers. Nucleic Acids Res. 18: 7213-7218.

Wilkinson GS (1985). The social organization of the common vampire bat. II. Mating system, genetic structure and relatedness. Behav. Ecol. Sociobiol. 17: 123-134.

Williams JGK, Kubelik AR, Livak KJ, Rafalski JA, Tingey SV (1990). DNA polymorphism amplified by arbitrary primers as useful genetic markers. Nucleic Acids Res. 18: 6531-6535.

Williams JGK, Hanafey MK, Rafalski JA, Tingey SV (1993). Genetic analysis using random amplified polymorphic DNA markers. Methods Enzymol. 218: 704-740.

Worthington Wilmer J, Barratt E (1996). A non-lethal method of tissue sampling for genetic studies of chiropterans. Bat Res. News 37: 1-3.

5

Evaluation of the potential of human mesenchymal stem cell engrafted after in utero transplantation in murine model

Groza I.[1], Daria Pop[2], Cenariu M.[1], Pall Emoke[1]

[1]Faculty of Veterinary Medicine, University of Agricultural Sciences and Veterinary Medicine Cluj-Napoca Romania.
[2]University of Medicine and Pharmacy Cluj-Napoca Romania.

Mesenchymal stem cells derived from human placenta (hPMSc) is an important alternative source of adult stem cells. These cells with multilinear capacity represent a biological material important for regenerative medicine. In this study, we established a mouse model for *in utero* transplantation of hPMCs to investigate if these cells would affect long-term, organ-specific engraftment. Murine model results can be extrapolated in human medicine in the treatment of various diseases.

Key words: Placenta, mesenchymal stem cells, engraftment, *in utero* transplantation.

INTRODUCTION

The human placenta is a fetomaternal organ, formed by both fetal and maternal tissue (Linju et al., 2005). Its successful formation is a critical process in embryogenesis, and the normal development and function of the placenta is crucial to the wellbeing of the fetus. This organ is discarded postpartum, after having performed its necessary function of supporting the embryo and fetus (Geordias et al., 2002; Linju et al., 2005). The intrauterine transplantation of stem cells provides in some instances a therapeutic option before definitive organ failure occurs (Shapiro et al., 2000). The early fetus is uniquely tolerant to foreign antigens, accepting allogeneic or xenogeneic cells without the need to match major histocompatibility complex (MHC) antigens or induce immunosuppression (Muench, 2005). Multiple clinical experiences show that certain diseases such as immune deficiencies and inborn errors of metabolism can be successfully treated using adult stem cells.

The major problem is the low level of engraftment. Some experiments in mice show similar early homing of allogeneic and xenogeneic stem cells and reasonable early engraftment of allogeneic murine fetal liver cells (17.1% donor cells in peripheral blood 4 weeks after transplantation) (Troegera et al., 2006). Multiple researches on animal models are designed to optimize engraftment and recipient microenvironment in order to increase levels of grafting. It is known that some diseases such as hemoglobinopathies (Fanconi´s anemia and thalassaemia), immunological defects (SCID), and certain inborn errors of metabolism can be treated by stem cell transplantation (Troegera et al., 2006). In this study, we established a mouse model for *in utero* transplantation of hPMCs stem cells to investigate if these cells would affect long-term, organ-specific engraftment.

*Corresponding author. E-mail: pallemoke@gmail.com.

Abbreviations: hPMSc, Human placental mesenchymal stem cells; **MHC,** major histocompatibility complex; **PBS,** phosphate-buffered saline; **EDTA,** ethylenediaminetetraacetic acid; **FITC,** fluorescein isothiocyanate; **FACS,** fluorescence-activated cell sorting; **MSCs,** mesenchymal stem cells; **SD,** standard deviation; **E13.5,** 13.5 days embryos; **E 20,** 20 days embryos.

MATERIALS AND METHODS

Harvest and preparation of placenta-derived cells

Biological material, clinically normal human term placentas (37 to 40 weeks of gestation, n =3) were collected after Cesarean section. Term placentas from healthy donor mothers were obtained with informed consent approved according to the procedures of the institutional review board. The harvested pieces of tissue were washed several times in phosphate-buffered saline (PBS) (Sigma) and then mechanically minced and enzymatically digested with 0.25% trypsin- ethylenediaminetetraacetic acid (EDTA) (Gibco) for 30 min at 37 °C. After centrifugation the cell suspension was filtered to eliminate undigested fragments. For erythrocytes lysis, cells suspensions were treated with fluorescence-activated cell sorting (FACS) Lysing Solution 10x (BD Biosciences) for 15 min. The suspension pelleted by centrifugation (1500 rpm/7 min) and suspended in propagation medium, which consist of Dulbecco's Modified Eagle's medium (DMEM) (Gibco) supplemented by 10% fetal calf serum (FCS), 100 U/ml penicillin-streptomycin (Gibco). Cultures were maintained in DMEM with 10% fetal bovine serum (FBS; Hyclone, USA) at 37 °C with 5% CO_2. Approximately 1 week later, some colonies consisting of fibroblast-like cells were observed. These cells were trypsinized and replated for expansion. In order to obtain single cell-derived hPMSc clones, cells were serially diluted in 96-well culture plates (BD Biosciences) at a final density of 60 cells/ plate. Colonies that grew with homogeneous bipolar morphology were expanded.

Flow cytometry

The cell surface phenotype of hPMCs was characterized after the second passage. The cells were trypsinised (0.25% trypsine EDTA), washed twice with PBS and stained according to the recommendation of the manufacturer with the monoclonal antibodies, FITC-CD44, examined with a FACS CantoII Apparatus (Becton–Dickinson).

In utero transplantation

For in utero transplantation of hPMCs, were prepared single cell suspensions. On day 13.5 after mating, pregnant mice were anesthetized with 4 mg/kg Kethamine and 40 mg/kg Xylazine cocktail administered by intraperitoneal injection. Under aseptic conditions, the uterine horns were exposed and donor cells were injected through a glass micropipette (inserted through the uterine wall and into the peritoneal cavity of each fetus under direct visualization. The injection consisted of 1×10^6 hPMCs in 5 μl of PBS. The abdominal incision was closed in two layers using 4-0 silk, and the mice were allowed to complete pregnancy to term.

Engraftment analysis

On E20, a low abdominal midline incision was made and the number of live fetuses in each uterine horn was recorded. Then, placenta, fetal blood and fetal organs including brain, heart, lung, liver, spleen and bone marrow were collected. To obtain single cell suspension as chopped tissues were processed by the Medimachine device. The percentage of cells of donor origin was evaluated by FACS using a flow cytometer (FACS Canto II). Red blood cells were lysed with whole-blood lysing solution (BD Biosciences). Cells were washed twice with cold PBS (Sigma) containing sodium azide (0.1%) and 0.5% bovine serum albumin (BSA) and incubated in the dark at room temperature for 30 min with 20 μl fluorescent antibody (anti - human CD45 PE-Cy5 antibody (PE-Cy5: phycoerythrin-Cy5), (FITC: fluorescein isothiocyanate), anti – human CD34-FITC antibody (FITC: fluorescein isothiocyanate) and anti- human CD44 antibody). Have prepared two samples for each antibody in the study: a sample and a sample labeled with antibody as blank unmarked. For positive control were used MSCs isolated from placenta (CD44 $^+$). Cells were then washed two more times with PBS/azide and analyzed. In order to perform serial sections of murine fetuses were embedded in paraffin. Sections (5 mm) were air-dried and fixed in ice-cold acetone or 4% paraformaldehyde for 10 to 20 min. To highlight

hPMCs were used antibody against anti-human CD44.

Statistical analysis

The data are described as mean±SD. Differences were assessed by using the independent-samples t-test, paired-samples t-test. A p-value of < 0.05 was considered significant.

RESULTS AND DISCUSSION

To isolate and determine the multipotent potential of hPMC in term placentas, we extracted 4.5 ± 10^7 nucleated cells from 3 (n=3) placentas delivered at a mean gestational age of 38.66 ± 1.52 weeks. After 7 to10 days, adherent cells with fibroblastic morphology were detected. The hPMCs were cultured for more than 7 passages without any spontaneous differentiation. In utero stem cell transplantation was performed in 10 (n=10) female carrying a total of 65 fetuses. To show that hPMCs injected in utero on E13.5 engrafted in fetal organs, we collected fetal organ samples at E20. Most fetal tissues had demonstrable hPMCs engraftment at E20. Although the distribution pattern and numbers of cells in individual fetuses varied, hPMCs were detectable in more than 60% of the fetus. Engraftment analysis was done using FACS Diva software and results are presented as histograms. We assessed the presence of hPMCs in various fetal mouse tissues (Figure 1). Grafting percentages shown ranged between 1.2 and 7.2% at a mean 4.04%±2.26 (93.37±4, 67 positive control), a low but consistent with published data in literature. Postmortem analysis of the organs from E 20 fetal mice confirmed that hPMCs engrafted in more than 60% of fetal organs after in utero transplantation (Figure 2).

Placenta derived mesenchymal stem cells are generally negative for CD34, CD45 and HLA-DR expression and positive for CD29, CD44, CD73, CD90, CD105 and CD166 (Barry et al., 1999, 2001; Parolini et al., 2008). HPMCs can proliferate in vitro, maintaining a homogenous morphology, consistent phenotype, and the capacity to differentiate into bone, cartilage, adipose tissue, hepatocytes or insulin secretion cells (Fukuchi et al., 2004; Yen et al., 2005; Chien et al., 2006; Zhang et al., 2006). Besides these capabilities they have a direct immunosuppressive effect on the proliferation of CD4$^+$ and CD8$^+$lymphocytes from human peripheral blood and umbilical cord blood in vitro, and are expected to have a potential application in allograft transplantation (Li et al., 2007). Cells of different origins have been used for in utero transplantation in a number of models. Tran's species animal models have been widely used in the study of stem cell migration and engraftment (Liechty et al., 2000; Saito et al., 2002). Human bone marrow-derived mesenchymal stem cells have been transplanted into fetal sheep and shown to persist for as long as 13 months with multilineage differentiation potential (Liechty et al., 2000).

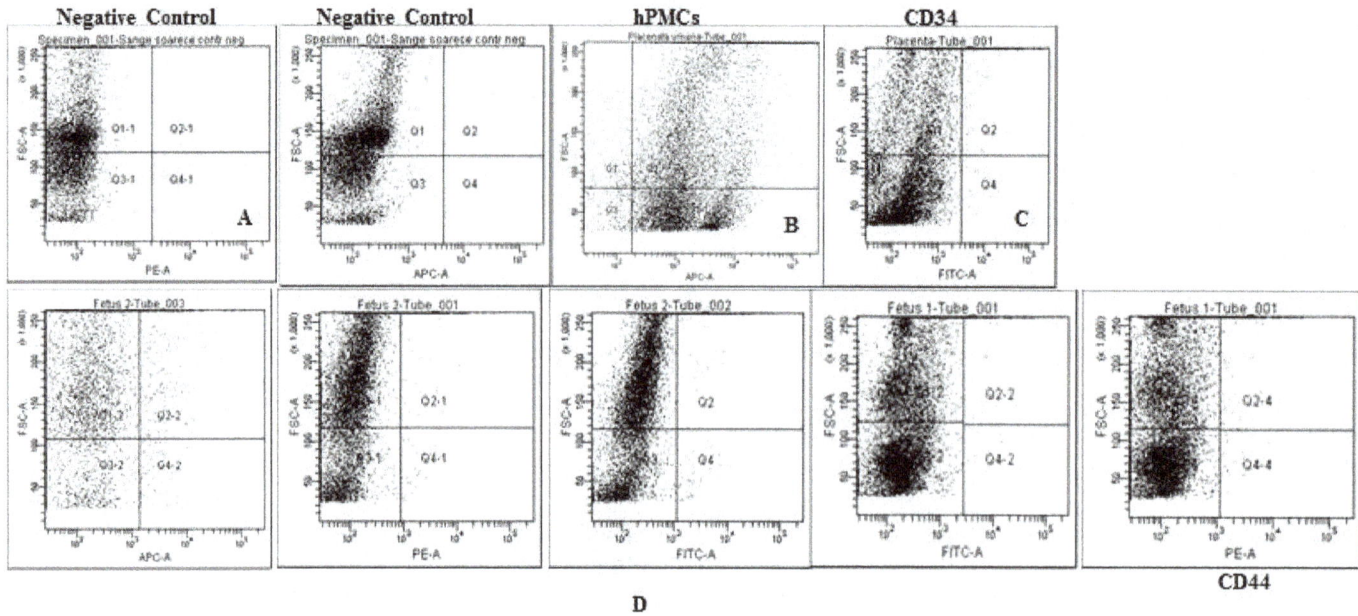

Figure 1. Flow cytometric analysis of CD34 positive hPMCs in the fetus and placenta after *in utero* transplantation of hPMCs. Representative data showing percentages of human CD44-positive hPMCs in placenta; **C** or fetus (organ mixture); **D** of recipient mice at E20; Control negative: mouse peripheral blood without hPMCs transplantation; **A** positive control hPMCs; **B**.

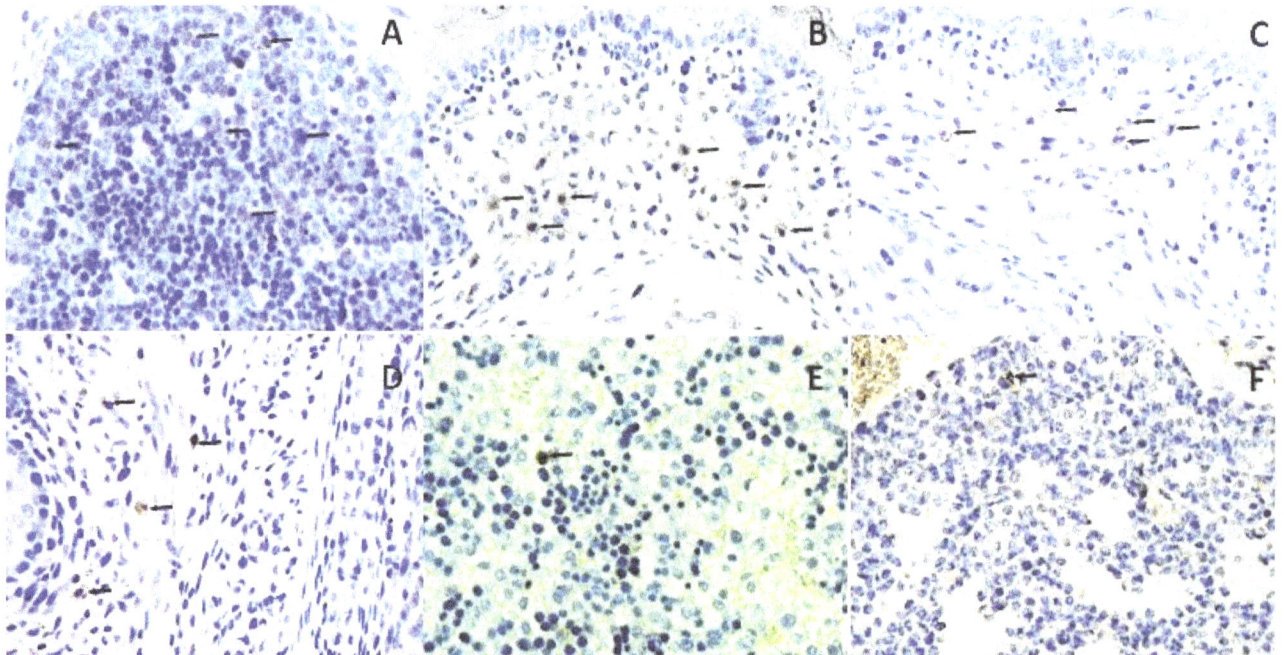

Figure 2. Hematoxylin-eosin (HE) stain and immunofluorescence of fetal mouse organs at embryonic day; **E** 20, were present in various fetal mouse tissues. Mouse tissues were immunostained using fluorescein isothiocyanate (FITC) - conjugated CD44 antibody (A,B,C,D,E,F); A- lymphatic node section with CD44[+] cells; **B**, **C** and **D**- murine fetal subcutaneous section with CD44[+] human cells, E,F -murine fetal liver section with CD44[+] (40x).

In utero transplantation of 1 x10[8]/kg CD34[+] paternal canine bone marrow-derived cells in a canine model achieved a low level of microchimerism (<1%) in various tissues (Blakemore et al., 2004). It has been shown that human cord blood-derived cells can differentiate into hepatocytes in the mouse liver without evidence of cellular fusion (Newsome et al., 2003). Human microchimerism was observed in various organs and tissues at 4 months after transplantation of human amnion and chorion mesenchymal progenitors in neo-natal swine and rats (Bailo et al., 2004). Differences observed in cell numbers may be due to colonization efficiency in different tissue environments or the rate of cell turnover in each organ (Krause et al., 2001). Our study adds to this body of work by establishing an *in utero* (E13.5) model of xenogeneic hPMC transplantation in mice.

Conclusions

In order to identify some sources of stem cell indispensable for regenerative therapy is imperative to identify new sources of mesenchymal stem cells, ethically acceptable, technically accessible and allows isolation of multipotent cells such proliferative potential and multilinear capacity at least similar to mesenchymal stem cells isolated from embryonic or other adult sources. MSCs are widely distributed in a variety of tissues in the adult human body (for example, bone marrow, kidney, lung and liver). These cells are also present in fetal environment (for example, blood, liver, bone marrow and kidney) but MSCs are a rare population in these tissues. The most well studied and accessible source of MSCs is bone marrow, although even in this tissue the cells are present in a low frequency. The human placenta is an attractive new source of MSCs, but the biological characteristics of placenta-derived MSCs have not yet been characterized. Our results show that mesenchymal stem cells are present in the human term placenta and may be a potential source of cells for transplantation therapy. Using routine cell culture techniques, placental derived mesenchymal stem cells can be successfully isolated and expanded *in vitro*. It appears that hPMCs from an allogeneic donor might constitute such a source. A further potential benefit is the exposure of the fetus to allogeneic cells, inducing tolerance such as future treatment.

ACKNOWLEDGEMENTS

This work was supported by National Council for Scientific Research in Higher Education – (CNCSIS) Human Resources Postdoctoral Project PD RU 278/2010.

REFERENCES

Bailo M, Soncini M, Vertua E, Signoroni PB, Sanzone S, Lombardi G, Arienti D, Calamani F, Zatti D, Paul P, Albertini A, Zorzi F, Cavagnini A, Candotti F, Wengler GS, Parolini O (2004). Engraftment potential of human amnion and chorion cells derived from term placenta. Transplantation.78: 1439-1448.

Barry F, Boynton R, Murphy M, Haynesworth S, Zaia J (2001). The SH-3 and SH-4 antibodies recognize distinct epitopes on CD73 from human mesenchymal stem cells. Biochem. Biophys. Res. Commun., 289: 519-524.

Blakemore K, Hattenburg C, Stetten G, Berg K, South S, Murphy K, Jones R (2004). In utero hematopoietic stem cell transplantation with haploidentical donor adult bone marrow in a canine model. Am. J. Obstet. Gynecol., 190: 960-973.

Chien CC, Yen BL, Lee FK, Lai TH, Chen YC, Chan SH, Huang HI (2006). In vitro differentiation of human placenta-derived multipotent cells into hepatocyte-like cells. Stem Cells. 24: 1759-1768.

Fukuchi Y, Nakajima H, Sugiyama D, Hirose I, Kitamura T, Tsuji K (2004). Human placenta-derived cells have mesenchymal stem/progenitor cell potential. Stem Cells. 22: 6.

Georgiades P, Ferguson-Smith AC, Burton GJ (2002). Comparative developmental anatomy of the murine and human definitive placentae. Placenta. 23: 3-19.

Krause DS, Theise ND, Collector MI, Henegariu O, Hwang S, Gardner R, Neutzel S, Sharkis SJ (2001). Multi-organ, multi-lineage engraftment by a single bone marrow-derived stem cell. Cell. 105: 369-377

Li C, Zhang W, Jiang X, Mao N (2007). Human-placenta-derived mesenchymal stem cells inhibit proliferation and function of allogeneic immune cells. Cell Tissue Res., 330: 437-446.

Liechty KW, MacKenzie TC, Shaaban AF, Radu A, Moseley AM, Deans R, Marshak DR, Flake AW (2000). Human mesenchymal stem cells engraft and demonstrate site-specific differentiation after in utero transplantation in sheep. Nat. Med., 6: 1282-1286.

Linju YB, Hsing-I H, Chih-Cheng C, Hsiang-Yiang J, Bor-Sheng K, Ming Y, Chia-Tung S, Men-luh Y, Meng-Chou L, Yao-Chang C (2005). Isolation of Multipotent Cells from Human Term Placenta, Stem Cells. 23: 3-9.

Muench MO (2005). In utero transplantation: baby steps towards an effective therapy. Bone Marrow Transplant. 35: 537-547.

Newsome PN, Johannessen I, Boyle S, Dalakas E, McAulay KA, Samuel K, Rae F, Forrester L, Turner ML, Hayes PC, Harrison DJ, Bickmore WA, Plevris JN (2003). Human cord blood-derived cells can differentiate into hepatocytes in the mouse liver with no evidence of cellular fusion. Gastroenterology. 124: 1891-1900.

Parolini O, Alviano F, Bagnara GP, Bilic G, Buhring HJ, Evangelista M, Hennerbichler S, Liu B, Magatti M, Mao N, Miki T, Mrongiu F, Nakajima H, Nikaido T, Portmann-Lanz CB, Sankar V, Soncini M, Stadler G, Surbek D, Takahashi TA, Redl H, Sakuragawa N, Wolbank S, Zeisberger S, Zisch A, Strom SC (2008). Concise review: isolation and characterization of cells from human term placenta: outcome of the first international Workshop on Placenta Derived Stem Cells. Stem Cells. 26: 300-311.

Saito T, Kuang JQ, Bittira B, Al-Khaldi A, Chiu RC (2002). Xenotransplant cardiac chimera: immune tolerance of adult stem cells. Ann. Thorac. Surg., 74(1): 19-24.

Shapiro E, Krivit W, Lockman L, Jambaqué I, Peters C, Cowan M, Harris R, Blanche S, Bordigoni P, Loes D, Ziegler R, Crittenden M, Ris D, Berg B, Cox C, Moser H, Fischer A, Aubourg P (2000). Long-term effect of bone-marrow transplantation for childhood-onset cerebral X- linked adreno-leukoldystrophy. Lancet. 356: 713-718.

Troegera C, Daniel S, Andreina S, Stephan S, Lisbeth D, Sinuhe H, Wolfgang H (2006). In utero haematopoietic stem cell transplantation, Experiences in mice, sheep and humans Swiss. Med. Wkly., 136: 498-503.

Yen BL, Huang HI, Chien CC, Jui HY, Ko BS, Yao M, Shun CT, Yen ML, Lee MC, Chen YC (2005). Isolation of multipotent cells from human term placenta. Stem Cells. 23: 3-9, 649-658.

Assessment of productive and reproductive performance of dairy cattle nexus with feed availability in selected peri-urban areas of Ethiopia

Zewdie Wondatir[1]*, Yoseph Mekasha[2] and Bram Wouters[3]

[1]Holetta Research Center, P.O. Box, 2003 Addis Ababa, Ethiopia.
[2]Haramaya University, P.O. Box 38 Dire Dawa, Ethiopia.
[3]Wageningen UR Livestock Research, P.O. Box 65, 8200 AB Lelystad, The Netherlands.

The study was conducted to assess the performance of dairy cattle in relation with feed availability and quality in selected peri-urban of Debre Birhan, Jimma and Sebeta areas of Ethiopia. Structured questionnaire, secondary data sources, field observations and laboratory analysis were employed to generate data. A total of 60 farmers (Debre Birhan=20, Jimma=20 and Sebeta=20) were randomly selected for the study. The overall estimated mean lactation length of cows was 296.5±8.7 days and was not different (*P>0.05*) among sites. The overall estimated mean age of heifers at first service was 27.5±1.0 months and age at first calving was 36.8±1.0 months and differed (*P<0.001*) considerably among the study sites. The result of the study indicated that grass hay was the main basal diet in all study areas. Laboratory analysis of major feed resources indicated that hay had Crude Protein (CP) content of 6.1% and crop residues varied from 3.1 to 6.7%. In addition, crop residues had lower digestibility (48%), its energy value ranged from 6.5 to 7.9 MJ/kg dry matter (DM). Wheat bran, and molasses had Metabolizable Energy (ME) content of 13.2 and 12.5 MJ/kg DM, respectively. Brewery wet grains had lower CP (27%) than cottonseed cake (42%) and enough seedcake (35%). Annual feed balance estimation revealed that the total estimated available feed supply met 83% of the maintenance DM requirement of livestock per farm per year while, the total estimated CP and ME were in accordance with the livestock requirement merely for maintenance. Therefore, from the current study it was concluded that the quality of available basal roughage feeds is generally low and strategic supplementation of protein and energy rich feeds should be required. Furthermore, optional feeds like brewery wet grains and other non-conventional feed resources should be further considered.

Key words: Age at first service, calving interval, crude protein, daily milk yield, days open, feed supply, feed quality, lactation length, metabolizable energy.

INTRODUCTION

Ethiopia is believed to have the largest livestock population in Africa (CSA, 2010). The livestock population census showed that Ethiopia has about 50.8 million heads of cattle, 25.9 million sheep, 21.9 million goats, 1.9 million horses, 5 million donkeys, 0.3 million mules, 0.8 million camels and 42 million poultry (CSA, 2010). This does not include livestock population of three zones of Afar and six zones of Somali regions. However, despite the large number of livestock resources in the country, its productivity is extremely low. In Ethiopia, annual milk production per cow is generally low due to reduced lactation length, extended calving interval, late age at first calving and poor genetic makeup (Alberro, 1983; Mukasa-Megerwa, 1989; Demke et al., 2000). Another major problem to such low livestock production and reproduction is shortage of livestock feeds both in quantity and in quality, especially during the dry season (Ahmed et al., 2010). Furthermore, quality of native pasture is very low especially in dry season due to their

*Corresponding author. E-mail: zewbt2006@yahoo.com.

low content of digestible energy and protein and high amount of fiber content (Zinash et al., 1995).

This is much worse for crop residues owing to their lower content of essential nutrients (protein, energy, minerals and vitamins) and lower digestibilities and intake. Despite, these problems, however, ruminants will continue to depend primarily on forages from natural pastures and crop residues (Zinash and Seyoum, 1989). Peri-urban dairy production systems have emerged around cities and towns, which heavily rely on purchased fodder (Vernooij, 2007). Commercialization of dairy production takes place around cities and towns where the demand for milk and milk products is high (medium and large towns) (Azage, 2004). However, the production system has been constrained by shortage of feed supply in dry season (quantity and quality) (Yoseph et al., 2003a). Few research works have been carried out with regard to feed availability in relation with dairy animals in urban and peri-urban dairy farms (Yitaye et al., 2009). Current and up-to-date baseline information is lacking in peri-urban areas on productive and reproductive performance of crossbred dairy cows in association with feed availability and quality under the prevailing situations. This study was therefore, aimed to look into the performance of cattle with respect to feed resources availability in selected areas of Ethiopia.

MATERIALS AND METHODS

In Jimma and Sebeta, there was no grazing land available and dairy cattle did not have access to grazing while in Debre Birhan dairy cattle are managed under indoor feeding system and have free access to graze for some hours a day. Milk supplied to Peri-urban and urban areas are only obtained from crossbred cows and hence crossbred cows with any exotic blood level inheritance were used for the study. Variables under productive and reproductive performance of cattle were estimated based on the farmer's estimation.

Description of the study areas

Debre Birhan is found in North Shoa administrative zone of the Amhara National Regional State and is located at 130 km north of the capital Addis Ababa, at 39°30' E longitude and 09°36' N latitude. It is a typical highland area with an elevation of 3360 masl. It receives an annual average rainfall of 731 to 1068 mm and has an annual temperature range of 6 to 20°C.

Sebeta is located 25 km Southwest of Addis Ababa and situated at a latitude and longitude of 8°55'N and 38°37'E, respectively. It has an elevation of 2356 m above sea level and has annual rainfall of about 1650 mm. The mean annual minimum and maximum temperature is 8°C and 19°C, respectively.

Jimma is located at 350 km away from the capital Addis Ababa. It is the largest city in the South Western Ethiopia. It lies between 36°10´ E longitude and 7°40´ N latitude. Its altitude is 2060 masl. Its climate is humid tropical with bimodal heavy annual rainfall, ranging from 1200 to 2800 mm. The mean annual minimum and maximum

temperature of the area is 11.3 and 26.2°C, respectively.

Sampling procedures

A reconnaissance survey was conducted in order to select specific dairy farmers and to get general picture of the study sites. Based on the record available from the respective district of the agricultural offices, there were about 200, 200 and 205 dairy farmers keeping crossbred cows in Debre Birhan, Jimma and Sebeta peri-urban areas, respectively. Based on the sample size to proportion technique 10% of the farms from each site were considered. Accordingly, a total of 60 dairy farms (20 from each site) were randomly selected from the peri-urban area of each study site. A structured questionnaire was prepared and pre-tested for its applicability before its administration. Interviews were carried out at the farmer's home to enable counterchecking of the farmer's response with respect to the availability of feed resources, livestock population, productive and reproductive performance of cows and the overall management system of the farm.

Feed quantity assessment

The quantity of feed dry matter obtainable from natural pastures were determined by multiplying the hectare with their respective estimated annual DM yield per hectare that is, 2.0 t/ha (FAO, 1987). The amount of purchased dry forages such as hay and straw was determined by estimating a single donkey load or lorry load and for baled hay by asking how many bales of hay would be purchased for a year. Whenever record was available, the quantity of purchased feeds was considered from the record. The quantity of available crop residues produced by farmers was estimated by applying grain to straw ratio as suggested by FAO (1987) and assuming 10% utilization wastage. The quantity of concentrates and non-conventional feed resources were estimated by interviewing the farm owners with regard to the frequency and quantity purchased per month. The grazing potential of crop stubbles was estimated using a mean of 0.5 ton per ha as reported by FAO (1987).

Chemical analysis of feed samples

Chemical analysis of feedstuffs was performed at Holetta Agricultural Research Center nutrition laboratory. DM and ash contents of feed samples were determined by oven drying at 105°C overnight and by igniting in a muffle furnace at 600°C for 6 h, respectively. Nitrogen (N) content was determined by Kjeldahl method and CP was calculated as N*6.25 (AOAC, 1995). Calcium (Ca) and phosphorous (P) contents were determined by atomic absorption spectrophotometry (Perkin, 1982). Acid Detergent Fiber (ADF), Acid Detergent Lignin (ADL), Neutral Detergent Fiber (NDF), were analyzed by the method of Van Soest et al. (1991). *In vitro* Organic Matter Digestibility (IVOMD) was determined by the modified Tilley and Terry method (Van Soest and Robertson, 1985). ME and Digestible Crude Protein (DCP) content of a particular feed were estimated from IVOMD and CP contents, respectively, as per the following equations.

ME (MJ/kg DM) =0.015*IVOMD (g/kg) (MAFF, 1984).

DCP (g) = 0.929*CP (g) -3.48 (Church and Pond, 1982).

Assessment of livestock feed requirement

Livestock populations were converted into Tropical Livestock Unit (TLU) as suggested by Gryseels (1988) for indigenous zebu cattle and Bekele (1991) for crossbreds. The DM requirements for maintenance were calculated based on daily DM requirements of 250 kg dual-purpose tropical cattle (an equivalent of one TLU). Nutrients supplied by each feed types were estimated from the total DM output and nutrients content of that feed on DM basis. The total nutrient requirements (DM, CP and ME) per day per livestock species were estimated based on the recommendations of Kearl (1982) and McCarthy (1986) for tropical livestock.

Statistical analysis

Data collected were analyzed using Statistical Analysis System software (SAS, 2002). Descriptive statistics were employed to describe qualitative variables. General Linear Model (GLM) procedure of SAS was employed to analyze the effect of classification variables. A completely randomized design was employed as per the following model.

$$y_{ij} = \mu + S_i + e_{ij},$$

Where, y_{ij}, Productive and reproductive performance of dairy cows, μ = overall mean, S_i = the effect of ith study sites, e_{ij} = random error

RESULTS AND DISCUSSION

Household characteristics

In the study areas (Debre Birhan, Sebta and Jimma), about 86.7% of the respondents were male dairy farmers while 13.3% were females (Table 1). Less number of female-headed households involved in livestock keeping in the current study could probably be due to cultural issues that force females to get married and/or for economic reason. The results of the current work differ from the report of Azage (2004) who reported 33% female-headed households and 67% male headed household livestock keepers in Addis Ababa Table 1.

Livestock herd structure

The average livestock holding per household in all study areas was (15.6±0.2 TLU) (Table 2). The number of sheep per household was higher (P<0.05) at Debre Birhan than the rest of study sites. This is because of suitable weather conditions and better grazing lands. The average number of horses per household was much larger in Debre Birhan (P<0.05) than in Jimma and Sebeta, which might be related to better adaptation to the environment and suitability of these animals for people to overcome transport problems associated with rugged

terrains. At Jimma and Sebeta, horses were rarely kept, but purchased from other areas for pulling carts. The average number of donkeys per household in Debre Birhan was higher (P<0.05) than in Sebeta. Donkeys are mainly used as pack animals in these areas Table 2.

Productive and reproductive performance of cows

Least squares means for daily milk yield, lactation length, age at first service, calving interval and days open are shown in Table 3. The estimated mean daily milk yield based on the farmers response varied significantly (P<0.001) among the study sites. In Sebeta area, the estimated daily milk yield (9.7±0.5 kg) was higher (P<0.001) than the rest of the study areas. The highest estimated daily milk yield observed for Sebeta area could possibly be the result of better access to brewery by-products, agro-industrial by-products and hay. The whole range of estimated daily milk yields (6.1 to 9.7 kg) reported in this study corresponds well with values reported earlier (Demeke et al., 2000). The current report also agreed with that reported by Mesfin et al. (2009) for crossbred dairy cows. Yoseph et al. (2003b) reported an average daily milk yield of (8.9 kg/day) for crossbred dairy cows in urban and peri-urban area of Addis Ababa, which is closer to the current finding. However, Moges and Baars (1998) reported higher average milk yields (9 to 12 kg/day). The difference could be attributed to differences in management conditions and the level of exotic gene inheritance in the crossbred animals. The overall estimated mean lactation length of cows in all study sites was 296 days and varied from 273 to 327 days and was not different (P>0.05) among sites.

The estimated lactation length was comparable to the ideal lactation length of 305 days as defined by Foley et al. (1972). However, farmers have the attitude that extended length of lactation favors growth of calves despite low milk yields. The overall estimated mean age of heifers at first service was 27.5 months and age at first calving was 36.8 months that differed (P<0.001) considerably among the study sites. Estimated mean ages of heifers at first service and calving were shortest at Sebeta (24.3 and 33.6 months) compared to other sites. The results are in accordance with the mean value of 25.6 months reported for age at first service and 36.2 months reported for age at first calving for dairy heifers under urban production systems (Emebet, 2006). Heifers maturing at younger ages are better milk producers and have lower rearing costs (Ruiz-Sanchez et al., 2007). Neither the age at first service nor the age at first calving in the present work meet the optimum age at first service (14.6 months) and calving (24 months) for milk yield under intensive management for exotic breeds as

Table 1. Demographic characteristic of the respondents in selected peri-urban dairy production system of Ethiopia.

Household variable	Study sites			
	Debre Birhan	Jimma	Sebeta	Total
Sex of household head	n=20	n=20	n=20	n=60
Male (%)	100.0	80.0	80.0	86.7
Female (%)	0.0	20.0	20.0	13.3
Overall (%)	100	100	100	100

n = number of respondents.

Table 2. Herd size and herd structure (Mean ± SE) per household in selected peri-urban dairy production system of Ethiopia.

Livestock species	Study sites				Study sites			
					TLU			
	DB	Jimma	Sebeta	Overall mean	DB	Jimma	Sebeta	Overall mean
Cattle	11.8±0.7	11.9±1.5	8.8±1.5	10.8±0.7	14.6±0.9	13.3±1.7	11.6±1.9	13.2±0.9
Cows	3.7±0.3	5.0±0.7	5.0±0.7	4.6±0.4	6.6±0.6	9.0±1.3	9.0±1.3	8.2±0.6
Oxen	2.8±0.3^a	0.2±0.1^b	0.6±0.3^b	1.2±0.2	4.2±0.5	0.2±0.1	0.7±0.3	1.7±0.3
Heifers	1.5±0.3	3.1±0.6	1.7±0.4	2.1±0.3	1.0±0.2	2.1±0.4	1.2±0.3	1.4±0.2
Bulls	1.0±0.2^a	0.7±0.2^a	0.1±0.1^b	0.6±0.1	1.6±0.4	0.8±0.2	0.2±0.1	0.9±0.2
Calves	3.0±0.3	3.0±0.5	1.4±0.3	2.4±0.2	1.2±0.1	1.2±0.2	0.6±0.1	1.0±0.1
Sheep	24.2±2.9^a	0.7±0.6^c	2.7±0.8^b	9.2±1.7	2.4±0.3	0.1±0.0	0.3±0.1	0.9±0.2
Goats	0.7±0.5	-	0.4±0.3	0.3±0.2	0.1±0.0	-	0.1±0.0	0.1±0.0
Horses	1.9±0.3^a	1.1±0.2^b	0.1±0.0^c	1.0±0.2	1.5±0.2	0.8±0.2	0.1±0.0	0.8±0.1
Donkeys	3.1±0.3^a	-	1.0±0.26^b	1.4±0.2	1.5±0.1	-	0.5±0.1	1.0±0.1
Total herd size					20.1±0.3	14.3±0.4	12.5±0.3	15.6±0.2

[a-b-c] means with different letters of superscripts in the same row differ significantly (*P<0.05*), TLU = tropical livestock unit. DB = Debre Birhan.

Table 3. Least square means (LSM ± SE) milk production and reproductive performance of crossbred dairy cows in selected peri-urban dairy production system of Ethiopia.

Variable	Study sites		
	DB	Jimma	Sebeta
MY (kg/day)	6.1±0.4^b	7.1±0.5^b	9.7±0.5^a
LL (days)	309±18.2	280.8±14.7	297.0±10.6
AFS (months)	32.5±1.7^a	25.7±1.4^b	24.3±1.7^b
AFC (months)	41.8±1.7^a	35.0±1.5^b	33.6±1.7^b
CI (days)	477.0±32.5	463.5±39.6	474.0±31.5
DO (days)	197.0±32.5	183.5±39.6	194.0±31.5

[a-b] means with different superscript in the same row for the same trait do significantly differ (P<0.05, MY = Milk Yield, LL = Lactation Length, AFS = Age at First Service, AFC = age at first calving, CI = calving interval, DB = Debre Birhan, DO = days open.

reported by Nilforooshan and Edriss (2004). The overall estimated mean calving interval and days open in the study sites were about 472 and 192 days, respectively. There was no marked difference (*P>0.05*) in length of calving interval and days open among the study sites. The length of days open was a bit more than 6 months in all study sites, which might affect the profitability and lifetime productivity of dairy cows Table 3.

Chemical composition and nutritive value of feeds

Chemical composition and nutritive value of the major feedstuffs in the study areas are shown in Table 4. All crop residues evaluated had lower CP contents than the minimum level of 7% CP required for optimum rumen microbial function (Van Soest, 1982). The mean IVOMD for cereal crop residues was about 48%, which is lower than the minimum level required for quality roughages (Seyoum and Fekede, 2008). The neutral detergent fiber (NDF) content of all crop residues was above 65%. Roughage feeds with NDF content of less than 45% are categorized as high quality, 45 to 65% as medium quality and those with more than 65% as low quality roughages (Singh and Oosting, 1992). All crop residues and stubbles in this study might be categorized as low quality roughages that may inflict limitations on animal performance. The ADF content of crop residues varied from 51.0% in field pea straw to 56.3% in wheat straw. The ADF content for both crop residues and stubbles was within the range reported by Ahmed (2006) and Solomon et al. (2008). However, Teklay (2008) reported a lower ADF values for barley and wheat straw, which could be attributed to differences in climate, crop management and soil fertility. Generally, Kellems and Church (1998) categorized roughages with less than 40% ADF as high quality and above 40% as low quality.

All crop residues and stubbles could be categorized as low quality roughages. In this study, the lignin content was high for both crop residues and grass hay as compared to the maximum level of 7%, which limits DM intake. Lignin is completely indigestible and forms lignin-cellulose/hemicelluloses complexes (Kellems and Church, 1998) due to physical encrustation of the plant fiber and reduces its availability to microbial enzymes (McDonald et al., 1995). The energy content of crop residues ranged from 6.5 MJ/kg DM (wheat) to 7.9 MJ/kg DM (barley) straw. The energy contents for crop residues in this study were within the range reported by Seyoum and Fekede (2008), but lower than the value of 10.3 MJ/kg DM reported by Teklay (2008). Differences might be due to differences in management practices, soil fertility and/or crop variety used (McDowell, 1988). Hay obtained from native grass had CP content of 6.1%. The value observed in the present study is lower than the minimum value required for optimum rumen microbial function reported by Van Soest (1982). It had also high NDF content. NDF content of hay reported in this study was within the range of the values reported by Dereje et al. (2010). The higher NDF content could be a limiting factor on feed intake, since voluntary feed intake and NDF content are negatively correlated (Ensminger et al., 1990). ME of commonly used energy supplements such as wheat bran, molasses and *Atela* varied from 12.5 to 13.2 MJ/kg DM.

Molasses had the lowest CP content as compared to wheat bran and *Atela*. The cell wall content of molasses was almost negligible, whereas wheat bran had relatively higher fiber contents. The nutritional values for the current feeds are compatible with that reported by Seyoum and Fekede (2008). Seyoum et al. (2007) proposed a standard for energy supplements as those feeds, which contain high CP (13.9%), IVOMD (82.2%) and ME (13.1 MJ/kg DM). With the exception of CP content of molasses, energy supplements (wheat bran,

Atela) evaluated in the present work closely matched to this standard. Among the protein supplements, brewery wet grains had slightly lower CP (26.8%) than cotton seed cake (42.0%) and noug seedcake (34.5%). This might be due to differences in the chemical composition and type of grains used as a raw material to produce these by-products (Negesse et al., 2009). The ME contents of protein supplements were not much different. The energy content, protein content and IVOMD in protein supplements were high though slightly lower than the reported thresholds (Seyoum et al., 2007) for good quality protein supplements of (CP = 32.6%), (IVOMD = 65.5%) and (ME = 10.2 MJ/kg DM). Ca and P concentrations of the major feedstuffs in the study areas except for barley straw were low as compared to the normal range of 2.0 to 3.5 g/kg DM reported for livestock feeds by McDonald et al. (1995) Kellems and Church (1998) Table 4.

Estimated annual feed availability

The total estimated feed DM, DCP and ME production per farm in the study areas is shown in Table 5. The major feed resources include hay, agro-industrial by-products and crop residues. Feed dry matter was commonly obtained from hay in all study sites. However, farmers at Debre Birhan heavily rely on crop residues compared to Jimma and Sebeta. Agro-industrial by-products and non-conventional feeds were important feed resources next to hay in both Jimma and Sebeta. Use of improved fodder trees as animal feed in the study sites was rare and the DM calculation did not account for these feed resources Table 5.

Estimated annual feed balance

The total annual nutrient intake, nutrient requirement and feed balances in the study areas are shown in Table 6. In all the study areas, the estimated available feed supply met about 83% of the maintenance DM requirement of livestock per farm per year while the total estimated DCP and ME for maintenance were 40 and 10% surplus per year per farm, respectively. At Debre Birhan the existing feed supply on a year round basis satisfies only 64% of the maintenance DM requirement of the animals per farm. Similarly, the total available DCP and ME in the same area satisfy only 66% and 81% of the total livestock requirement per farm on a yearly basis. In Jimma, total annual DM requirement was 11.5% less than the annual DM requirement for maintenance. On the other hand, the total DCP and ME were 51 and 25% per farm, respectively, above the total annual maintenance requirement. In Sebeta, the total annual DM requirement for maintenance was 3% less than the requirement for maintenance while total DCP and ME were 102 and 26%

Table 4. Chemical composition and nutritive value of major feedstuffs in the study areas.

Feedstuff	DM (%)	Chemical composition (% DM)						Nutritive values				
		Ash	NDF	ADF	Lignin	CP	DCP (g/kg DM)	IVOMD%	ME (MJ/kg DM)	Ca (g/kg)	P (g/kg)	
Crop residue												
Wheat straw	93.4	9.5	80.3	56.3	13.1	3.1	25.7	43.2	6.5	0.2	0.9	
Barley straw	91.6	8.5	76.8	52.8	12.1	3.6	29.5	52.6	7.9	3.3	0.8	
Oats straw	92.4	7.1	75.3	54.5	15.0	3.1	24.9	48.8	7.3	0.4	1.0	
Faba bean straw	92.6	6.6	73.4	51.0	9.9	6.1	53.5	47.1	7.1	1.5	0.8	
Field pea straw	91.8	6.5	72.7	52.3	11.1	6.7	59.1	48.4	7.3	1.4	1.0	
Grass												
Hay	92.4	13.7	76.0	49.2	10.6	6.1	53.5	48.7	7.3	0.4	1.3	
Non-conventional feeds												
Coffee pulp	90.3	9.0	55.5	48.6	6.7	11.1	99.9	49.0	7.4	0.5	1.1	
Bean hull	90.9	3.1	72.7	61.4	8.2	6.5	57.3	55.9	8.4	0.6	3.0	
Pea hull	91.0	3.6	58.6	40.8	7.5	16.4	148.7	63.7	9.6	0.4	2.0	
Atela	21.8	5.8	60.2	22.5	11.0	21.0	167.3	87.8	13.2	0.2	0.6	
AIBP												
Brewery wet grain	22.2	4.7	78.6	29.9	10.7	26.8	245.7	60.3	9.1	0.3	1.7	
Wheat bran	86.5	4.4	52.8	8.1	-	16.9	153.2	83.0	12.5	0.2	0.8	
Cotton seedcake	92.3	7.6	47.2	20.8	6.3	42.0	386.7	60.2	9.0	0.2	1.1	
Noug seedcake	93.4	10.9	33.1	27.2	7.1	34.5	317.0	68.1	10.2	1.1	0.2	
Molasses	72.4	18.5	-	-	-	4.0	29.0	99.6	14.9	0.8	0.2	
Crop stubbles												
Barley stubble	92.5	6.2	80.3	68.5	7.5	2.2	17.0	53.5	8.0	0.9	0.3	
Wheat stubble	93.0	6.4	81.7	69.7	8.1	2.1	15.9	48.3	7.2	0.4	0.7	
Faba bean stubble	92.7	4.2	76.0	62.4	10.2	3.1	24.9	44.3	6.6	0.8	0.3	
Field pea stubble	92.5	3.8	77.8	58.7	12.9	3.8	31.4	41.4	6.2	0.5	0.4	
Oats stubble	93.2	7.3	79.8	71.5	7.7	1.9	14.6	50.2	7.5	0.3	0.2	

AIBP = Agro-Industrial By-products, Atela = a by-product of local beverage.

above the total annual requirement per farm. Surplus DCP and ME above the maintenance requirement in Jimma and Sebeta could probably be attributed to the use of better energy and protein supplements. The larger deficit observed mostly under Debre Birhan area may be associated

Table 5. Estimated available dry matter production, DCP and ME supply per annum per farm in selected peri-urban dairy production system of Ethiopia.

Feedstuff	Study sites								
	Debre Birhan			Jimma			Sebeta		
	DM (t)	DCP (kg)	ME (MJ)	DM (t)	DCP (kg)	ME (MJ)	DM (t)	DCP (kg)	ME (MJ)
Crop residues	11.2	330	86100	-	-	-	4.6	120	30000
Hay	10.9	600	81600	14.4	770	105200	10.2	540	74000
AIBP	4.2	770	50400	8.8	1300	110200	10.2	2100	108700
Non-conventional feeds	-	-	-	4.4	500	43100	0.2	20	1600
Total	26.3	1700	218100	27.6	2570	258500	25.2	2780	214300

- Not available, Atela = a by-product of local beverage, AIBP = Agro-industrial by-products.

Table 6. Estimated annual feed dry matter and nutrient balance of livestock per farm per annum in selected peri-urban dairy production system of Ethiopia.

Study site	Annual nutrient supply			Estimated annual nutrient requirement (for maintenance)			Balance of supply and requirements		
	TDM (t)	TDCP (kg)	TME (MJ)	TDM (t)	TDCP (kg)	TME (MJ)	TDM	TDCP	TME
DB TLU=20.1	26.3	1700	218100	41.4	2600	270900	-15.1	-900	-52800
Jimma TLU=14.3	27.6	2570	258500	31.2	1700	206900	-3.6	+870	+51600
Sebeta TLU=12.5	25.2	2780	214300	26.0	1400	174100	-0.8	+1380	+40200
Average	26.3	2350	230300	32.9	1900	217300	-6.5	+450	+13000

DB = Debre Birhan, TDM = total dry matter, TDCP = total digestible crude protein, TME = total metabolizable energy.

with poor quality of roughages and absence of supplements.

CONCLUSION AND RECOMMENDATIONS

The quality of available basal roughage feeds for dairy cattle in peri-urban areas of Ethiopia is generally low. Better milk yield and reproductive performance observed at Sebeta area could be a point of interest to further study on the biological and economic efficiency of feeding agro-industrial by-products such as brewery wet grain for dairy cattle kept close to brewery factories. Alternative means of feed production and supply particularly in dry season should be in place with the involvement of all stakeholders and development actors Table 6.

ACKNOWLEDGEMENTS

The authors acknowledge Wageningen University of The Netherlands for its financial support to this research work.

REFERENCES

Ahmed H (2006). Assessment and Utilization Practice of Feed Resources in Basona Worana Wereda of North Shoa, unpublished MSc. Thesis, Haramaya University, Dire Dawa, Ethiopia, P. 131.
Ahmed H, Abule E, Mohammed K, Treydte AC (2010). Livestock feed resources utilization and management as influenced by altitude in the Central Highlands of Ethiopia. http://www.lrrd.org/lrrd22/12/cont2212.html.
Alberro M (1983). Comparative performance of F1 Friesian X Zebu heifers in Ethiopia. Anim. Prod. Sci., 37: 247-252.
AOAC (Association of Official Analytical Chemists) (1995). Official Methods of Analysis. PP.5-13. (16th edition), Washington DC.
Azage T (2004). Urban livestock production and gender in Addis Ababa. PP.3. Urban Agriculture Megazine, number 12, MEI, 2004, International Livestock Research Institute. Addis Ababa, Ethiopia.
Bekele S (1991). Crop livestock interactions in the Ethiopian highlands and effects on sustainability of mixed farming: a case study from Ada district, Debrezeit. An MSc. Thesis Agricultural University of Norway, Oslow, Norway, P. 163.
Church DC, Pond WC (1982). Basic Animal Nutrition and Feeding Record. John Wiley and Sons, U.S.A. p. 1135.
CSA (Central Statistical Agency) (2010). Livestock and Livestock Characteristics, Agricultural Sample Survey. Volume II, Stat. Bull., 468: 107.
Demeke S, Neser FWC, Schoeman SJ, Erasmus GJ, Van Wyk JB, Gebrewolde A (2000). Crossbreeding Holstein-Friesian with

Ethiopian Boran Cattle in A Tropical Highland Environment: Preliminary Estimates of Additive and Heterotic Effects on Milk Production Traits. S. Afr. J. Anim. Sci., 30(1): 32-33.

Dereje F, Seyoum B, Aemiro K, Tadesse D, Getu K, Getnet A (2010). Near Infrared Reflectance Spectroscopy for Determination of Chemical Entities of Natural Pasture from Ethiopia. Agric. Biol. J. N. Am., retrieved from http://www.scihub.org/ABJNA.

Emebet M (2006). Reproductive Performance of dairy Cows Under urban Dairy Production Systems in Dire-Dawa. MSc. Thesis, Alemaya University, Dire Dawa, Ethiopia, P. 82.

Ensminger RE, Oldfield JE, Heineman WW (1990). Feed and Nutrition. (2nd edition). The Ensminger publishing company, P. 1151.

FAO (Food and Agriculture Organization of the United Nations) (1987). Land use, production regions, and farming systems inventory. Technical report 3 vol. 1. FAO project ETH/78/003, Addis Ababa, Ethiopia, P. 98.

Foley RC, Bath DL, Dickinson FN, Tucker HA (1972). Dairy cattle principles, practices, problems, profits, Philadelphia, USA. P. 669.

Gryseels G (1988). Role of Livestock on a Mixed Smallholder Farms in Debre Berhan, PhD Dissertation, Agricultural University of Wageningen, The Netherlands, p. 249.

Kearl LC (1982). Nutrient Requirement of Ruminants in Developing Countries International Feed stuffs Institute, Utah Agricultural Experiment Station, Utah State University, Longman 84322. USA, p. 381.

Kellems RO, Church DC (1998). Livestock Feeds and Feeding.(4th edition.). Prentice-Hall, Inc., New Jersey, USA, P. 573.

MAFF (Ministry of Agriculture Fisheries and Food) (1984). Energy allowances and feeding systems for ruminants. Reference Book 413 HMOs, London, p. 85.

McCarthy G (1986). Donkey Nutrition. In: Reed JD and Capper BS and Neate JH (eds.) The professional Hand book of the Donkey (Compiled for the donkey sanctuary).Sid mouth (UK), P. 248.

McDonald P, Edwards RA, Greenhalgh JFD, Morgan CA (1995). Animal Nutrition. (Fifth Edition). Longman Group, Harlow, United Kingdom, P. 607.

McDowell RE (1988). Improvement of Crop Residues for Feeding Livestock in Smallholder Farming Systems. PP. 3-27. In: J.D. Reed, B.S. Capper and P.J.H., Neate (eds.). Plant Breeding and Nutritive Value of Crop Residues. Proceedings of a Workshop. Held at ILCA, Addis Ababa, Ethiopia, 7-10 Dec. 1987. ILCA, Addis Ababa.

Mesfin D, Seyoum B Aemiro K Getu K, Kedir N (2009). On-farm evaluation of lactating crossbred (Bos taurus x Bos indicus) dairy cows fed a basal diet of urea treated teff (Eragrostis tef) straw supplemented with escape protein source during the dry season in crop-livestock production system of north Shoa, Ethiopia. Livestock Res. Rural Dev., http://www.lrrd.org/lrrd 21/5/cont2105.htm.

Moges D, Robert Baars (1998). Long-Term Evaluation of Milk Production and Reproductive Performance of Dairy Cattle at Alemaya. In: proceedings of the 6th annual conference of the Ethiopian Society of Animal Production. Addis Ababa, Ethiopia, October 14-15 May 1998, PP. 176-183.

Mukasa-Mugerwa E (1989). A review of reproductive performance of female Bos indicus (Zebu) cattle. ILCA, Monograph No. 6. International Livestock Center for Africa, Addis Ababa, Ethiopia, P. 134.

Negesse T, Makkar HPS, Becker K (2009). Nutritive value of some non-conventional feed resources of Ethiopia determined by chemical analyses and an in vitro gas method. Anim. Feed Sci. Technol., 154: 204-217.

Nilforooshan MA, Edriss MA (2004). Effect of age at first calving on some productive and longevity traits in Iranian Holsteins of the Isfahan Province. American Dairy Science Association. J. Dairy Sci., 87: 2130-2135.

Perkin E (1982). Analytical Methods for Atomic Absorption Spectrophotometry. Perkin Elmer Corporation, Norwalk, Connecticut, USA.

Ruiz-Sanchez R, Blake RW, Castro-Gamez HMA, Sanchez F, Montaldo

HH, Castillo-Juarez H (2007). Changes in the association with between milk yield and age at first in Holstein cows with herd environment level for milk yield. J. Dairy Sci., American Dairy Science Association, 90: 4830-4834.

SAS (2002). Statistical Analysis System software, Version 9.0, SAS Institute, Inc., Cary, NC, USA.

Seyoum B, Zinash S, Dereje F (2007). Chemical Composition and Nutritive Values of Ethiopian Feeds. Research Report 73, Ethiopian Institute of Agricultural Research, Addis Ababa, Ethiopia, P. 24.

Seyoum B, Fekede F (2008). The status of animal feeds and nutrition in the West Shewa Zone of Oromiya, Ethiopia, PP. 27-49. In: Proceedings of the Workshop 'Indigenous Tree and Shrub Species for Environmental Protection and Agricultural Productivity', November 7-9, 2006, Holetta Agricultural Research Centre, Ethiopia. Series on Conference and Workshop Proceedings of KEF (Commission for Development Studies at the Austrian Academy of Sciences): 2008/1.

Singh GP, Oosting SJ (1992). A Model for Describing the Energy Value of Straws. Indian Dairyman XLIV, PP. 322-327.

Solomon B, Solomon M, Alemu Y (2008). Potential Use of Crop Residues as Livestock Feed Resources Under Smallholder Farmers Conditions in Bale Highlands of Ethiopia. J. Trop. Subtrop. Agroecosyst., 8: 107-114.

Teklay A (2008). Assessment of the feeding systems and feed resources of dairy cattle in Lemu-Bilbilo Wereda dairy products-processing cooperatives, Arsi Zone of Oromia Regional State, Ethiopia, unpublished MSc. Thesis, Addis Ababa University, Addis Ababa, Ethiopia, P. 77.

Van Soest PJ (1982). Nutritional Ecology of the Ruminants: Ruminant metabolism, Nutritional strategies, the cellulolytic Fermentation and the Chemistry of Forages and Plant Fibers. Ithaca, New York. P. 373.

Van Soest PJ, Robertson JB (1985). Analysis of Forages and Fibrous Foods. A Laboratory Manual for Animal Science 613. Cornel University, Ithaca. New York, USA, p. 202.

Van Soest PJ, Robertson JB, Lewis BAL (1991). Methods for Dietary fiber, Neutral Detergent Fiber and Non-Starch Polysaccharides in relation to Animal Nutrition. J. Dairy Sci., 74: 3583-359.

Vernooij AG (2007). Report Ethiopia Mission, 22-29 September, 2007. Internal Report 200706. Animal Sciences Group, Wageningen University.

Yitaye A, Maria W, Azage T, Werner Z (2009). Performance and limitation of two dairy production systems in the Northwestern Ethiopian Highlands. Trop. Anim. Health Prod., 41(7): 1143-1150.

Yoseph M, Azage T, Alemu Y, Ummuna NN (2003a). Variations in nutrient intake of dairy cows and feed balance in urban and peri-urban dairy production systems in Ethiopia. In: Proceedings of the 10th annual conference of the Ethiopian Society of Animal Production (ESAP) held in Addis Ababa, Ethiopia, August 22-24, 2002, PP. 177-184.

Yoseph M, Azage T, Alemu Y, Ummuna NN (2003b). Milk Production, milk composition and body weight change of crossbred dairy cows in urban and peri-urban dairy production systems in Ethiopia. In: Proceedings of the 12th annual conference of the Ethiopian Society of Animal Production (ESAP) held in Addis Ababa, Ethiopia, August 22-24, 2002, pp. 185-192.

Zinash S, Seyoum B (1989). Utilization of Feed Resources and Feeding Systems in the Central zone of Ethiopia, pp. 129-132. In: Proceedings of the Third National Livestock Improvement Conference. Addis Ababa, Ethiopia, May 1989. IAR, pp. 24-26.

Zinash S, Seyoum B, Lulseged G, Tadesse T (1995). Effect of harvesting stage on yield and quality of natural pasture in the central highlands of Ethiopia. In : proceedings of the Ethiopian Society of Animal Production (ESAP); Third National Conference 27-29 April 1995. IAR, Addis Ababa, Ethiopia, pp. 316-322.

7

Discovery of soft bodied metazoans and microphytofossils from the Mesoproterozoic sediments of Vindhyan Supergroup, India

Veeru Kant Singh[1]*, Rupendra Babu[1], Prabhat Kumar[2] and Manoj Shukla[1†]

[1]Birbal Sahni Institute of Palaeobotany, 53-University Road, Lucknow-226007, India.
[2]Department of Zoology, University of Lucknow-226007, India.

An assemblage of microbial remains consisting two types of preservations are being recorded for the first time from the phosphoritic chert intercalated in Tirohan Limestone (Rohtas Limestone Formation) of Semri Group, Vindhyan Supergroup, exposed in Chitrakoot, Madhya Pradesh. The uniquely preserved, cryptic forms are multicellular, bilaterally symmetrical, monoecious, cylindrical elongated coiled body with clitellum and ornamented structures (setae). This assemblage shows possible biomineralized novel post embryonic developmental stages of the juvenile and young forms of the annelids. The dark brown organic-walled microfossils (prokaryote and protists) are sparse, cellularly preserved, distorted, small sized, simple sphaeromorphs. The present explored preliminary results are based on the Mesoproterozoic stray fossils of two bioentities in limited material, and are the authentic key events to consider remarkable potential value in understanding the evolution and divergence of the metazoan in India along with status of Indian subcontinents (geographic position) in geological past of world history. It is a team effort of the earth scientists (geologists, botanist, and zoologist) based on the extinct and extant annelids in present brief communication.

Key words: Soft bodied metazoan, annelid, clitellum, Mesoproterozoic, Tirohan limestone, Vindhyan.

INTRODUCTION

The great Vindhyan is one of the largest and thickest Precambrian basin of peninsular India and is tectonically least disturbed and unmetamorphosed. Its sediments are about 5000 m thick and belong to a supergroup of heterogeneous lithofacies that are areno-argillaceous, siliciclastic, and carbonate, typified in the Son Valley in Uttar Pradesh (Sastri and Moitra, 1984) (Figure 1). Since 1997, the most authentic data on some animal faunas have come from the Vindhyan. Any discovery from Precambrian sediments is always a matter of interest to earth scientists of the world. The present discovery of Mesoproterozoic metazoan fossils is one of the most important, for the phosphoritic chert in which they occur belongs to the Semri group, the oldest of the Vindhyan. There is no earlier report of fossils possessing essential annelid characters; hence, this finding is of great importance in considering the origin of annelids as a taxon.

The best molecular clock estimation for the last common bilaterian ancestor varies from 1200 to 600 Ma, focusing on the Ediacaran sediments and divergence of the metazoan m cells with molecules (Wray et al., 1996; Ayala et al., 1998; Rasmussen et al., 2002; Tang et al., 2006). However, the origin and evolution of animals are based on morphology as well as chemo- and molecular-fossil data (Jermilin et al., 2005). This find of morphological data thus helps us understand how the annelid fauna related to the coeval Ediacaran fauna. The Chitrakoot region from where materials of the present study has been taken, is an area confined to an unevenly condensed succession of Semri group strata resting unconformably upon Bundelkhand granites (Narain, 1960). The Tirohan limestone is the youngest lithostratigraphic unit of this group (Table 1), containing phosphoritic chert stromatolites, *Jurussania* (Krylov), which are widely exposed in isolated hillocks, including Kamtanath Hill in the Paisuni River Valley of the Chitrakoot area in Uttar

---

*Corresponding author. E-mail: veerukantsingh@gmail.com.
[†]Deceased on 06/06/2006.

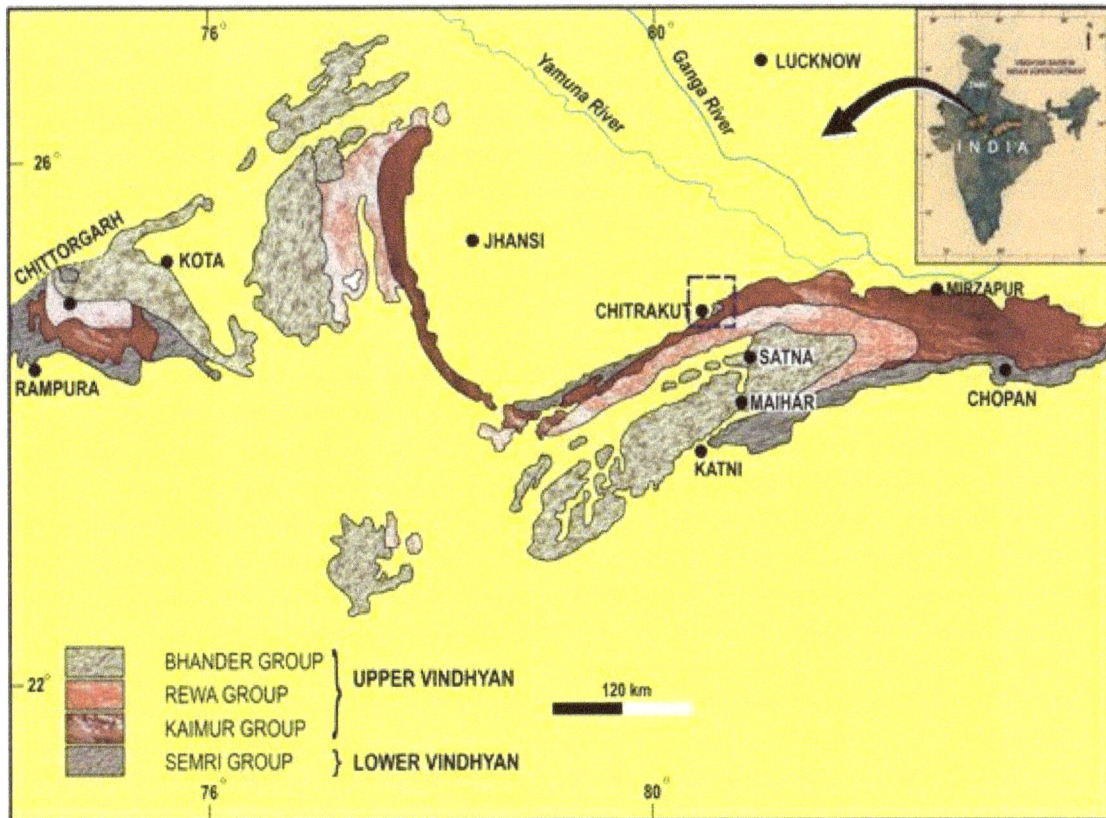

Figure 1. Generalized geological map of the Vindhyan Basin showing the position of study area (Krishnan and Swaminathan, 1959).

Table 1. Lithostratigraphic succession of Vindhyan Supergroup in Chitrakut area (Safaya, 1975).

Supergroup	Group	Formation	Lithologies/Members	Age (Ma)
Vindhyan	Rewa	Panna Formation	Calcareous shale, glauconitic shale, sandstone, siliceous and cherty shale	
	Kaimur	Baghain Sandstone	Light pinkish brown to buff colored medium to coarse grained, cross bedded, indurated sandstone, conglomerate bed at base	
		Bhounri Formation	Pitted flaggy fine to medium grained sandstone	
		Khoh Formation	Silicified chert, agate and breccia	
	Semri	Chitrakut Formation	Tirohan Limestone Member Upper Green Sandstone Member Pellet Limestone Member Lower Green Sandstone Glauconitic Limestone Member	1504-1409 Ma (based on Rb-Sr dating)[1]Kumar et al., 2001
Bundelkhand Gneissic Complex				2500 Ma (Pb-Pb dating)[2] Singh and Kumar, 1978

Pradesh and Madhya Pradesh, India. The fossiliferous localities lie at Latitude 25° 09'44.9"N and Longitude 80° 52' 05.33" E (Figure 3) and Latitude 25° 10' 18.8" N and Longitude 80°50' 37. 5"E (Figure 2). Because of the relative scarcity of fossils prior to ~543 Ma and with the explosion and radiation of the better-known, morphologically complex populations, the earliest appearance of animal remains is yet to be understood.

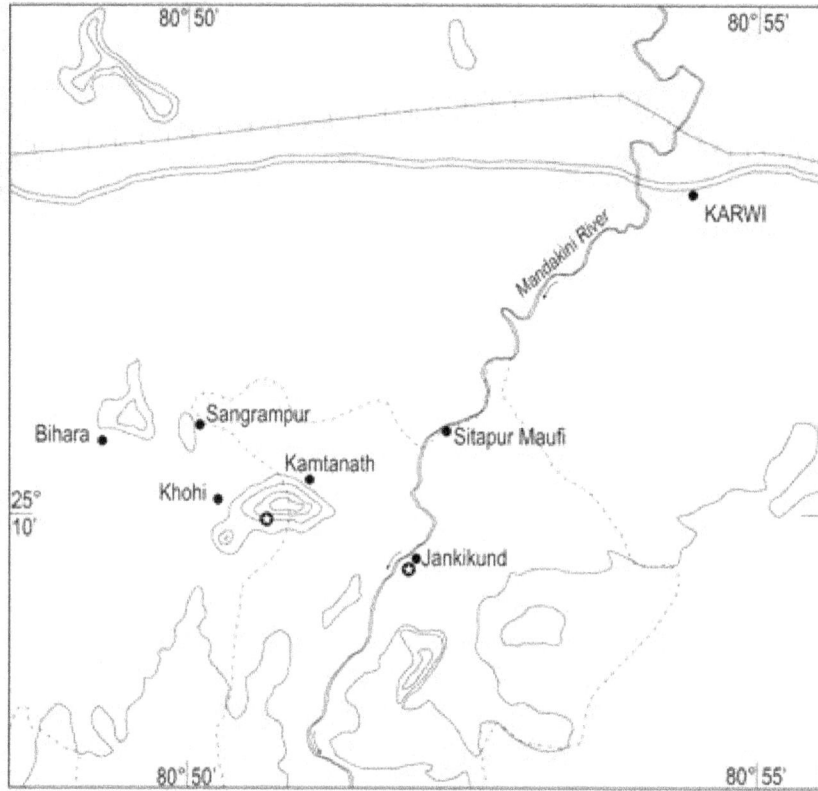

Figure 2. Map showing the position of sample collection sites in study area.

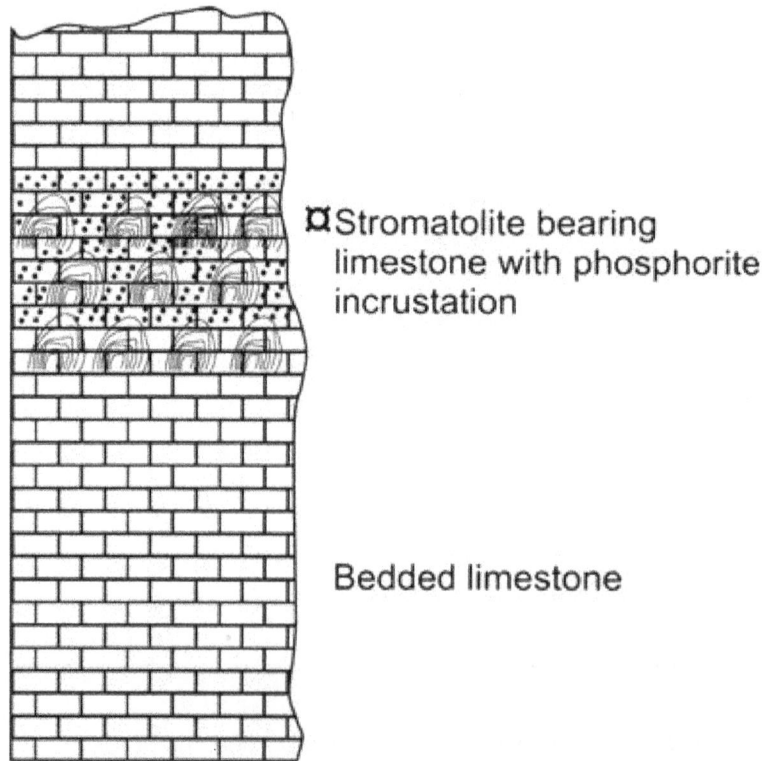

¤ Stromatolite bearing limestone with phosphorite incrustation

Bedded limestone

Figure 3. Lithocolumn of Jankikund section (Anabrasu, 2001). Star (¤) showing the position of fossilliferous unit.

There is, as found so far, no other coeval occurrence of metazoan and algal information from the Proterozoic deposits in the world. The OWMs recorded data have been summarized from the Vindhyan Supergroup (Prasad et al., 2005). The aim of the present study is to disseminate and add new information for the findings of triploblastic animals belonging to the lower metazoans group and their origin including habit and habitat on the land nearly Ediacaran period earlier than the evisences previously suggested by pioneer earth scientists.

MATERIALS AND METHODS

Petrographic thin sections were prepared both in perpendicular aspect as well as on the bedding planes of the surface phosphoritic chert samples associated with columnar stromatolites, Jurussania (Krylov), in bedded limestone. The preservation of the microfossils varies from section to section due to cutting of samples from different angles (Figure 4b). Variation is even more obvious in sections from the same sample and depends upon the thinness of the sections. The slides have been examined and microphotographed under immersion oil from transmitted optical light in an Olympus BH2 microscope under 40X and 100X. The studied materials, slides, and photographic negatives have been deposited in the museum of Birbal Sahni Institute of Palaeobotany, Lucknow, India (Statement No. BSIP 1168).

OBSERVATIONS AND RESULTS

The present findings—that is, with two kinds of preserved microfossil remains —have been recorded for the first time from the phosphoritic chert intercalated in Tirohan limestone (equivalent to the Rohtas Limestone Formation) of the Semri Group of the Vindhyan Supergroup exposed in Chitrakoot, Madhya Pradesh. The uniquely preserved cryptic forms are multicelluar, bilaterally symmetrical, monocious, cylindrical and elongated coiled bodies with clitella, ornamented structures, and setae. The dark brown organic-walled microfossils—prokaryotes, and simple sphaeromorphs (protists) are sparse, cellularly preserved, distorted, small in size. These specimens, recovered as scattered Mesoproterozoic fossils, are of remarkable potential value for understanding the evolution and divergence of metazoans of India.

Figure 4a represents a fossil earthworm specimen in cross section, measuring 165.0 μm long and 10.0 μm broad, showing a simple cylindrical and elongated body, somewhat flattened, annulated, with pointed anterior and blunt posterior. The anterior end possesses a small, fleshy, preoral lobe, a prostomium under which lies the mouth. There is no well-demarcated head, but there appears to be a single oral segment, a peristomium behind the prostomium. The body is divided into a series of 30 to 70 circular metameres, or ring-like segments, with prominent grooves. A groove between segments is about 1.0 μm thick and bears two or three dark spots, possibly nodules, on the body. Each segment, which is

separated by a groove, bears four pairs of "S" shaped chitinoids, locomotor appendages, and setae measuring ±3.0 to 5.0 μm, situated on the ventral side of the body; the first and last segments, however, lack them. A prominent saddle shaped ± 8.0 μm long clitellum or cingulum is situated between metameres or segments 33 and 36.

There are no parapodia, and the last segment of the posterior end, the pygidium, bears an oval ventral anus, which is invisible here due to the angle of cutting of the material. The size of these fossil forms compares to extant forms in a ratio of 1: 15 to 30. The reported fossil form most resembles the extant *Lumbricus terrestris*, which lives in moist soil and makes deep burrows. It is found in Europe and North America, as well as in the eastern ghat of India (Bhal, 1950). Figure 4c is of a fragment of fossil earthworm showing some metameres measuring 30.0 μm long and 10.0 μm broad. Figure 4h represents fragmentary small prokaryotes, aseptate trichomes, Siphonophycus (upper arrow), and simple protists, Leiosphaeridia (lower arrow), resembling known forms from Palaeoproterozoic and Mesoproterozoic transitory sediments (Prasad et al., 2005; Knoll et al., 2006). The food matter consumed by an earthworm is discharged as excrement or fecal matter in casts on the ground surface during the action of digging and boring in the earth, as shown in Figure 4e (enlarged round pellets) and 4i. This material interpreted as fecal matter, resembles the fecal matter of extant forms. Figure elements 4d, f, and g show layers of the body with engulfed materials.

DISCUSSION

We discuss the present findings from chert, intercalated on the top of the limestone during the formation of the Semri group, in light of the available data on other fossils of the lower Vindhyan. The calcified metazoan tubes associated with stromatolites are also known from the Nama group of Namibia (Grotzinger et al., 2000). It is assumed that the origin and development of metazoans depended on the availability of essential requirements such as oxygen, salicic acid and phosphate minerals in carbonate host rocks in suitable oceanic water conditions from the Proterozoic to the present. The oxygen-producing carbonaceous mega-remains of Eukaryotes were reported from Palaeoproterozoic and Mesoproterozoic transitory sediments (Han and Runnegar, 1992; Rai and Singh, 2006). Atmospheric oxygen rose to a level that allowed aerobic metabolism (Knoll, 1996). The development of biocommunties is dependent on the phosphatic minerals which are essential for survival (Sly et al., 2003). The oxygen isotope has reversible properties as toxin and atoxin based on the composition of surface and oceanic water, which may differ in the same lithofacies, particularly carbonate host rocks usually having salicic properties in ion exchange that affect water

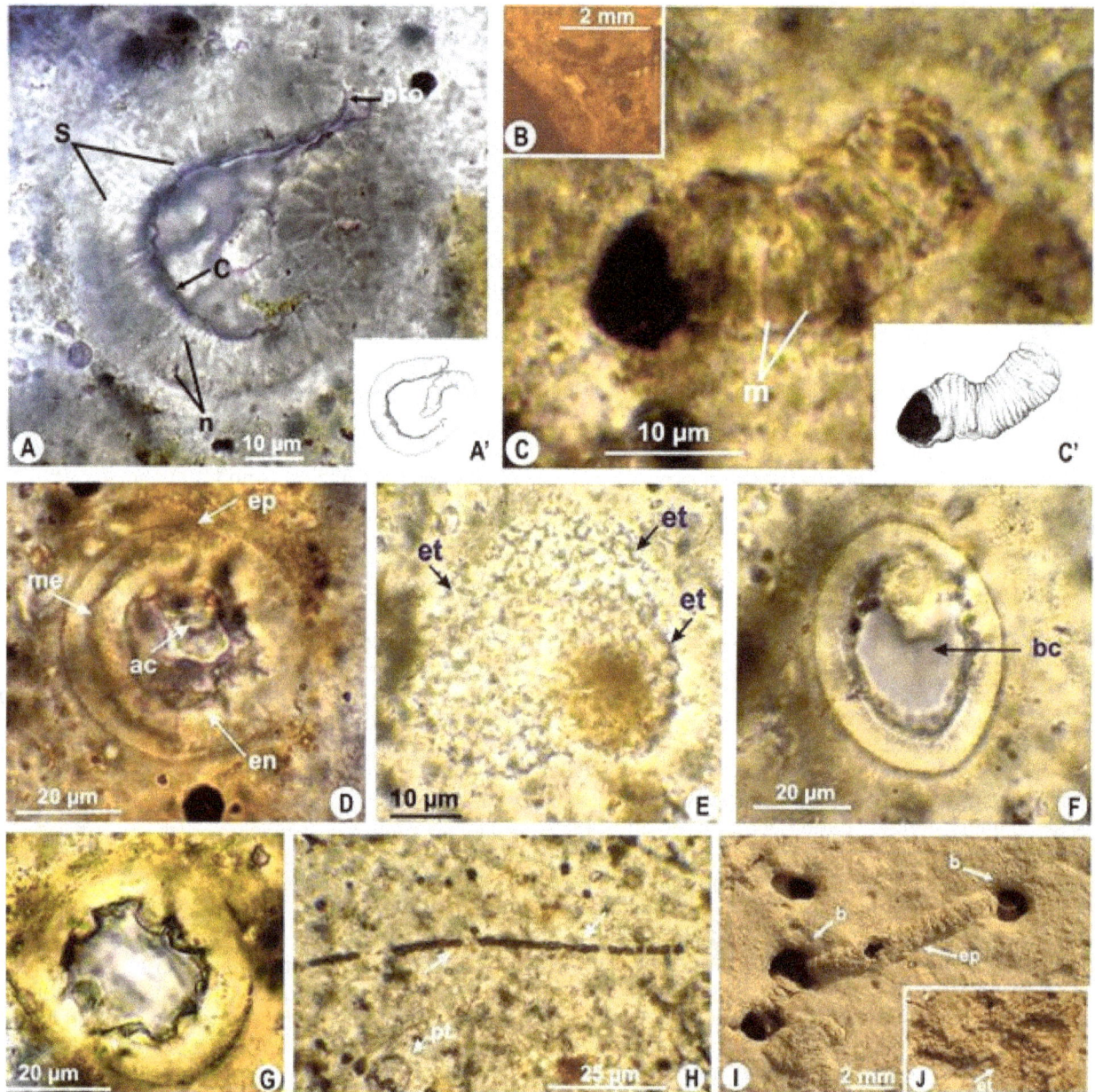

Figure 4. (a) A complete crescent or sickle shaped fossil annelid earthworm. (a') line diagram of 'a' (BSIP Slide no. 13317); (b) A view of fossil-bearing chert section; (c) A cylindrical fragment of fossil annelid earthworm; (c') line diagram of 'c' (BISP Slide no. 13319); (d, f, g) Transverse sections of annelid earthworm fossils body in different angled position showing body layers (BSIP Slide no.13317-13319); (e) excrement or fecal material (BSIP Slide no. 13319); (h) *Siphonophycus* sp. , upper arrow ; *Leiosphaeridia* sp. lower arrow (BSIP Slide no. 13319); (i) A specimen of fossil annelid earth worm and burrows in which earthworms take shelter; (j) excrement or fecal matter of fossil annelid earthworm (BSIP Slide no. 13319). Abbreviations b-*burrow*; bc- *body cavity* c- *clitellum* ac –*alimentary canal*; ep-*epidermis;* en- *endodermis;* et-*excreta* ; m-*metameres or body* segments; me-*mesodermis;* n-*nephridiopore;* pro-*prosternium;* pt- *protist*s; s-*setae.* (Bar in each figure showing the measurements of the forms).

quality for worms.

Preserved forms such as compressions, impressions, molds and biorelics stromatolites is always a matter of debate (Walter, 1972; Fendonkin, 2003; Narbonne, 2005; Prasad et al., 2005; Knoll et al., 2006). The importance of essential minerals in lithofacies belonging from the

Proterozoic–Lower Cambrian age has been debated since their discovery (Cook and Shergold, 1984; Gehling, 1999). The body plan of invertebrate fossil fauna from Neoproterozoic sediments is radial in parazoa and sponges and bilateral in metazoans such as mollusks, annelids, and arthropods (Fedonkin, 2003; Narbonne,

2005; Maithy and Babu, 1997; Chen et al., 2000). The earliest macroscopic fossils representing worms are traces formed on the bedding planes of Mesoproterozoic sandy facies from Australia and India (Seilacher et al., 1998; Maithy and Babu, 1988).

The oldest invertebrate fossil is a parazoan in the Porifera. A new body fossil and oldest tissue-grade colonial eukaryotes have been reported from India and Canada (Maithy and Babu, 1986; Fedonkin and Yochelson, 2002). Probable Proterozoic fungi and the origin of the bilaterian body plan with the evolution of a developmental regulatory mechanism have been discussed by Davidson et al. (1995) and Butterfield (2005).

The present findings are added to earlier records from Pre-Ediacaran up to Mesoproterozoic sediments in the world. The discovery of triploblastic fossils (Seilacher et al., 1998) and small shelly faunas (Azmi, 1998) a decade ago generated controversy regarding the age and nature of metazoans from lower Vindhyan sediments. In an attempt to resolve this issue, micropaleontological information on prokaryotes and protists from overlain and underlain glauconitic limestone of the Vindhyan Supergroup in the Paisuni and Son River valleys has been studied (Prasad et al., 2005). Gregory et al. (2006) recently summarized their age as indicated by various methods. Recently published report (Bengston et al., 2009) proofs more than a billion years older origin for the "Cambrian fossils" reported by Azmi (1998). In some parts of the world, earlier recorded annelid fossils are preserved in bedding planes of Mesoproterozoic rocks; these are actual annelid fossils rather than uninformative artifacts of inorganic origin or pseudo-fossils.

The body wall of invertebrates including annelids may fail to fossilize due to microbial action and post-mortem disintegration. Earlier reported worms from Neoproterozoic sequences are cellularly preserved organisms without the well-marked essential affinities of worms and possibly have most decay-resistant reproductive parts of some prokaryotes in persisting biocommunties. The paucity of multicellular epibiont and endolithic microbiota from this transitional zone suggests that they were either destroyed or survived due to changes of habit and habitat, such as adoption of hibernation, decrease of size, or distorted OWM with certain ornamentation. The extant microbiocommunties appear to be the most sensitive and tolerant. According to Darwinian law, reproductive success would be enhanced by changing their lifestyle and physiology to better fit a prevailing ecosystem that was created by abrupt changes due to natural disaster during deposition, probably by mucoci-liary creeping or peristaltic crawling on sandy facies, and hibernating in soil for their existence.

The reported annelid fossils appear to be true tube dwellers, the tubes of which may be composed of mucus and hardened to a parchment-like consistency and upon which particles of sand or shell stick together. A secretion of lime is then laid down on this mucous framework. The

worms reported were up to 4470 m deep sediments and the extant form found in sand and mud was up to 5550 m deep. The nature has provided appropriate strength to persist during the Columbia and Rodinia supercontinent assembly, and its convergent range is ca. 19000 to 1100 Ma (Karlstorm et al., 2001; Rogers et al., 2003). It is our assumption that the third law of Newton could be most appropriate to this view.

Metazoan findings in the present study are the most significant for Precambrian sediments. They also speak to a hypothesis based upon a Paleogene floral assemblage, eucalyptus and their geographical distribution in three continents (Mehrorta, 2003); extant annelid earthworms are mostly found in the duff beneath these trees (Bhal, 1950). The comprehensive analyses of geographical distributions of dinosaurs with natural calamities (such as craters) in time strongly support the same view based on other data.

Conclusion

The composition and distribution of the present findings, published data on palaeobiology as well as on extant forms from the Vindhyan and elsewhere in the world, indicate the Calymian age and complex geodynamics, stable and unstable environments of the upper part of the Semri group, and it also reflects the divergence of this animal phylum in the terminal Proterozoic followed by a monophyletic process. One can confidently say that most probably India, Australia and South America were together as one supercontinent, Gondwana, at that time based on comprehensive analysis of synergetic data.

ACKNOWLEDGEMENTS

The authors are thankful to Dr. N. C. Mehrotra, Director, Birbal Sahni Institute of Palaeobotany, Lucknow, India for providing necessary facilities and permission to publish research work in prestigious journal. The authors are highly indebted to Prof. Steve Westrop, and Dr. Sandy Dengler, University of Oklahoma for critical review and admirable suggestions of the manuscript.

REFERENCES

Ayala FJ, Rzhetsky A, Ayala FJ (1998). Origin of metazoan phyla, molecular clocks confirm paleontological estimates. PNAS USA, 95: 606-611.
Azmi RJ (1998). Discovery of lower Cambrian small shelly fossils and brachiopods from the lower Vindhyan of Son valley, Central India. J. Geol. Soc. India, 52: 381-389.
Bengston S, Belivanova V, Rasmussen B, Whitehouse M (2009). The controversial "Cambrian" fossils of the Vindhyan are real but more than a billion years older. PNAS USA, 106(19): 7729-7734.
Bhal NK (1950). The Indian Zoological memoirs, on Indian animal types:, Lucknow Publishing House, Lucknow, P. 84.
Butterfield NJ (2005). Probable Proterozoic fungi. Palaeobiology, 31(1):

165-182.

Chen JY, Oliveri P, Li C W, Zhou GQ, Gao F, Hagadorn JW, Peterson KJ, Davidson EH (2000). Precambrian Diversity, Putative phosphotised embryos from Doushantuo Formation, China. PNAS USA, 97: 4457-4462.

Cook PJ, Shergold JH (1984). Phosphorus, phosphorites and skeletal evolution at the Precambrian-Cambrian boundary. Nature, 308: 231-236.

Davidson EH, Peterson KJ, Cameron RA (1995). Origin of bilaterian body plans, evolution of developmental regulatory mechanisms. Science, 270: 1319-1325.

Fedonkin MA (2003). The origin of the metazoan in the light of the Proterozoic fossil record. Paleontol. Res., 7(1): 9-14.

Fedonkin MA, Yochelson EL (2002). Middle Proterozoic (1.5 Ga) Horodyskia moniliformis Yochelson and Fedonkin, the oldest known tissue grade colonial Eukaryote. Smithsonian Contribution to Palaeobiology, 94: 29.

Gehling JG (1999). Microbial mats in terminal Proterozoic siliciclastic, Ediacaran death marks. Palaios, 14: 40-57.

Gregory LC, Meert JG, Tamrat E, Malone S, Pandit MK, Pradhan V (2006). A palaeomagnetic and geochronologic study of the Majhgawan Kimberlite India, implication for the age of the Upper Vindhyan Supergroup. Precamb. Res., 149: 65-75.

Grotzinger JP, Watters WA, Knoll AH (2000). Calcified metazoans in thrombolite-stromatolite reefs of the Terminal Proterozoic Nama Group, Namibia. Paleobiology, 26: 334-359.

Han TM, Runnegar B (1992). Megascopic Eukaryotic algae from the 2.1 Billion year old Negaunee Iron Formation, Michigan. Science, 257: 232-235.

Jermilin LS, Poladian L, Charleston MA (2005). Is the "Big Bang" in animal evolution real? Science, 310: 1910.

Karlstrom KE, Áhäll KI, Harlan SS, Williams ML, Mclelland J, Geissman JW (2001). Long-Lived (1.8-1.0 Ga) Convergent Orogen In Southern 12 Laurentia, Its Extensions to Australia and Baltica, and Implications for Refining Rodinia. Precam. Res., 111: 5-30.

Knoll AH (1996). Breathing rooms for early animals. Nature, 382: 111-112.

Knoll AH, Javaux E, Hewitt D, Cohen P (2006). Eukaryotic organisms In Proterozoic oceans. Philos. Trans. Royal Society of London 'B', 361(1470): l023-1038.

Kumar A, Gopalan K, Rajagopalan G (2001). Age of Lower Vindhyan sediments, Central India. Curr. Sci., 81(7): 806-809.

Maithy PK, Babu R (1986). Misraea, a new body fossil from the lower Vindhyan Supergroup (Late Precambrian) around Chopan, Mirzapur district, U.P. Geophytology, 16(2): 223-226.

Maithy PK, Babu R (1988). The Mid-Proterozoic Vindhyan macrobiota from Chopan, South east Uttar Pradesh. J. Geol. Soc. India, 31(6): 584-590.

Maithy PK, Babu R (1997). Upper Vindhyan biota and the Precambrian/Cambrian Boundary. Palaeobotanist, 46: 1-6.

Mehrotra RC (2003). Status of plant mega fossils during Early Paleogene in India. Special J. Geol. Soc. Am., 369: 415-422.

Narain K (1960). Vindhyan sedimentation in the Karwi area. Record Geol. Surv. India, 98(2): 91-106.

Narbonne GM (2005). The Ediacaran biota: Neoproterozoic origin of animal and their ecosystems. Ann. Rev. Earth Planet Sci., 33: 421-442.

Prasad B, Uniyal SN, Asher R (2005). Organic walled microfossils from the Proterozoic Vindhyan Supergroup of Son valley, Madhya Pradesh, India. Palaeobotanist, 54: 13-60.

Rai V, Singh VK (2006). Discovery of megascopic multicellularity in deep time, New evidences from the ~ 1.63 Billion years old Lower Vindhyan succession, Vindhyan Supergroup, Uttar Pradesh, India. J. Appl. Biosci., 32(2): 196-203.

Rasmussen B, Bengtson S, Fletcher IR, Mcnaughton NJ (2002). Discoidal impressions and trace-like fossils more than 1200 Million years old. Science, 296: 1112-1115.

Rogers M, John JW, Santosh M (2003). Supercontinents in Earth History. Gondwana Res., 6(3): 357-368.

Sastri MVA, Moitra AK (1984). Vindhyan stratigraphy-A review. Memoir. Geol. Surv. India, 116: 1-37.

Seilacher A, Bose PK, Pfluger F (1998). Triploblastic animals more than 1 billion years old, Tracefossil evidence from India. Science, 282: 80-83.

Singh IB, Kumar S (1978). On the stratigraphy and sedimentation of the Vindhyan sediments in Chitrakoou area, Banda district (U. P.) – Satna district (M. P.) J. Geol. Soc. India, 19: 359-367.

Sly BJ, Snoke MS, Raff RA (2003). Who came first - larvae or adults? Origins of bilaterian metazoan larvae. Int. J. Dev. Biol., 47: 623-632.

Tang C, Twahara J, Glore GM (2006). Molecules after work together in complex. Nature, 444: 383-386.

Walter MR (1972). Stromatolites and the biostratigraphy of Australian Precambrian and Cambrian. Special Paper, Paleontol. Assoc. London, 11: 190.

Wray GA, Levinton JS, Shapiro LR (1996). Molecular evidences for deep Precambrian divergences among metazoan phyla. Science, 274: 568-573.

8

Comparative study of *β-lactoglobulin* gene polymorphism in Holstein and Iranian native cattle

Abbas Doosti[1*], Asghar Arshi[1,2], Mehdi Yaraghi[1] and Mehdi Dayani-Nia[1]

[1]Biotechnology Research Center, Islamic Azad University, Shahrekord Branch, Shahrekord, Iran.
[2]Islamic Azad University, Shahrekord Branch, Young Researchers Club, Shahrekord, Iran.

β-Lactoglobulin (β-LG) is the major protein in whey, making up about 50 to 60% of the total protein. A biological task for β-LG has been the subject of much speculation, but no specific function has been proven. About eleven protein variants of β-LG are known, of which variants A and B are common in most cattle breeds. The purpose of this study was to identify genotypes of the *β-LG* gene in Holstein and Iranian native cattle using polymerase chain reaction-restriction fragment length polymorphism (PCR-RFLP) technique. Blood samples were obtained from 278 Holstein and 210 Iranian native cattle. The primers β-LG F and β-LG R amplified a 247 bp fragment. The *Hae*III enzyme was used for restriction of the polymerase chain reaction (PCR) products. The frequency of allele *A* and *B* in Holstein and native cattle were 0.53, 0.47 and 0.23, 0.77, respectively. According to previous studies *AA* genotypes of β-LG gene is an important factor in increasing the milk production and the results of this study showed that high frequency of *AA* genotypes in Holstein and Iranian native cattle and could be use in creating the next generation for increasing the milk production.

Key words: Polymorphism, *β-lactoglobulin* gene, polymerase chain reaction-restriction fragment length polymorphism, Holstein, native cattle.

INTRODUCTION

β-Lactoglobulin (β-LG) is the major whey protein in the milk of ruminants, horses, pigs, cats, dogs, dolphins, whales, and kangaroos but not in the milk of humans, rodents, and lagomorphs (Perez et al., 1995). β-LG is amphiphatic and an extremely acid stable protein which exists at the normal pH of bovine milk as a dimer with a molecular weight of 36,000 daltons. It is a single chain polypeptide of 18 kDa comprising of 162 amino acid residues (Karimi et al., 2009).

The quantitative effects of the variants on milk composition and cheese-making properties have been functions of this protein are still not known. It could have a role in metabolism of phosphate in the mammary gland reported frequently (Lum et al., 1997). The biological and the transport of retinol and fatty acids in the gut (Hill et al., 1997). The *β-LG* gene is situated on bovine chromosome 11 and encodes the main protein of whey. *β-LG* loci affect the milk production parameters and quality of milk protein. Their polymorphisms partly explain the genetic variance and improve the estimation of breeding value. Such loci can be taken into account as suitable supplements to conventional breeding procedures (Karimi et al., 2009). Polymorphism of this gene was discovered in 1955 and more than ten alleles are known so far (Matejicek et al., 2007). Two most common allelic variants of β- lactoglobulin (β-LG) are *β-LG A* and *β-LG B* (Lum et al., 1997). The point mutations in exon IV of the bovine *β-LG* gene determine two allelic variants A and B.

The bovine *β-LG A* variant differs from the B variant by two amino acids only, for example aspartate-64 and valine-118. These amino acids are substituted by glycine and alanine, respectively in the B variant (Rachagani et al., 2006). Heterozygous cattle have significantly more allelic production of *β-LG A* than of *β-LG B* (Elmaci et al., 2006). The *β-LG* locus affects mainly milk composition and milk quality and the *B* allele has been especially

*Corresponding author. E-mail: biotechshk@yahoo.com.

Abbreviations: *β-LG*, Beta-lactoglobulin; RFLP, restriction fragment length polymorphism.

recognized as superior for milk quality in cattle breeds, whereas allele *A* is associated rather with yield parameters (Strzalkowska et al., 2002). The aim of this study was to determine the frequencies of alleles and genotypes of *β-LG* gene in Holstein and Iranian native cattle using polymerase chain reaction-restriction fragment length polymorphism (PCR-RFLP) method (Table 1).

MATERIALS AND METHODS

Samples collection and DNA extraction

528 blood samples were collected in vacutainers containing sodium EDTA as an anticoagulant from 278 Holstein and 250 native cattle. The tubes were maintained at -20°C until used for DNA extraction. Genomic DNA was extracted from 100 µl blood samples. The gel monitoring was used for determination of the DNA quality and quantity.

β-lactoglobulin gene amplification and genotyping analysis

Genotyping of *β-LG* was performed using PCR-RFLP. The 247 bp *β-LG* promoter fragment was PCR amplified from genomic DNA. The sequences of primers used for amplification of exon IV of *β-LG* gene containing polymorphic sites for *A* and *B* alleles were: *β-LG* F: 5′TGTGCTGGACACCGACTACAAAAAG-3′ and *β-LG* R: 5′ GCTCCCGGTATATGACCACCCTCT- 3′. Amplification reactions were done in a final volume of 25 µl, containing 100 µl DNA, 0.2 µM of each primer, 1X PCR buffer, 1.5 mM MgCl$_2$, 200 mM dNTPs and 1U of *Taq* polymerase. Thermal cycling conditions included: an initial denaturation step at 95°C for 5 min followed by 30 cycles of 94°C for 1 min, 60°C for 1 min, 72°C for 1 min and a final extension for 5 min at 72°C. PCR products were recognized by electrophoresis on 2% agarose gel. The restriction digestion of the PCR products was carried out with *Hae*III enzyme. The PCR products were subjected to digestion by restriction enzymes in a total volume of 20 µl (10 µl PCR product, 2 µl enzyme buffers, 0.2 µl enzymes, and 7.8 µl distilled water) and placed in the incubator at 37°C for 4 h. The digested products were analyzed on 2% agarose gel.

Statistical analysis

The genotype and allele frequencies of the *β-LG* locus were estimated by direct counting. The χ^2 test was performed on the basis of the Hardy–Weinberg law for determining genetic equilibrium.

$$x^2 = \frac{\sum (0-e)^2}{e} \qquad G^2 = -2(LnLo - LnL1)$$

RESULTS

Digestion of 247 bp fragment of *β-LG* gene by *Hae*III restriction endonuclease generated two types of restriction pattern: two fragments at 99, 74 bp sizes (*BB*-genotype) and three fragments 148, 99 and 74 bp (*AB*-genotype).

The restriction pattern that includes two fragments of 148, 99 bp was *AA*-genotype. In the analyzed popula-tions of cattle, the *BB* genotype was most frequent. Frequencies of *AA*, *BB* and *AB* genotypes were 0.280, 0.220, 0.498 and 0.052, 0.592, 0.354 in Holstein and native cattle, respectively. The frequencies of the *A* and *B* alleles were 0.53, 0.47 in Holstein and 0.23, 0.77 in native cattle, respectively. A representative PCR-RFLP pattern is depicted in Figure 1. Genotype frequencies were in accordance with the Hardy-Weinberg equilibrium (P-value >0.001).

DISCUSSION

β-Lactoglobulin, a globular protein of 162 amino acid residues, is the most abundant whey protein in bovine milk (Ariyaratnea et al., 2001). A large part of the variation in amount of *β-LG* protein is associated with *β-LG* protein variants (Ganai et al., 2008). Certain genetic variants of *β-LG* are correlated with the presence of a high percentage of milk fat. So far, 11 genetic variants which encode different forms of the *β-LG* protein have been discovered, thus influencing the quality of the milk: A, B, C, D, E, F, G, H, I, J and W. Out of these, the A and B forms are of the highest interest since they are associated with milk production performances. The *BB* homozygote individuals supply milk rich in fat and protein, very valuable in the process of cheese making, while the *AA* homozygote ones supply milk with a low percentage of fat (Balcan et al., 2007). Variant *A* has a higher *β-LG* protein concentration than variant *B* (Hill et al., 1997). It is likely that this difference in amount of *β-LG* protein is not caused by the amino acid substitutions, but rather by different levels of expression of the corresponding *A* and *B* alleles of the *β-LG* gene (Ganai et al., 2008). Variants *A* and *B* differ at two sites. Asp64 in *A* is changed to Gly in *B*, and Val118 in *A* is changed to Ala in *B* (Ariyaratnea et al., 2001). Milk from cows with the *β-LG BB* genotype had a greater fat percentage, while the milk yield and protein percentages of the *β-LG AB* genotype cows were higher, but this difference was not significant (Karimi et al., 2009).

The study of Kumar et al. (2009) on *β-LG* gene in Indian goats showed that the amplified product was of 426 bp and the restriction digestion with *Sac*II revealed three genotypes, namely S_1S_1, S_1S_2 and S_2S_2 at the *β-LG* locus (Kumar et al., 2006). Mele et al. (2007) on dairy ewes showed a significant relationship among hetero-zygous *β-LG* and TFA milk content that was more than 20% higher compared to the two homozygote *β-LG*. Patel et al. (2007) in India were showed that *B*-allele frequency (0.72) of beta-lactoglobulin was much higher compared to *A*-allele (0.27) in crossbred dairy bulls.

Karimi et al. (2009) showed that allele frequency of *β-LG* with regard to the *B*-allele (0.9125) was higher than that of the *A*-allele (0.0875). Frequencies of *AB* and *BB* genotypes were 0.175 and 0.825, respectively, while the

Table 1. The frequency of alleles in Holstein and Iranian native cattle.

Population	Frequency A (%)	Frequency B (%)	S.e	Interval confidence A	Interval confidence B
Holstein	0.53	0.47	0.03	0.75 - 0.87	0.13 - 0.25
Native	0.23	0.77	0.03	0.53 - 0.65	0.35 - 0.47

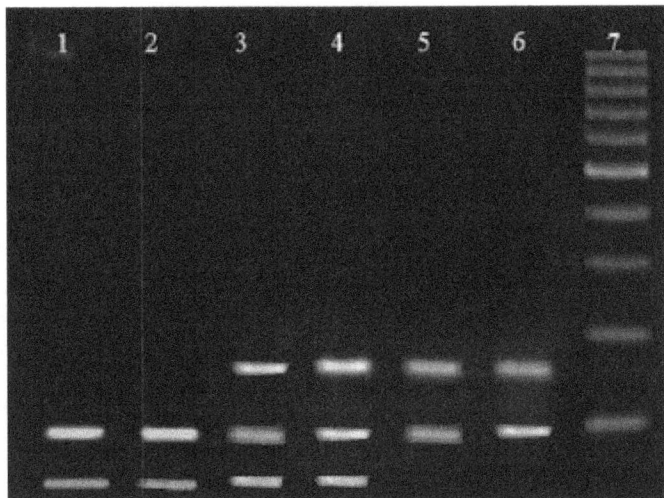

Figure 1. Gel electrophoresis of PCR product after digestion with HaeIII restriction enzyme. Line 1 and 2: BB genotype, line 3 and 4: AB genotype, Line 5 and 6: AA genotype, and line 7: 100 bp DNA marker (Fermentas, Germany).

AA genotype was found to be absent (Karimi et al., 2009).

To test differences between observed and expected frequencies, χ2 analysis was performed on the basis of the Hardy-Weinberg law (Recio et al., 1997). In conclusion, these genotypes appear to be obvious candidates for selection aiming at improving milk production traits. Although the allele frequency of B is high in native population, the A allele (favorable allele) frequency is not too high. It is suggested that cross-breeding should be done between these populations or with exotic cattle to increase the frequency of the favorable allele.

ACKNOWLEDGEMENTS

The authors would like to thank all the staff of the Biotechnology Research Center of Islamic Azad University of Shahrekord Branch in Iran for their sincere support.

REFERENCES

Ariyaratnea K, Browna R, Dasguptaa A, Jongea J, Jamesona GB, Loob TS, Weinbergb C, Norrisb GE (2001). Expression of bovine b-lactoglobulin as a fusion protein in Escherichia coli:a tool for investigating how structure affects function. Int. Dairy J., 12: 311-318.

Balcan RA, Georgescu SE, Adina M, Anca D, Tesio CD, Marieta C (2007). Identification of beta-lactoglobulin and kappa-casein genotypes in cattle. Zool. Biotechnol., 40(1): 211-216.

Elmaci C, Oner Y, Balcioglu M (2006). Genetic Polymorphism of β-Lactoglobulin Gene in Native Turkish Sheep Breeds. Biochem. Genet., 44: 379-384.

Ganai NA, Bovenhuis H, Arendonk J, Visker M (2008). Novel polymorphisms in the bovine b-lactoglobulin gene and their effects on b-lactoglobulin protein concentration in milk. Anim. Genet., 40: 127-133.

Hill JP (1997). The relationship between β-lactoglobulin phenotype and milk composition in New Zealand dairy cattle. J. Dairy Sci., 76: 282-286.

Karimi K, Beigi Nassiri MT, Mirzadeh Kh, Ashayerizadeh A, Roushanfekr H, Fayyazi J (2009). Polymorphism of the β-lactoglobulin gene and its association with milk production traits in Iranian Najdi cattle. Int. J. Biotechnol., 7(2): 82-85.

Kumar A, Rout P, Roy R (2006). Polymorphism of β-lactoglobulin gene in Indian goats and its effect on milk yield. J. Appl. Genet., 47(1): 1-5.

Lum LS, Dovc P, Medrano JF (1997). Polymorphisms of Bovine b-Lactoglobulin Promoter and Differences in the Binding Affinity of Activator Protein-2 Transcription Factor. J. Dairy Sci., 80: 1389-1397.

Matejicek A, Matejickova J, Nemcova E, Frelich J (2007). Joint effect of CSN3 and LGB genotypes and their relation to breeding values of milk production parameters in Czech Fleckvieh. Czech. J. Anim. Sci., 52: 83-87.

Melea M, Conte G, Serra A, Buccion A, Secchiari P (2007). Relationship between beta-lactoglobulin polymorphism and milk fatty acid composition in milk of Massese dairy ewes. Small Ruminant Res., 73: 37-44.

Patel R, Chauhan JB, Singh KM, Soni KJ (2007). Allelic Frequency of Kappa-Casein and Beta-Lactoglobulin in Indian Crossbred (Bos taurus × Bos indicus) Dairy Bulls. J. Vet. Anim. Sci., 31(6): 399-402.

Perez MD, Calvo M (1995). Interaction of β-lactoglobulin with retinol and fatty acids and its role as a possible biological function for this protein. J. Dairy Sci., 78: 978-988.

Rachagani S, Gupta ID, Gupta N, Gupta SC (2006). Genotyping of beta-lactoglobulin gene by PCR-RFLP in Sahiwal and Tharparkar cattle breeds. BMC. Genet., 7: 31-34.

Recio I, Fernandez A, Alvarez M, Ramos M (1997). β-Lactoglobulin polymorphism in ovine breeds: Influence on cheesemaking properties and milk composition. Lait, 77: 259-265.

Strzalkowska N, Krzyzewski J, Zwierzchowski L, Ryniewicz Z (2002). Effect of kappa-casein and beta-lactoglobulin polymorphism cows age, stage of lactation and somatic cell count on daily milk yield and milk composition in Polish Black and white cattle. Anim. Sci. Pap. Rep., 20: 21-35.

9

A new species of the genus *Hilethera* Uvarov (Oedipodinae: Acrididae: Orthoptera) from Pakistan

Barkat Ali Bughio, Riffat Sultana* and M. Saeed Wagan

Department of Zoology, University of Sindh, Jamshoro, Pakistan.

The subfamily Oedipodinae has considerable economic importance in Pakistan. It poses constant threat to pastures, orchards and variety of crops in both irrigated and rain-fed areas. Amongst its members, the genus *Hilethera* has not been studied extensively and it was described by Uvarov in 1925. It comprised two species: *Hilethera hierichonica* Uvarov and *Hilethera aeolopoides* Uvarov, but at present, we have added one new species, *Hilethera balucha* to this genus. This new species is closely related to *Trilophidia annulata* Thunberg because it has a refined body form but can easily be separated from the same on the following bases: it has the structure of pronotum; its anterior projections are conical, extending outwardly. It has lateral plates with convex median process, has shorter ancorae placed angularly, and has hyaline wings that are transparent with a smoky marginal band, and slightly greenish at the base. This insect has been collected from rocky areas with scattered vegetation that consists of grasses and herbs. *H. balocha* new species is described from Balochistan Province of Pakistan.

Key words: Oedipodinae, new species, *Hilethera Balocha*, Balochistan.

INTRODUCTION

The fauna of grasshopper insects belonging to the subfamily Oedipodinae are of considerable economic importance in Pakistan. Their representatives pose constant threat to pastures and variety of crops in both irrigated and rain-fed areas. Some species of Oedipodinae can reach high densities, concentrate their feeding on valued plants, and thus damage the agriculture value of both range and crop land and cause economic loss to mankind. They are commonly known as band-winged grasshoppers. Oedipodinae is distributed in wide range throughout the world; it contains185 genera throughout the world. Amongst the members of Oedipodinae, *Locusta migratoria* Linnaeus causes devastating swarm all over the old world (Vickery and Kevan, 1983). Members of Oedipodinae occur throughout Pakistan due to their diversity of habitats such as agricultural crops, hilly areas and desert like plain.

Mostly, they are known as geophiles (living in open grounds) and phytophyles (found at vegetation, grasses, herbs and shrubs). Literature review showed that the band- winged grasshoppers are being reported as important pests of agriculture.

Earlier, Cotes (1893) recorded a serious damage of *Aiolopus* species in upper Sindh. Mooed (1966) reported the damaging status of *Locusta migratoria* at agricultural fields of Larkana District. Ahmed (1980) surveyed the fauna of grasshoppers of Pakistan and reported that some of the Oedipodinae grasshoppers are severe pests of orchards. Wagan and Solangi (1990) recorded heavy damage of some Oedipodinae species on cultivated crops in different areas of Sindh Province. But their work was restricted to specific region and did not provide sufficient information about its current status. It is therefore essential to identify them accurately; so that diagnosis of an economic problem could be properly made. In this present study, observation has been made of the genus *Hilethera*, that contains three species; and one new species has been described as well.

*Corresponding author. E-mail: riffatumer@hotmail.com.

Figure 1. *Hilethera balucha* Male n sp.

Hilethera balocha n.sp.

Diagnostic characters

Its body is medium size; antennae, filiform with about 20 to 21 segments, and slightly longer than the head and pronotum together. Its head is sub-globular, thickly constricted with light brown spots; having sparse hairs, smaller than the pronotum. It has round prominent eyes.

Description of holotype ♂

Its body is medium size; antennae, filiform with about 20 to 21 segments, and slightly longer than the head and pronotum together. Its head is sub-globular, thickly constricted with light brown spots; having sparse hairs, smaller than the pronotum. It has round prominent eyes (Figure 1). Fastigium of vertex is raised at the anterior; is pentagonal, sub-angular, and concave with undulated lateral carinulae; fastigial foveolae is somewhat larger and triangular; frontal ridge is flat and slopes down. Pronotum has saddled shape, and densely punctured; anterior margin is obtusely round, while posterior is rectangular. Pronotum is highly constricted in prozona; median carina, visible, sharp, and slightly deep in prozona. Tegmina and wings well developed with obtuse rounded apices. Tegmina have a v- shape mark at the base. Hind femur is stout, smaller but wide a little; finally, dorsal carina is entire, and dorsal genicular lobes are smoothly round. Hind tibia is slender, with 9 inner and 8 black tipped spines. Claws are shorter and arolium, smaller.

General coloration

Generally, it is light yellowish brown in color. Antennae is dark

brownish with white spots near the base. Fastigium of vertex has smaller black spots. Tegmina are semi-transparent with two smoky dark bands; apex has light brown speckles. Hyaline wings are transparent, have a slightly smoky marginal band and greenish a little at the base. Hind femur is internally black except for a pale pre-apical band. Hind tibia is dirty whitish at the base; and has a pre-median and pre-apical ring.

Description of allotype ♀

It is similar to that of the male but larger in size. Antennae is filiform with about 23 to 24 segments, slightly longer than the head and pronotum together. Head is sub-globular, thickly constricted with light brown spots; having sparse hairs, smaller than the pronotum. It has round prominent eyes. Fastigium of vertex is raised at the anterior; is pentagonal, sub-angular, concave with undulated lateral carinulae; fastigial foveolae is slightly larger, and triangular; frontal ridge is flat and slopes down. Pronotum has saddled shape, densely punctured; anterior margin is obtusely round, while posterior is rectangular. Pronotum is highly constricted in prozona; median carina, visible, sharp, and slightly deep in prozona. Tegmina and wings well developed with obtuse rounded apices. Tegmina have v- shape mark at the base. Hind femur is stout, and smaller but wide a little; dorsal carina is entire, and dorsal genicular lobes are smoothly round. Hind tibia is slender, with 10 inner and 9 black tipped spines. Claws are shorter and arolium, smaller. Supra-anal plate is long, flat, with marked but narrow transverse carina; margin is obtusely round. Cerci is short and conical; and has sub acute rounded apices. Ovipositor has curved valves, and pads of ventral valves in tubercles. Sub-genital plate is long, conical, with margin obtusely round.

General coloration

Generally, it is light yellowish brown in color. Antennae is dark brownish. Fastigium of vertex has smaller black spots. Tegmina are semi-transparent with two dark bands; apex has smoky brown speckles. Hyaline wings are transparent with a smoky marginal band; and slightly greenish at the base. Hind femur is internally black except for a pale pre –apical band. Hind tibia is dirty whitish at the base, and has a pre-median and pre- apical black ring.

Phallic complex

Apical valve of penis is like vertical plough at the dorsal part and shorter than the valve of cingulum; valve of penis tapers towards the apex with rounded sub acute apices (Figure 2a to d). Valve of cingulum is finger like form, straight upwardly, and larger than that of penis valve; has rounded acute apices at apex, and slightly deep at the base. Arch of cingulum is flat, curved inwardly and well developed. Basal bridges are folded slightly thin. Apodemes are smaller, narrower, straight; anterior has rounded acute apices. Zygoma is remarkable, smaller and is up ward. Rami is irregular, dorsally extending like lobe and denticulate at margins. Gonopore is like a straight, thin rod, having angular apices. Ejaculatory duct is shorter and broad, directed anteriorly. The epiphallus is attached to the ninth sternite and to the zygoma by muscular tissues; epiphallus has bridge shape; bridge is straight, narrow and thick a little. Anterior projections are conical, protruding outwardly with obtuse rounded apices; lateral plates emarginate marginally, and have incurved processes. Posterior projections have externo–lateral expansions at the base. Ancorae is shorter and placed angularly, is sub acute round at the apex, with lightly wider median processes, and round at the base. Lophi is moderate with acute apices, curves upwardly; is convex; anteriorly directed with slightly wide apical

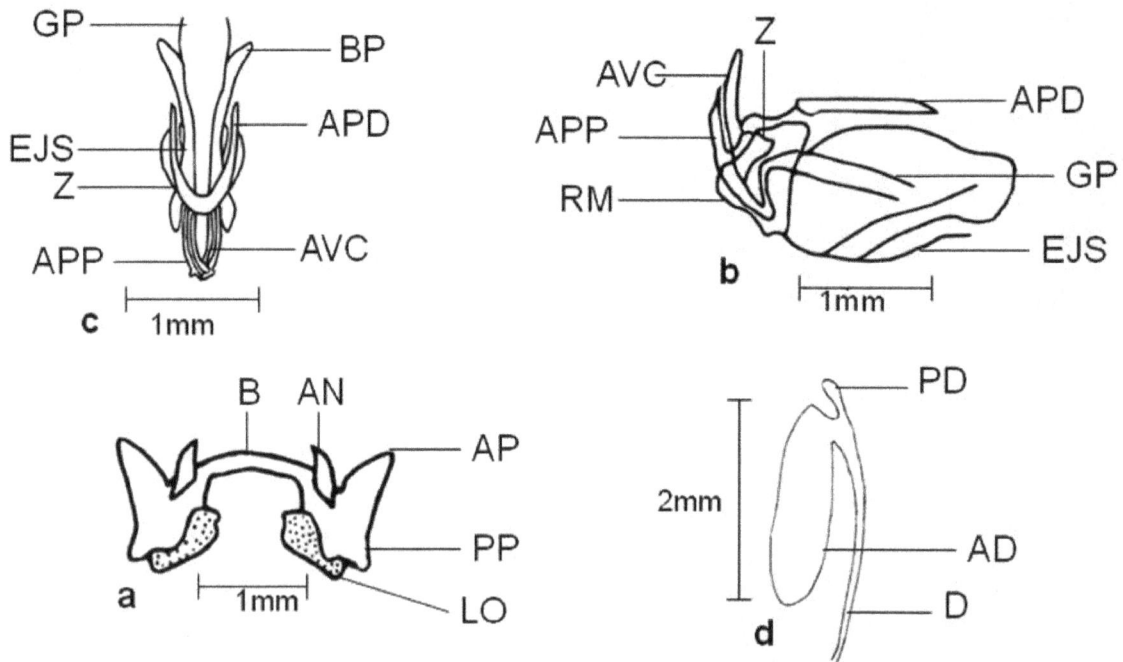

Figure 2. *Hilethera balucha,* a) Epiphallus, b) Endophallus and Cingulum lateral view, c) same dorsal view, d) Spermatheca.

lobes, ending into small rounded terminal apices. Lophi is set with enormous smaller spines, besides the lateral plates, with oval circular sclerites.

Female: Cerci is short, conical and hairy. Ovipositor has medium length, valves are curved and robust. The spermatheca usually resembles a long tube and opens on the dorsal wall of the genital cavity, opposite the genital opening. Pre – apical diverticulum is laterally placed, broad at apex with sub acute rounded apices. Apical diverticulum has tube like shape, elongated with rounded processes at the base. Measurement of various body parts is shown in Tables 1 and 2

Materials examined

Balochistan: Loralai: Shabozai Holotype (2♂♂) 15.x.2001, Allotype (3♀♀ Female), 15.x.2001, Paratype (16♂♂, 18 ♀♀) 22.x.2001 (Wagan, Barkat and Riffat)

Etymology

This species has been named after the locality.

Repository

The material has been deposited in the Museum of Entomology, Department of Zoology, University of Sindh Jamshoro, Pakistan.

DISCUSSION

Walker (1870) first raised the status of Oedipodinae family level. Since then it has been considered as a family or subfamily. Kirby (1914) and Bei-Bienko and Mishchenko (1951) considered it as a subfamily. Dirsh (1956) included it in subfamily Acridinae.

However, Uvarov (1966) clearly separated this subfamily from Acridinae since it has been regarded as subfamily by Dirsh (1975), Vickery and Kevan (1983) and Otte (1995). This subfamily differs from all other subfamilies because it has the presence of strong and serrated intercalary vein of median area of tegmen and the mesosternal interspace is about twice wider than long. Uvarov, in his series of publications (1921, 1929, 1942a, b), gave the comprehensive description of various species of Oedipodinae and in 1925 he also revised the genera *Hilethera* with addition of two new species *Hilethera hierichonica* and *Hilethera aeolopoides*. In the present, we added one new species *Hilethera balucha*, making it now catalogued as three species.

Tokhai (1997) recorded 17 species of band-winged grasshoppers from Balochistan Province of Pakistan. Baloch (2000) described 20 species including 2 new species Oedipodine grasshoppers from Punjab Province of Pakistan. Garai (2001) studies the grasshoppers of Pakistan and listed 19 species of Oedipodine grasshoppers mostly from N.W.F.P. (now Khyber Punkhton Khawa). Now the status of Oedipodinae (= locustinae) as a subfamily is accepted by Roberts (1940), Kevan and Knipper (1961), Uvarov (1966), Dirsh (1975), Vickery and Kevan (1983), Eades (2000) and Eades and Otte (2010). As the present study is based on a small number of species, we strongly recommend that if more

Table 1. The measurement of various body parts of *Hilethera balocha* (holotype).

Parameter	Holotype ♂	Allotype♀
Length of body	15.0	18.0
Length of antennae	6.5	7.0
Length of pronotum	4.0	4.2
Length of tegmina	17.2	19.0
Maximum width of tegmina	2.5	4.1
Length of hind femur	9.1	10.0
Maximum width of hind femur	3.6	4.0
Length of hind tibia	8.0	9.0

Table 2. The measurement of various body parts of *Hilethera balocha* (paratype).

Parameter	Paratype ♂♂		Paratype ♀♀	
	(Mean ± Sd)	(Range)	(Mean ± Sd)	(Range)
Length of body	15.33±0.89	15-16	18.8±3.60	17-21
Length of antennae	6.66±1.62	6-8	6.2 ±1.49	6-7
Length of pronotum	3.33±3.69	3-4	4.0±00	4-4
Length of tegmina	14.66±1.63	14-16	18.22 ±3.5	17-21
Max. width of tegmina	2.66±0.38	2.5-3	4.0±00	4-4
Length of hind femur	8.33±0.81	8-9	10.0±1.41	9-11
Max. width of hind femur	3.66±0.80	3-4	3.84±0.69	3.2-4
Length of hind tibia	7.33± 0.81	7-8	8.22±1.59	7.1-9

extensive survey is carried out throughout the country it would be helpful to find out new diversity in the collection of genus *Hilethera* from this region.

REFERENCES

Ahmed FU (1980). Survey of Grasshoppers in Arid and semi Arid region of Pakistan Final Rep. Pl-480 No.P.K-ARS-20 (FG-Pa-21), p. 500.

Bei-Beinko GY, Mishchenko LL (1951). Locust and Grasshoppers of U.S.S. R and adjacent countries. 1 & 2 Monson, Jerusalem pp. 691.

Baloch N (2000). Survey and taxonomy of grasshoppers belonging to family Acrididae (Orthoptera) of the Punjab. Ph. D. Thesis, University of Sindh, pp. 1-197.

Cotes EC (1893). A conspectus of the insects which affects crop in India. Indian Mus. Notes, 2: 145-176.

Dirsh VM (1956). The Phallic complex in Acridoidea (Orthoptsa) in relation to taxonomy. Trans. R. Ent. Soc. Lond., 108(66): 223-256.

Dirsh VM (1975). Classification of the Acridomorphoid Insects. Farringdon E.W. Classey Ltd., 8: 171.

Eades DC (2000). Evolutionary relationships of phallic Structures. Acridomarpha (Orthoptera). J. Orthoptera Res., 9: 18-40.

Eades DC, Otte D (2010). Orthoptera species file on line. Version 2.0/3.5.<http://Orthoptera Species File'Org>.

Garai A (2001). Orthopteroid Insects from Pakistan. Esperiana, Bd, 9: 431-447.

Kirby WF (1914). Orthoptera (Acrididae)The fauna of British India including Ceylon and Burma London Taylor and Francis, p. 276.

Kevan DKMcE, Kniper H (1961). Garad fluglerous Qstafrika (Orthopteriods. Dermapteroida and Blattopteroda) *Beitr. Ent.*, 3: 356-413.

Mooed A (1966). Taxonomy of Tetrigidae (Tetrigoidea, Orthoptera) and Acridinae and Oedipodinae (Acrididae, Acridoidea, orthoptera) of Hyderabad region. M.Sc Thesis, University of Sindh, p. 260.

Roberts HR (1940). A comparativ study of the subfamilies of Acrididae (Orthoptera) primarily on the Basis of their phallic structure. Proc. Acad. Nat. Sci. Phild., 93: 201-246. 90 figs. Acridiae (Orthoptera). J. Morph., 71: 523-576.

Tokhai S (1997). Survey and taxanomy of Orthoptera of Zhob Division (Balochistan) and adjoining areas. M.Phil thesis. University of Sindh, pp. 1-201.

Uvarov BP (1921). Notes on the Orthoptera in the British Museum. 1. The group of Euprepocnemini. Trans. Ent. Soc. Lond., 1(2): 106-143.

Uvarov BP (1925). The genus *Hilethera* Uv and its species (Orth: Acrid). Eos, 1: 33-42.

Uvarov BP (1941). Geographical variation in *Scintharista notabilis* (Walker, 1870) (Orthoptera, Acrididae). Proc. R. Ent. Soc. Lond., (B) 10: 91-97.

Uvarov BP (1942a). New Acrididae from India and Burma. Ann. Mag. Nat. Histol., 9: 587-607.

Uvarov BP (1942b). Palaerctic Acrididae new to the Indian Fauna. Eos., 18: 97-103.

Uvarov BP (1966). Grasshoppers and Locusts. A hand book of general acridology Cambridge Univ. Press (London), 1: Xii + 481.

Wagan MS, Solangi SM (1990). Distribution and incidenceof grasshoppers (acrididae) of Sind. Biol. San. Veg. Plagas. (Fuera de serie), 20: 125-129.

Walker F (1870). Catalogue of specimens of *Dermaptera saltatoria* in the collection of the British Museum . London, 3: i- iv + 425-604, 4: i-iv +605-809.

10

Application of premetamorphic oral cavity electron micrographs for Egyptian toads' taxonomy

Gamal Mahmoud Bekhet

[1]Department of Zoology, Faculty of Science, Alexandria University, Alexandria 21511, Egypt.
[2]Department of Biological Sciences, College of Science, King Faisal University, Al-Hassa 31982, Saudi Arabia. E-mail: dr_gamal_bekhet@yahoo.com.

In the present study, the microanatomy of both the oral disc and buccal cavity of the tadpole of *Bufo regularis* was described. Tadpoles of 32, 38 and 40 stages were dissected and analyzed using scanning electron microscope. In all the stages, the mouth was ventral and the oral disk width was large, that is, equal to about 44% of the greatest width of the body. The disk was provided with a broad gap on the lower lip; the rest of the mouth was bordered by a large number of papillae. The papillae were arranged in a single row on the dorsolateral part of the mouth; the ventrolateral and ventral lip was surrounded by a double row of papillae. The number of papillae increased with larvae growth, from zero in stage 32 to about 150 in stage 40. The tooth row formula is 2(1)/3(2). The upper and lower beaks were pigmented and serrated. While the upper beak was broadly arched and formed a smooth arc, the lower beak had V-shape. Premetamorphic papillae were observed during the early metamorphic stages, and these degenerated rapidly at about late metamorphic stage. Metamorphic atrophy of the oral structures occurred roughly in the reverse order of development, although the procedure was rapid and more haphazard than the development. We suggested that the oral flaps and the roof papillae play a significant role in the capture of food particles by establishing the inflow of "alimentary water", and aggregating food particles and mucus inside the buccopharyngeal cavity, which may reflect ecological and functional constraints that are relative to the morphology of other suspension feeding anuran larvae. Herein, we described the oral features of the tadpoles of *B. regularis*.

Key words: Anuran tadpoles, scanning electron microscope, oral disc, oral papillae, jaws, gap.

INTRODUCTION

The number and arrangement of tooth rows on the oral disc of tadpoles is specific. The labial tooth row formula (LTRF) is a synaptic representation of this arrangement. A number of systems have been devised for numbering labial tooth rows and designating the rows with medial gaps. Accordingly, we used the fractional designation (Altig, 1970) to specify the number and gross morphology of tooth row. This system accommodates all row configuration easily, does not use Roman numerals, has been used for quite a long time, is familiar to many workers and takes less space than formulas that are written in spatial order.

Rows on the anterior labium are numbered from distal (labial margin) to proximal (mouth). The notation "A-1" denotes the first anterior (most distal from mouth) row; also A-n denotes the row adjacent to the mouth. Rows on the posterior labium are numbered proximal to distal. The first row to the mouth is p-1, and more rows are numbered sequentially through P-n. Rows with medial gaps are designated with parentheses, and rows that vary between individuals (gap present or absent) are placed within brackets. A gap in a tooth row is a physical break in the tooth ridge and therefore expressed in the tooth row. The functional and developmental considerations of gaps in tooth rows have not been examined. These gaps allow larger excursions of the jaw sheaths during feeding.

In summary, a labial tooth row formula (LTRF) of 5 (2-5) / 3[1] indicates a tadpole with 5 upper tooth rows with medial gaps in rows A-2 through A-5 and 3 lower rows with or without a gap in P-1. Some tadpoles, particularly of Ranids species and Pelotids species have accessory rows situated in the lateral areas of the oral disc. So the formulation of Altig (1970) was modified by Webb and Korky (1977) by placing the number of accessory rows between solid, 5 (2-5)4/3[1].

Variations in the size, density and shape of the oral disc, the papillae at the margins of the oral disc, the shape of the jaws, the numbers of denticle rows and any gaps in those rows are all important features in identifying tadpoles of different species. Even among closely related taxa and in many cases, they seem to reflect lineage and habitats (Duelman and Trueb, 1983, 1986; Grandison, 1981; Channing, 2001). McDiarmid and Ronald (1999), Nascimento et al. (2005), and Rossa-Feres and Nomura (2006) showed that in Bufonidae, both oral disc and keratinized mouth parts were present and oral disc emerged; teeth formula was 2/2 or 2/3. In Ranidae, the formula was smaller (3/3) or larger (5/3, 2/4, 3/4 or 6-7/6); in Hylidae, Pipidae "Xenopus laevis" and Rhinophrynidae "Rhinophrynus dorsalis", oral disc did not emerge and the formula was 2/2, 2/3 or 2/4. Tadpoles of Hylorina sylvatica have unique characteristics (Echeverría et al., 2001; de Sá and Langone, 2002; Formas and Brieva, 2004; Alcalde and Blotto, 2006; Altig, 2007; Vieira et al., 2007).

The Central American tree frog, Hyla microcephala, which feeds on large food particles in ponds, has small jaws set back in an oral tube and has no labial teeth. Tadpoles of a certain tropical stream-dwelling Hyla, in contrast, have the highest number of rows: 17 upper and 21 lower rows. The tadpole of Rhamphophryne proboscidea is characterized by a small size (17.8 mm), with tooth row formulae of 2(2)/3, oral papillae only on lateral margins and jaw sheath of V-shape (Menin, 2006; Gomes et al., 2010; Zimkus et al., 2010). Albertina (2007) shows that seven papillae of oral disc of Colostethus marchesianus tadpole occur on the right side of the disc and six on the left, probably resulting from the fusion of two papillae. A few labial teeth are missing from A-1 in this specimen. Tadpoles of Eupsophus emiliopugini have four lingual and four infralabial papillae and third lower labial ridge is absent. Savage (2002) showed that in Bufo coniferus, the mouth is moderate and anteroventral and the oral disc emerges with the papillae of the upper labium confined to the corners of the mouth; on the right side, there are 10 to 13 small papillae in the outer row and three to six inner, and on the left side there are 10 to 11 outer and 2 to 4 inner. The lower labium is also free of papillae except at the corners; on each side, there are 9 to 15 papillae. The incidence of oral deformities could be high in natural populations. For instances Batrachochytrium dendrobatidis infection exerts a strong influence on the occurrence and type of oral deformities in tadpoles (Dana et al., 2007; Frost, 2009; Matthew et al., 2010; Marion et. al., 2010).

Matthew et al. (2010) suggested that tadpoles with missing teeth compensate for inferior feeding kinematics during mouth closing in each gape cycle by increasing the number of gape cycles per unit time. The ways in which these structures actually function have received little study. However, it is clear that large oral discs with many denticle rows are common among stream-dwelling tadpoles exposed to water currents. The larvae use these structures to hold on to surfaces and resist being swept downstream (Van and McCollum, 2000; Tolledo et al., 2009). Jaw sheaths that have sharp edges are characteristic to many tadpoles that feed on active prey. High-speed video of feeding North American bullfrog (Rana catesbeiana) larvae, which have the common pattern of two upper and three lower tooth rows, showed that tadpoles use their labial teeth to anchor the oral disc to surfaces while their jaws bite at the substrate.

Feeding behaviour changes drastically during metamorphosis as larval suction feeders become adult lingual feeders (Sanderson and Sarah, 1999; Relyea, 2001). In order to understand this transition, the general morphological development of the floor of the buccal cavity in Egyptian anurans species Bufo regularis larvae was studied up to the completion of metamorphosis by scanning electron microscope.

MATERIALS AND METHODS

Ribbons of fertilized eggs from couples of the available B. regularis were collected from breeding sites. After hatching, the larvae were daily fed on a meal of boiled spinach daily, until they reached the desired stages needed for experimental work, according to the normal table of Sedra and Michael (1961). Three larval stages were used for the present study namely, stages number 32, 38 and 40. Specimens were dissected and subsequently fixed in a 2 to 3% glutaraldehyde solution for 3 to 4 h at room temperature, then washed in 0.1 M phosphate buffer for 15 min each. Next, specimens were dehydrated in a graded ethanol series as follows: 35, 50, 70, 80, 95%; three changes at 100% for 15 min each, and a final wash in acetone for 5 min. Specimens were dried in CO_2, mounted on aluminium stubs and sputter coated with gold. Features of dorsal and ventral internal oral anatomy were examined and photographed using a scanning electron microscope (Jeol) attached to a computer. Terms used to describe features of the oral cavity are derived from Wassersug (1997, 1980) and Wassersug and Heyer (1988).

RESULTS

In the early stages, particularly stage 32, it is shown that the oral disc structures were not developed yet in the mouth (Figure 1) while during development of the tadpoles in stage 36, it is observed that the oral structures started to appear gradually. In this, the upper jaw sheath was formed first followed by the 87 appearance of the ventrolateral margins of the lower labium, which were considered the first soft tissues of the oral disc that materialized from the surrounding body surfaces. Nascent marginal papillae subsequently appeared in these areas before they did on the other margins of the disc. After that the lateral emergency

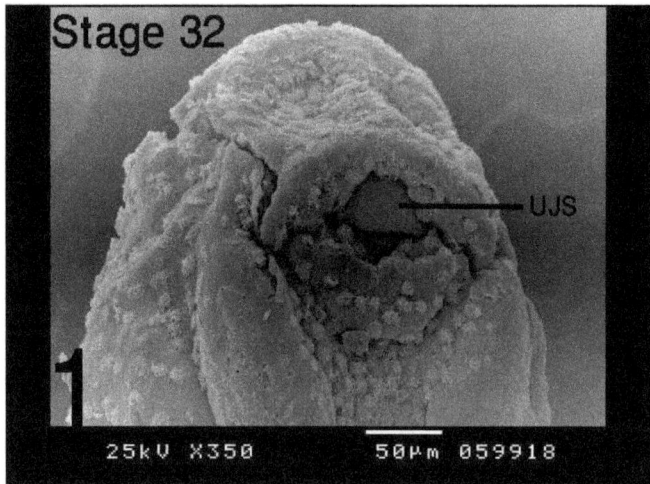

Figure 1. Scanning electron micrograph of oral apparatus of *B. regularis* tadpole (stage 32). UJS- upper jaw sheath.

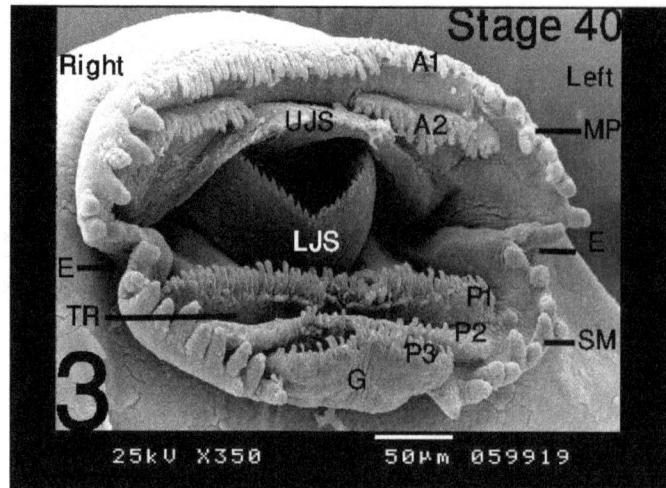

Figure 2. Scanning electron micrograph of oral apparatus of *B. regularis* tadpole (stage 36). A-1 and A-2, First and second anterior tooth rows; E, lateral emargination of oral disc; LJS, lower jaw sheaths; MP, marginal papillae; P-2-3, second and third posterior tooth rows; SM, submarginal papillae; UJS, upper jaw sheath.

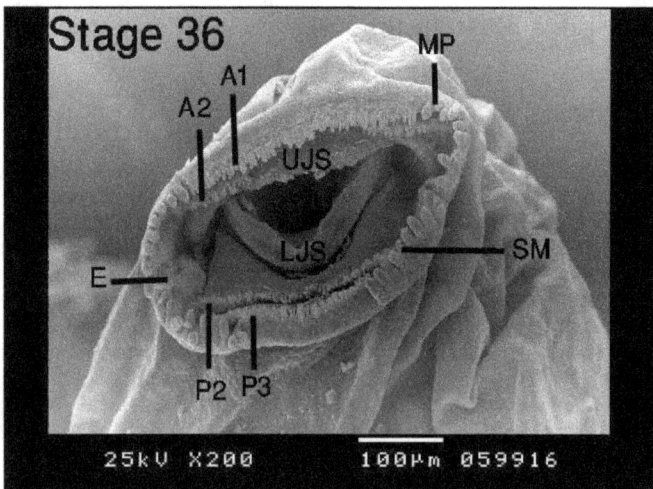

Figure 3. Scanning electron micrograph of oral apparatus of *B. regularis* tadpole (stage 40) with 2(1)/3(2) tooth rows formula. A-1 and A-2, first and second anterior tooth rows; E, lateral emargination of oral disc; LJS, lower jaw sheaths; MP, marginal papillae; P-1-3, first through third posterior tooth rows; SM, submarginal papillae; TR, tooth ridge for tooth row P-2; UJS, upper jaw sheath.

appeared with small notch on one side only. In addition, there were only two rows of oral papillae which appeared in the posterior side, named (p2 and p3) (Figure 2). The fully developed oral disc of the tadpole of *B. regularis* (Figure 3; stage 40) consisted of various structures (although some tadpoles lack all of these), two keratinized jaw sheaths for grasping and shearing food, one above and the other below the mouth; tooth row ridges that bear the keratinized teeth (these existed as rows above, or anterior to the jaw sheaths named the A rows, and below or posterior to the jaw sheaths named the P rows), and one or more rows of papillae along the borders of the oral disc (these can completely surround

the oral disc or be interrupted at the anterior or both anterior and posterior sides).

Oral disc located ventrally, emerged on both sides (Figure 3); border of disc was surrounded with 38 marginal papillae: 16 located antero-laterally (eight on right side and eight on left side); and 22 post-laterally papillae (11 on right side and 11 on left side). Also, there was a single dorsal gap in papillae while the submarginal papillae were absent, and all these papillae were the last structures to be atrophied during metamorphosis. Furthermore, it was noted that the ventral (posterior) gap was clearly discernable and there was no dorsal (anterior) gap formed. Concerning the jaw sheaths in the present result, the upper jaw sheath barely had a concave medial shape while the lower jaw sheath took V-shape; both upper and lower jaw sheaths had serrated edges; serrations extended on the entire lengths of sheaths. Labial tooth row formula LTRF of the tadpoles of *B. regularis* was: 2(2)/3(2). The corners of the mouth of a tadpole did not extend backwards to form a toad mouth until tissue atrophies.

DISCUSSION

Our observations indicate that the oral morphology of *B. regularis* could serve as a specialized feeding mode. The functional roles that have been proposed for oral papillae fall into the following basic categories: chemosensory, tactile receptors and structures that control water flow (Van Dijk, 1981), enhance attachment to substrates (Altig and Brodie, 1972), modify the shape of the oral disc during feeding and manipulate food and substrate particles.

The number and prominence of marginal papillae varies concordantly with observed reductions in the size of the oral disc among tadpoles within the same lineage. For example Ranid tadpoles have larger papillae arranged more sparsely than Hylids; and stream Hylids have smaller, more densely arranged papillae than pond Hylids. For these and other reasons, oral structures have been used infrequently in systematic analyses (Grandison, 1981, Duelman and Trueb, 1983, 1986; Rdel, 2000; Anstis, 2002). As Eterovick et al. (2002) suggested that, the tadpoles of the species in the *Hypsiboas polytaenius* clade may be distinguished by their tooth row formula. *H. polytaenius* has LTRF 2(2)/3(1,2) (Heyer et al., 1990); *Hypsiboas cipoensis,* 2(2)/3(1) and *H. goianus,* 2(1,2)/3(1) (Eterovick et al., 2002).

Although the tadpole of *B. regularis* presented the same LTRF as *H. cipoensis,* 2(2)/3(1), the position of the marginal papillae was different. *H. cipoensis* has a single row of marginal papillae on the upper and lower lips, presenting a rostral gap, and two rows laterally, while *H. leptolineatus* has a row of marginal papillae on the upper lip, also with a rostral gap, and two rows of papillae laterally and on the lower lip. Our results on the internal oral anatomy of the tadpole of *B. Regularis* tadpole are in agreement with previous description (Wassersug and Heyer, 1988), but we are adding two new characters that were observed in the present study, viewed by means of a scanning electron microscope. The first character is that, from the roof of the oral cavity there was a posterior gap instead of anterior one as in the other species and the second one is: the oral discs of toad tadpoles showed a caudad double row of papillae.

Tadpoles of these species possibly do not aggregate when predators are present compared to *Boraras maculatus* that does (Channing, 2001). Tadpoles of *B. regularis* differ from other species by having (i) numerous papillae surrounding the oral disc compared to 18 papillae in back-riding tadpoles in stages 25 to 62 papillae; 12 to 16 papillae in *Colostethus marchesianus* and 13 to 16 papillae in *C. caeruleodactylus,* in stages 25 to 42 (Caldwell et al., 2002; Castillo, 2004 Trenn, 2004); (ii) small and numerous papillae arranged in a double row surrounding the oral disc, except for a medial gap in the posterior disc. These characters may be useful in elucidating the phylogeny of *B. regularis* within its genus and even to provide evidence of relationships at the species level. These variables likely reflect an nterplay of evolutionary history and functional demands (for example, stream vs. pond dwellers). Species with rapid developmental times (for example, *Bufo*) seemingly start and complete develop-mental sequences slightly earlier than species with longer developmental times. Species with labial tooth row formula greater than 2/3 start development earlier (Marinelli and Vagnetti, 1988), complete it later and retain mouth parts longer into metamorphosis than species with tooth formula equal to or lesser than 2/3 tooth rows.

Lentic species tend to form the oral structures later and faster and atrophy the oral apparatus earlier and faster than lotic forms. For these reasons, oral development is often discordant with features of limb development used in staging (Tubas et al., 1993; Liu et al., 1997).

The mechanisms that account for the formation of marginal and submarginal papillae are unknown, although apoptosis is surely an important mechanism. As development progresses toward metamorphosis, these structures gradually regress until they disappear.

REFERENCES

Albertina P, Diego E, Jesus R (2007). A New Amazonian Species of the Frog Genus *Colostethus* (Dendrobatidae) that Lays its Eggs on Undersides of Leaves. Copeia, 1: 114-122.

Altig R (2007). A primer for the morphology of anuran tadpoles. Herp. Boil., 2: 71-74.

Anstis M (2002). Tadpoles of South-eastern Australia: A Guide with Keys. Sydney, Australia: New Holland Publishers.

Antonella B, Elvira B, Emilio S , Sandro T (2008). The oral apparatus of tadpoles of *Rana dalmatina, Bombina variegata, Bufo bufo,* and *Bufo viridis* (Anura). Zoologischer Anzeiger. J. Comp. Zool., 247: 47-54.

Castillo-Trenn P (2004). Description of the Tadpole of *Colostethus kingsburyi* (Anura: Dendrobatidae) from Ecuador. J. Herp. 38(4): 600-606.

Channing A (2001). Amphibians of Central and Southern Africa. Cornell University Press.

Caldwell JP, Lima AP, Biavati GM (2002). Descriptions of tadpoles of *Colostethus marchesianus* and *Colostethus caeruleodactylus* (Anura: Dendrobatidae) from their type localities. Copeia, 2002: 166-172.

Dana L, Altig R, James B, Susan C (2007). Occurrence of Oral Deformities in Larval Anurans. Copeia, 2: 449-458.

Erik H, Jan C, Rene P, Cristina F (2010). Descriptions of the Tadpoles of Two Poison Frogs, *Ameerega parvula* and *Ameerega bilinguis* (Anura: Dendrobatidae) from Ecuador. J. Herp. 44(3): 409-417.

Frost D (2009). Amphibian Species of the World: An online reference. Version 5.3. American Museum of Natural.

Gomes F, Provete D, Martins I (2010). The tadpole of *Physalaemus jordanensis* Bokermann, 1967 (Anura, Leiuperidae) from Campos do Jordo, Serra da Mantiqueira, Southeastern Brazil. Zootaxa 2327: 65-68.

Liu H, Wassersug RJ, Keiji K (1997). The Three Dimensional Hydrodynamics of Tadpole Locomotion. J. Exp. Biol., 20: 2807-2819.

Marinelli M, Vagnetti D (1988). Morphology of the oral disc of *Bufo bufo* (Salientia: Bufonidae) tadpoles. J. Morphol., 1951: 71-81.

Marion A, Michael J, Roberts J, Luke C, Paul D (2010). A new species of *Litoria* (Anura: Hylidae) with a highly distinctive tadpole from the north-western Kimberley region of Western Australia. Zootaxa 2550: 39-57.

Matthew DV, Wassersug RJ, Matthew JP (2010). How Does a Change in Labial Tooth Row Number Affect Feeding Kinematics and Foraging Performance of a Ranid Tadpole (*Lithobates sphenocephalus*)? Biol. Bull., 218: 160-168.

McDiarmid R, Altig R (1999). Tadpoles: The Biology of Anuran Larvae. Chicago: University of Chicago Press, Ltd., London.

Menin M, Rodriguez D, Lima A (2006). The Tadpole of *Rhinella Proboscidea* (Anura: Bufonidae) with notes on adult reproductive behavior. Zootaxa, 1258: 47-56.

Nascimento L, Caramaschi U, Cruz C (2005). Taxonomic review of the species groups of the genus *Physalaemus* Fitzinger, 1826 with revalidation of the genera *Engystomops* Jiménez-de-la-Espada, 1872 and *Eupemphix* Steindachner, 1863 (Amphibia, Anura, Leptodactylidae). Arquivos do Museu Nacional, 63: 297-320.

Rdel M (2000). Herpetofauna of West Africa, Vol. I. Amphibians of the West African Savanna. Edition Chimaira, Frankfurt, Germany.

Relyea R (2001). Morphological and Behavioral Plasticity of Larval

Anurans in Response to Different Predators. Ecology. 2: 523-540.

Rossa-Feres D, Nomura F (2006). Characterization and taxonomic key for tadpoles (Amphibia: Anura) from the northwestern region of São Paulo State, Brazil. Biota Neotropic., 6: 1.

Sanderson S, Sarah J (1999). Development and Evolution of Aquatic Larval Feeding Mechanisms. In The Origin and Evolution of Larval Forms, edited by Brian K. Hall and Marvalee H. Wake. San Diego: Academic Press.

Savage J (2002). The Amphibians and Reptiles of Costa Rica. University of Chicago Press, Chicago and London.

Sedra SN, Michael MI (1961). Normal table of the Egyptian toad, *Bufo regularis* Reuss, with an addendum on the standardization of the stages considered in previous publication. Cesk. Morf. 9: 333-351.

Tolledo J, Oliveira E, Feio R, Weber L (2009). Distribution extension and geographic distribution map. Amphibia, Anura 5: 422-424.

Trenn P (2004). Description of the Tadpole of Colostethus kingsburyi (Anura: Dendrobatidae) from Ecuador. J. Herp., 38: 600-606.

Tubas L, Stevens R, Altig M (1993). Ontogeny of the oral apparatus of the tadpole of *Bufo americanus*. Amphibia-Reptilia., 14: 333-340.

Van B, McCollum S (2000). Functional Mechanisms of an Inducible Defence in Tadpoles: Morphology and Behavior Influence Mortality Risk from Predation. J. Evol. Biol., 13: 336-347.

Wassersug R (1997). Assessing and Controlling Amphibian Populations from the Larval Perspective. In Amphibians in Decline: Canadian Studies of a Global Problem, edited by David Green. Herpetological Conservation, St. Louis: Society for the Study of Amphibians and Reptiles Publications. Vol. 1.

Webb RG, Korky JK (1977). Variation in Tadpoles of Frogs of the *Rana tarahumarae* Group in Western Mexico (Anura: Ranidae). Herp. 33: 73-82.

Zimkus B, Rödel M, Hillers A (2010). Complex patterns of speciation and diversity among African frogs (genus *Phrynobatrachus*). Mole. Phylo. Evol., 55: 883-900.

11

Changes in the haematological profile of the West African hinge-backed Tortoise (Kinixys erosa) anaesthetized with ether or thiopentone sodium

Saba A. B.* and Oridupa O. A.

Department of Veterinary Physiology, Biochemistry and Pharmacology, University of Ibadan, Ibadan Nigeria.

The effect of ether or thiopentone sodium on haematological parameters of tortoise was determined by evaluating the Packed Cell Volume (PCV), Red Blood Cell (RBC) count, Haemoglobin (Hb) concentration, Mean Corpuscular Haemoglobin Concentration (MCHC), Mean Corpuscular Volume (MCV), Mean Corpuscular Haemoglobin (MCH) and White Blood Cell (WBC) count of ether or thiopentone-anaesthetized West African Hinge-Backed Tortoise (Kinixys erosa). The blood clotting and sleeping time were also determined. Fifteen tortoises were randomly but equally divided into three groups. Tortoises in Group I were administered with 0.9% Physiological saline, while tortoises in Group II and III were administered with ether and thiopentone sodium respectively. Ether was administered by inhalation while thiopentone sodium was administered intramuscularly. The sleeping time was significantly longer for ether than for thiopentone sodium. The difference in the sleeping time is ascribed to the differences in the physicochemical, pharmacokinetic or pharmacodynamic properties of the two anaesthetics. Blood clotting was delayed in tortoises anaesthetized with ether compared with thiopentone-anaesthetized tortoises, which makes thiopentone a more reliable anaesthetic agent for invasive surgical procedures in the tortoise than ether. The two anaesthetics elicited depression of the haematological parameters of the tortoises with significant (P<0.05) decreases in the PCV and RBC values. Hb concentration and MCV were significantly decreased for ether-anaesthetized tortoises, while MCH was significantly decreased for thiopentone-anaesthetized tortoises. The WBC count was elevated in ether-administered tortoises while the value decreased in thiopentone-administered tortoises. The elevation of WBC count was attributed to the irritant effect of ether. It was concluded that ether or thiopentone caused depression of haematological parameters in West African Hinge-Backed tortoise which should be taken into consideration when interpreting values of blood parameters obtained from anaesthetized subjects.

Key words: Ether, thiopentone sodium, haematology, sleeping time, West African hinge-backed tortoise.

INTRODUCTION

The characteristic shell sets the tortoise distinctly apart as a separate order that can not be confused with other animals. They are probably the only reptiles that most humans view with prejudice and the number of people who like the tortoise is surprisingly large. Both terrestrial tortoise and aquatic turtles are often kept as pets in gardens, terrarium and aquarium. As pets, these animals may perio-

dically require evasive or invasive handling for routine Veterinary checks, administration of drugs, exami-nation of the oral or anal orifice, endoscopy, repair and dressing of wounds and fractures, minor and major surgeries, amongst others (Balcombe et al., 2004). Anaesthetic drugs (local or general) may be required to achieve these purposes, several of which have been employed in the tortoise. These include methohexital sodium (Gaztelu et al., 1991; Jackson et al., 2000), atipamezole (Sleeman and Gaynor, 2000; Dennis and Heard, 2002), medetomidine (Sleeman and Gaynor, 2000; Dennis and Heard, 2002), isoflurane and sevoflurane (Heard, 2001),

*Corresponding author. sabadee200@yahoo.com, ab.saba@mail.ui.edu.ng.

ketamine (Holz and Holz, 1994; Greer et al., 2001) and propofol (Heard, 2001; Bertelsen, 2007).

Ether, an inhalant anaesthetic, is known to be fast-acting and cheap, causing minimal physiological effects, but irritating to the respiratory tract (Brunson, 1997) and stressful during the induction period (Van Herck et al., 2001). Thiopentone sodium is an intravenous, short-acting barbiturate, with a rapid onset. It is known to reduce intracranial pressure more effectively than pentobarbital, which is a desirable effect in comatose animals (Cole et al., 2001; Pérez-Bárcena et al., 2005). Reports on the effects of ether or thiopentone on haematological parameters of tortoise is very scarce. The need for this is quite pressing when viewed against the background that sample collection procedures and anaesthesia have been shown to exhibit profound effect on biochemical and haematological effects in animals (Frolich et al., 1996; Heard and Huft, 1998; Dressen et al., 1999), which should be considered when interpreting blood values obtained from anaesthetized subjects. This study was therefore aimed at evaluating haematological changes accompanying anaesthesia in West African Hinge-Backed tortoise (Kinixys erosa) using ether and sodium thiopentone as anaesthetic agents.

MATERIALS AND METHODS

Experimental animals

Fifteen tortoises were purchased and kept at the animal house of Department of Veterinary Physiology and Pharmacology, University of Ibadan. They were fed ripe pawpaw, banana, green vegetables and cooked potato, and had access to clean drinking water *ad libitum*, with normal light: dark timing unaltered. A shallow dish about 10 cm deep was placed in the house to allow the animals to submerge their heads into the water.

Measurement of weight

The body mass of the tortoises was measured to the nearest 5 g with 2 kg Soehnle spring balance.

Induction of anaesthesia and evaluation of sleeping time

Anaesthesia was induced using 20mls of ether by placing the tortoises in anaesthetic chamber containing ether-soaked cotton wool. The time taken from placing of the tortoise in the chamber to when the tortoise began to show first signs of drowsiness was taken as induction time. The sleeping time was taken from the first sign of drowsiness to first sign of recovery from anaesthesia, such as movement of the head or limbs. The tortoises designated for the intravenous anaesthetic agent were weighed and 3 mg/kg of thiopentone was administered via the intramuscular route using the Quadriceps muscle as reported by Bouts and Gasthuys (2002). The sleeping time was recorded as done for the inhalant group.

Blood sample collection and evaluation of blood clotting time

Blood samples were collected by sub-carapacial venipuncture from

the tortoises in the three groups following evidence of induction of anaesthesia. A large drop of blood from anaesthetized tortoises was placed on a clean glass slide and mixed continuously with a pin. The time of appearance of the first strand of fibrin was recorded as the blood clotting time. 4 mls of blood was also collected into Lithium-heparinized bottles for haematological analysis. The Packed cell volume (PCV), red blood cell (RBC) count, haemoglobin (Hb) concentration and white blood cell (WBC) count were determined by Cole's method (1986), while the mean corpuscular haemoglobin (MCH), mean corpuscular haemoglobin concentration (MCHC) and mean corpuscular volume (MCV) were calculated.

Statistical analysis

Student t-test was used to analyze the data (Steel and Torrie, 1996). The differences of the means were considered significant at $P < 0.05$.

RESULTS

Sleeping and blood clotting time

The mean sleeping time in the tortoises anaesthetized with ether was 17.03 ± 2.06 min with average blood clotting time of 4.13 ± 0.15 min. Those anaesthetized with sodium thiopentone had mean sleeping time of 4.30 ± 0.05 min and blood clotting time of 3.05 ± 0.21 min. This was a significant ($p < 0.05$) difference between the mean sleeping times for both groups. Tortoises in the control group had a mean blood clotting time of 3.05 ± 0.06 min, and this was significantly ($p < 0.05$) lesser than that observed for tortoises anaesthetized with ether (Table 1).

Haematological parameters

Red blood cell indices

There was a significant reduction in the mean PCV values of tortoises anaesthetized with ether ($18.00 \pm 3.54\%$) and sodium thiopentone ($16.13 \pm 1.04\%$) respectively compared with the unanaesthetized tortoises ($26.54 \pm 0.98\%$). No significant difference was observed between the mean RBC of unanaesthetized tortoises ($0.46 \pm 0.02 \times 10^6/\mu l$) compared with those anaesthetized with ether ($0.44 \pm 0.03 \times 10^6/\mu l$), but there was a significant decrease in those anaesthetized with sodium thiopentone ($0.37 \pm 0.02 \times 10^6/\mu l$) (Table 2).

The mean Hb of tortoises anaesthetized with ether (8.25 ± 0.44 g/dl) were significantly lower than the unanaesthetized tortoises (11.05 ± 0.51 g/dl), while there was no significant difference between the unanaesthetized tortoises and those anaesthetized with sodium thiopentone (8.95 ± 1.19 g/dl) (Table 2). The mean MCH of tortoises anaesthetized with ether (192.35 ± 22.73 pg) was significantly lower than that of the unanaesthetized tortoises (246.10 ± 11.83 pg), while there was no significant difference compared with the mean MCH of

Table 1. Sleeping and blood clotting time for tortoises anaesthetized with ether or sodium thiopentone.

	Ether-anaesthetized tortoise (n=5)	Thiopentone anaesthetized tortoise (n=5)	Unanaesthetized tortoise (n=5)
Weight (Kg)	0.49±0.26	0.49±0.04	0.49±0.16
Sleeping time (min)	17.03±2.06 a	4.30±0.05a	NA
Clotting time (min)	4.13±0.15a	3.05±0.21	3.05±0.06a

Same superscripts on the same row are significantly (P < 0.05) different; NA - Not applicable.

Table 2. Haematological parameters determined anaesthetized and unanaesthetized tortoises.

Blood parameter	Unanaesthetized group (n=5)	Ether group (n=5)	Thiopentone group (n=5)
PCV %	26.54±0.98ab	18.00±3.54a	16.13±1.04b
RBC (X106/μl)	0.46±0.02a	0.44±0.03	0.37±0.02a
Hb (g/dl)	11.05±0.51a	8.25±0.44a	8.95±1.19
MCH (pg)	246.10±11.83a	192.35±22.73a	238.87±22.68
MCHC (g/dl)	41.68±1.32	52.18±11.82	56.93±9.80
MCV (fl)	596.0±33.27a	422.73±104.69	437.53±28.50a
WBC (X103/μl)	5500±204.39ab	6550±131.39ac	5225±194.05bc

Superscripts on the same row are significantly (P < 0.05) different from each other.

sodium thiopentone (238.87 ± 22.68 pg) (Table 2). The mean MCHC of the unanaesthetized and anaesthetized tortoises was also not significantly (p>0.05) different (Table 2). The MCV of unanaesthetized tortoises (596.0 ± 33.27fl) was significantly higher than that of tortoises anaesthetized with sodium thiopentone (437.53 ± 28.50 fl), but there was no significant difference compared with those anaesthetized with ether (422.73 ± 104.69 fl) (Table 2).

White blood cell count (WBC)

The mean WBC of unanaesthetized tortoises (5500 ± 204.39X103/μl) was significantly lower than that of those anaesthetized with ether (6550 ± 131.39X103/μl) and significantly higher than that observed for those anaesthetized with sodium thiopentone (5225 ± 194.05X103/μl). There was a significant difference between the tortoises anaesthetized with ether and sodium thiopentone (Table 2).

DISCUSSION

In this study, the sleeping time was significantly longer for ether, which is an inhalant anaesthetic agent, than for thiopentone sodium, an intravenous anaesthetic. The difference in the sleeping time is unarguably ascribable to differences in the physicochemical, pharmacokinetic or pharmacodynamic properties of ether and thiopentone. Certain considerations needed to be taken when choosing

anaesthetic as chemical restraint in reptiles. Trkova et al. (2008) submitted that the induction and recovery periods should be as short as possible and Heard (2001) recommended inhalation anaesthesia as the technique of choice with respect to minimizing side effects. These considerations place ether as drug of choice as oppose to thiopentone. However, Girling and Raiti (2004) strongly recommended injectable rather than inhalant anaesthetic on the ground that inhalation anaesthesia being administered by either mask or tracheal intubation, or a combination of both often leads to animals struggling or difficulties with restraint. The only reservation raised by Schumacher and Yelen (2006) about injectable anaesthetic is the prolonged recovery phase, which is really not applicable to thiopentone sodium being an ultra short barbiturate (Hung et al., 1992); a fact further confirmed in West African Hinge-Backed Tortoise in this study.

Blood clotting was delayed in tortoises anaesthetized with ether-compared with thiopentone-anaesthetized tortoises. The mean body weight of the tortoises used in this study though lower than what was reported by Oyewale et al. (1998) was consistent for the three groups, therefore the difference in the blood clotting time may not be due to such factors like weight or age as reported by Chaloupka and Musick (1996) or Bradley et al. (1998) but to varying individual effects of the two anaesthetics used. Dordoni et al. (2004) had actually reported that thiopentone sodium reduced platelet function *in vivo* and *in vitro*, consequently prolonging blood clotting time. The findings in this study however showed that ether prolonged blood clothing process much more than thiopentone in tortoises. This therefore makes thiopentone a much more reliable anaesthetic agent for invasive surgical procedures in the tortoise because of its faster blood clotting time, which is an important consideration in such type of surgical interventions (Furie and Furie, 2005).

The two anaesthetics used in this study generally elicited depression of the haematological parameters of the tortoises with significant decreases in the pack cell volume in the test animals. The depression of RBC values was not significant but haemoglobin concentration and mean corpuscular volume were significantly decreased for ether-anaesthetized tortoises, while mean corpuscular haemoglobin was significantly decreased for thiopentone-anaesthetized tortoises.

Anaesthetics-induced depression of the haematological parameters has similarly been reported in other reptiles like Iguana (Knotkova et al., 2006) or in mammals such as rats (Deckardt et al., 2007), boar (Golemanov et al., 1986) or sheep (Edjtehadi, 1978). These changes occur within 15 min and begin to return to normal by 45 min after induction (Dressen et al., 1999); and are caused by anaesthetic-induced splenic vasodilatation resulting in sequestration of blood cells (Marini et al., 1994).

In this study, the erythrocytes were found to be microcytic, a condition thought to be responsible for relative elevation of the values of the mean corpuscular haemoglobin concentration of the tortoises. Every of the parameter responded in the same direction for the ether- or thiopentone-administered tortoises except the white blood cell count which was elevated in ether-administered tortoises while it decreased in thiopentone-administered tortoises. It is difficult to establish the reason for this difference which was also demonstrated by Golemanov et al. (1986) with thiopentone increasing white blood cell count in boar and on the other hand causing a decrease in sheep (Edjtehadi, 1978). It is generally admissible that anaesthetics lower RBC, WBC, PCV, Hb concentration due to splenic and capillary sequestration (Niezgoda et al, 1987; Heard and Huft, 1998; Apple et al., 1993; Knotkova et al., 2006). It has also been shown specifically that anaesthetics exhibit anti-inflammatory effects (Kenyon et al., 1985; O' Donnel et al., 1992; Singh, 2003), but the added component of stress attendant to the strenuous process of administration of anaesthetics especially the inhalant agent, serve to counteract the lowering of white cell count in anaesthetized subjects (Wall, 1985). More specific is the fact that the irritant effect of ether on the respiratory tract (Brunson, 1997) and its stressful effect during induction period (Van Herck et al., 2001) are capable of elevating white blood cell count during anaesthesia as observed in this study.

REFERENCES

Apple J, Minton E, Parsons KM, Unruh JA (1993). Influence of repeated restraint and isolation stress and electrolytes administration on pituitary-adrenal secretions electrolytes and other blood components of sheep. J. Anim. Sci., 71: 71-77

Balcombe JP, Barnard ND, Sandusky C (2004). Laboratory routines cause animal stress. Comtemp. Top. Lab. Anim. Sci., 43: 42-51.

Bertelsen MF (2007). Squamates. In: Zoo animal and wildlife immobilization and anaesthesia. West G, Heard D and N Caulket (eds). Blackwell Publishing Professionals, Ames, Iowa, pp. 233-243.

Bouts T, Gasthuys F (2002). Anaesthesia in reptiles. Part 1: Injection anaesthesia. Vlaams Diergenees.Tijds, 71: 183-194

Bradley TA, Norton TM, Latimer KS (1998). Hemogram values, morphological characteristics of blood cells and morphometric study of loggerhead sea turtles, Caretta caretta in the first year of life. Bull. Assoc. Rept. Amph. Vet., 8: 8-16.

Chaloupka MY, Musick JA (1996). Age, growth and population dynamics. In: The biology of sea turtles. Lutz PL, Musick JA (eds), New York, CRC Press Inc., pp. 233-276.

Brunson DB (1997). Pharmacology of inhalation anaesthetics. In: Anaesthesia and Analgesia for laboratory animals. DF Kohn, SK Wixson, WJ White and GJ Benson (eds), Academic Press, San Diego, California, pp. 29-41.

Cole EH (1986). Veterinary clinical pathology, 4th ed W.B Saunders Publishers.

Cole DJ, Cross LM, Drummond JC, Patel PM, Jacobsen WK (2001). Thiopentone and methohexital, but not pentobarbitones, reduce early focal cerebral ischemic injury in rats. Canad. J. Anaesthes., 48(8): 807-814.

Deckardt K, Weber I, Kaspers U, Hellwig J, Tennekes H, van Ravenzwaay B (2007). The effects of inhalation anaesthetics on common clinical pathology parameters in laboratory rats. Food Chem. Toxicol., 45(9): 1709-1718.

Dennis PM, Heard DJ (2002). Cardiopulmonary effects of a medetomidine-ketamine combination administration intravenously in gopher tortoises. J. Americ. Vet. Med. Assoc., 220: 1516-1519.

Dordoni PL, Frassanito L, Bruno MF, Rodolfo P, De Cristofaro R, Ciabattoni G, Ardito G, Crocchiolo R, Landolfi R, Rocca B (2004). In vivo and in vitro effects of different anaesthetics on platelet function. Br J. Haematol., 125(1): 79-82.

Dressen PJ, Wimsalt J, Burkhard (1999). The effects of Isoflurane anaesthesia on haematologic and plasma biochemical values of American kestrels. J. Avian Med. Surg., 13(3): 173-179.

Edjtehadi M (1978). Effects of thiopentone sodium, methoxyflurane and halothane on haematological parameters in sheep during prolonged anaesthesia. Clin. Exp. Pharmacol. Physiol., 5(1): 31-40.

Frolich D, Rothe G, Schwall B, Schmitz J, Taeger K (1996). Thiopentone and propofol, but not methhexitone nor midazolam, inhibit neutrophils oxidative response to the bacterial peptide FMLP. Eur. J. Anaesthesiol, 13(6): 582-588.

Furie B, Furie BC (2005). Thrombus formation in vivo. J. Clin. Invest., 115(12): 3355- 3362.

Gaztelu JM, Gracia-Aust E, Bullock TH (1991). Electrocortigrams of hippocampal and dorsal cortex of two reptiles: Comparison with possible mammalian homologs. Brain Behav. Evol., 37: 144-160.

Golemanov D, Aminkov B, Ianeva V (1986). Hematologic and biochemical changes in the blood of boars undergoing potentiated anaesthesia with droperidol, fentanyl and thiopental. Vet. Med. Nauki, 23(7): 53-60.

Greer LL, Jenne KJ, Diggs HE (2001). Medetomidine-ketamine anaesthesia in red-eared slider turtles (Trachemys scripta elegans). Contemp. Top. Lab. Anim. Sci., 40: 9-11.

Heard DJ, Huft V (1998). Effect of short term physical and isoflurane restraint on hematological and plasma biochemical values in the island flying fox (Pteropus hypomelanus). J. Zool. Wildl. Med., 29(1): 14-17.

Heard DJ (2001). Reptile anaesthesia. Vet. Clin. North Am. Exot. Anim. Pract., 4: 83-117.

Holz P, Holz RM (1994). Evaluation of ketamine, ketamine/xylazine and ketamine/midazolam anaesthesia in red-eared sliders (Trachemys scripta elegans). J. Zool Wildl. Med., 25: 531-537.

Hung OR, Varvel JR, Shafer SL, Stanski DR (1992). Thiopental pharmacodynamics II. Quantitation of clinical and electroencephalographic depth of anaesthesia. Anaesthesiol, 77:

237-244.

Jackson DC, Ramsey AL, Paulson JM, Crocker GE, Ultsch GR (2000). Lactic acid buffering by bone and shell in anoxic softshell and painted turtles. Physiol. Biochem. Zool., 73: 290-297.

Kenyon CJ, McNeil LM, Fraser R (1985). Comparison of the effects of etomidate, thiopentone and propofol on cortisol synthesis. Br. J. Anaesthes., 57(5): 509-511.

Knotkova Z, Knotek Z, Trnkova S, Mikulcova P (2006). Blood profile in green iguanas after short-term anaesthesia with propofol. Vet. Med., 51(10): 491-496.

Marini RP, Jackson LR, Esteves MI, Andrutis KA, Goslant CM, Fox JG (1994). Effect of isoflurane on hematologic variables in ferrets. Am. J. Vet. Res., 55(10): 1479-83

Niezgoda J, Wronka D, Pierzchala K, Bobek S, Kahl S (1987). Lack of adaptation to repeated emotional stress evoked by isolation of sheep from the flock. Zentr. für Veterinärmed, 34: 734-739

O' Donnell NG, McSharry CP, Wilkinson PC, Asbury JA (1992). Comparison of the inhibitory effect of propofol thiopentone and midazolam on neutrophils polarization *in vitro* in the presence or absence of human serum. Br. J. Anaesthes., 69(1): 70-74

Oyewale JO, Ebute CP, Ogunsanmi AO, Olayemi FO Durotoye LA (1998). Weights and blood profiles of the West African Hinge-Backed tortoise, Kinixys erosa and the Desert Tortoise, Gopherus agassizii. J. Vet. Med. A, 45: 599-605.

Pérez-Bárcena J, Barceló B, Homar J, Abadal JM, Molina FJ, de la Peña A, Sahuquillo J, Ibáñez J (2005). Comparison of the effectiveness of pentobarbital and thiopental in patients with refractory intracranial hypertension. Preliminary report of 20 patients. Neurocirugia, 16(1): 5-12

Singh M (2003). Stress response and anaesthesia: Altering the peri nd post-operative management. India J. Anaesthesiol, 47(6): 427-464.

Sleeman JM, Gaynor J (2000). Sedative and cardiopulmonary effects of medetomidine and reversal with atipamezole in desert tortoises (Gopherus agassizii). J. Zool Wildl. Med., 31: 28-35.

Steel RGD, Torrie JII (1996). Principles and procedure of Statistics. A biometric approach (2nd Edn.), McGraw-Hill, New York, pp. 6-15.

Schumacher J, Yelen T (2006). Anaesthesia and analgesia. In: Mader DR (Ed.): Reptile Medicine and Surgery. Saunders Elsevier, St. Louis, pp. 442-452.

Girling S, Raiti P (2004). BSAVA Manual of Reptiles. Simon Girling and Paul Raiti (ed) 2nd Revised Edition. British Small Animal Veterinary Association, London, UK.

Trnková Š, Knotková Z, Knotek Z (2008). Effect of butorphanol on anaesthesia induction by Isoflurane in the Green Iguana (Iguana iguana) Acta Vet. Brno, 77: 245-249.

Van Herck K, Baumans V, Brandt CJWM, Boere HAG, Hesp APM, van Lith HA, Schurink M, Beynen AC (2001). Blood sampling from the retro-orbital plexus, the saphenous vein and the tail vein in rats: comparative effects on selected behavioural and blood variables. Lab. Anim., 35: 131-139.

Wall HS, Worthman C, Else JG (1985). Effects of ketamine anaesthesia, stress and repeated bleeding on haematology of vervet monkeys. Lab. Anim., 19: 138-144.

cDNA, genomic sequence cloning and over-expression of ribosomal protein P2 gene (*RPLP2*) from the Giant Panda (*Ailuropoda melanoleuca*)

Sinan Zhang, Yiling Hou, Wanru Hou*, Xiang Ding and Chunlian Wu

Key Laboratory of Southwest China Wildlife Resources Conservation, College of Life Science, China West Normal University, 1# Shida Road, 637009, Nanchong, P. R. China.

RPLP2 is a component of the 60S large ribosomal subunit encoded by *RPLP2* gene and directly participate in protein synthesis, which is located in the cytoplasm. The cDNA and genomic sequence of *RPLP2* were cloned successfully from the Giant Panda using RT-PCR and Touchdown-PCR technology, respectively. The results showed that the cDNA fragment cloned is 394 bp in size, containing an open reading frame of 348 bp. The length of the genomic sequence is 1838 bp, with four exons and three introns. The deduced protein is composed of 115 amino acids with 11.66 KD of estimated molecular weight and 4.14 of pI. Alignment analysis indicated that the nucleotide sequence and the deduced amino acid sequence are highly conserved to other four species reported, including *Homo sapiens*, *Bos taurus*, *Mus musculus* and *Rattus norvegicus*. The homologies for nucleotide sequences of Giant Panda *PRLP2* are 90.80, 87.64, 88.79 and 87.64% to those of the four species, while the homologies for amino acid sequences are 88.27, 85.20, 86.31 and 85.20%. The *RPLP2* gene was overexpressed in *Escherichia coli* BL21, and the result indicated that fusing RPLP2 with the N-terminally His-tagged form gave rise to the accumulation of an expected 17.5 KD polypeptide, in accordance with the predicted protein, which could be used to purify and investigate the function of this protein.

Key words: cDNA cloning, overexpression, ribosomal protein P2 gene (*RPLP2*), Giant Panda (*Ailuropoda melanoleuca*).

INTRODUCTION

Ribosome, a compact ribonucleoprotein (RNP) grain that catalyzes protein synthesis, consists of 4 RNA species and approximately 80 structurally distinct proteins (Yoshihama et al., 2002; Hwang et al., 2004). It can be dissociated into a small subunit and a large one, whose shape and structure are irregular and asymmetric. The large ribosomal subunit has a distinct lateral protuberance called the stalk, which is an important and essential structure involved directly in the interaction of the elongation factors with ribosome during protein synthesis (Rodriguez-Gbriel et al., 1999; Krokowski et al.,

2006). Acidic ribosomal phosphoprotein with pH 3 to 5 of isoelectric point, named P protein, including RPLP0, RPLP1 and RPLP2 in large ribosomal subunit in eukaryotic cell, has been reported directly participate in protein synthesis (Remacha et al., 1995; Qiu et al., 2006). RPLP2, together with RPLP0, RPLP1 and the conserved domain of 28S rRNA, constitutes a major part of the GTPase-associated center in eukaryotic ribosomes (Hagiya et al., 2005). In addition, it can be attached to ubiquitin and assist the latter in regulating numerous important cellular processes including apoptosis, transcription, and the progression of cell cycle (Archibald et al., 2003).

Giant Panda is a rare species currently found only in China. For many years, studies on the Giant Panda have been mainly concentrated on fields such as breeding and

---

*Corresponding author. E-mail: hwr168@yahoo.com.cn.

propagation, ecology, morphology, taxology, physiology and pathological biochemistry (Hou et al., 2007a, b, 2008, Du et al., 2007, 2008). And recently, functional gene analysis is one of the hot issues in current Giant Panda research (Wu et al., 1990; Liao et al., 2003; Zhang et al., 2009; Hou et al., 2009; Sun et al., 2011). There are few reports about *RPLP2* gene of Giant Panda. This study was conducted to amplify the cDNA and genomic sequence of *RPLP2* gene from Giant Panda, and then analyzed the sequence characteristics of the cDNA and the deduced protein. This gene was over-expressed in *Escherichia coli* using pET28a plasmids.

MATERIALS AND METHODS

The skeletal muscle tissue were collected from a dead Giant Panda at the Wolong Conservation Center of the Giant Panda, Sichuan, China, and kept in liquid nitrogen. A total of 500 mg muscle tissue from Giant Panda was ground in liquid nitrogen to a fine powder, and the powder was suspended completely in 15 ml lysis buffer containing 10 mM Tris-HCl, pH 8.0, 100 mM EDTA and 0.5% SDS. After treatment with proteinase K (100 mg/ml, final concentration) at 55°C for 3 h, the mixture was then cooled to room temperature and mixed with an equal volume of saturated phenol (pH 8) before being centrifuged at 5000 g at 4°C for 20 min. The supernatant was pooled and then mixed with an equal volume of 1:1 (v:v) phenol-chloroform and then centrifuged as above and the supernatant collected, from which the DNA was precipitated by ethanol. The DNA obtained was then dissolved in TE buffer and kept at -20°C.

Total RNA was isolated using the Total Tissue/Cell RNA Extraction Kits, dissolved in DEPC water, and kept at -70°C. The quality and quantity of the products were detected by electrophoresis and spectrophotometry.

The PCR primers were designed by Primer Premier 5.0, according to the mRNA sequence of *RPLP2* gene from *Homo sapiens* (NC_000011), *Bos Taurus* (AC_000186), *Mus musculus* (NC_000073) and *Rattus norvegicus* (NC_005100), as following:

Forward: 5'-GAGGCTTCTCCGCCGCCGAG-3';
Reverse: 5'-CAGGGGAGCAGGAACCTAAT-3'.

The cDNAs was synthesized using a reverse transcription kit, and the genomic sequence was amplified using Touchdown-PCR. The PCR products were analyzed by electrophoresis and purified using a DNA harvesting kit. The products were ligated into vector plasmid pUC18 which has been digested by restriction endonuclease *SmaI* and then transformed into *E. coli* competent cells (JM109). The insert size was verified by digesting with *PstI* and *ScaII*. Recombinant pUC18 was sequenced by Huada Zhongsheng Scientific Corporation. The sequence data were analyzed by GenScan software, Blast 2.1, ORF finder software, DNAMAN 6.0, Predict Protein software. Software ExPASy Proteomics Server and SWISS-MODEL software respectively.

The genomic sequence was amplified using Touchdown-PCR, and the primers were the same as following:

Forward: 5'-ATGCGCTACG TTGCCTCCTA-3';
Reverse: 5'-CTAATCGAACA AGCCGAATC-3'.

The amplification conditions were: 94°C for 30 s, 62°C for 45 s, 72°C for 4 min in the first cycle and the anneal temperature deceased 0.5°C per cycle; after 20 cycles conditions changed to 94°C for 30 s, 52°C for 45 s, 72°C for 4 min for another 20 cycles. The fragment amplified was also purified, ligated into the clone

vector and transformed into the *E. coli* compence cells. Finally, the recombinant fragment was sequenced by Sangon (Shanghai, China).

PCR fragment corresponding to the RPLP2 polypeptide was amplified from the *RPLP2* cDNA clone with the forward primer, 5'-GAGGAATTC ATGCGCTACG TTG-3' (*EcoRI*) and reverse primer, 5'-AACCTCGAG CTAATCGAAC AAG-3'(*XhoI*), respectively. The amplified PCR product was cut and ligated into corresponding site of pET28a vector (Stratagen). The resulting construct was transformed into *E. coli* BL21 (DE3) strain (Novagen) and used for the induction by adding isopropyl-b-D-thiogalactopyranoside (IPTG) at an OD600 of 0.6 and culturing further for 4 h at 37°C, using the empty vector transformed BL21 (DE3) as a control. The recombinant protein production was induced after 0, 0.5, 1, 1.5, 2, 2.5, 3 and 4 h and then protein bands were separated by SDS-PAGE and stained with Commassie blue R250.

RESULTS

About 400 bp of cDNA fragment was amplified from the Giant Panda (Figure 1). The exact length of the Cloned cDNA was 394 bp (accession number: HQ318036) by sequencing analysis. Blast research showed that the cDNA sequence shares a high homology with the *RPLP2* gene from other mammals reported, including *H. sapiens*, *B. taurus*, *M. musculus* and *R. norvegicus*. On the basis of the high identity, it could be concluded that the cDNA isolated is just the cDNA encoding *RPLP2* protein. The *rpLP2* CDS sequence contains an ORF of 348 bp encoding 115 amino acids, 31 bp of 5'-untranslated sequence and the 15 bp of 3'-untranslated region in length.

A DNA fragment of about 2000 bp was amplified and the sequenced length was1838 bp (accession number: HQ318037). Comparison of the cDNA with the amplified genomic DNA fragment indicated that the cDNA sequence is in full accord with four fragments in the genomic DNA fragment, which manifests that the DNA fragment amplified is the genomic sequence of the *rpLP2* gene from Giant Panda (Figure 2).

Primary structure analysis revealed that the molecular weight of the putative *RPLP2* protein is 11.66 KD with a theoretical pI of 4.14. Topology prediction showed there are 3 different patterns of functional sites in the *RPLP2* protein of Giant Panda: one protein kinase C phosphorylation site, two casein kinase II phosphorylation sites, and two N-myristoylation sites (Figure 3). Further study found that the protein is composed of 21 negatively charged residues (Asp and Glu), 12 positively charged residues (Arg and Lys) and 82 uncharged residues.

The secondary structure prediction of the RPLP2 protein sequence indicated that 8.7% of the protein sequence is strand, 49.57% is helix and 41.74% is coil for Giant Panda. Further comparison was conducted to understand the secondary structure of protein subunit RPLP2 in five mammals is consistent or not, the results showed that there is very little difference in the secondary structure of protein subunit RPLP2 between the five mammalian species (Table1).

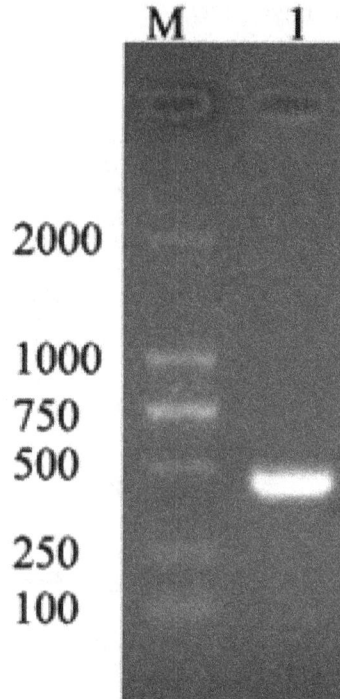

Figure 1. Reverse transcription polymerase chain reaction products of the Giant Panda *RPLP2*. M: Molecular ladder DL2000; 1: The amplified *RPLP2*.

Figure 2. Nucleotide sequence of cDNA of Giant Panda *RPLP2* gene and the amino acid sequence deduced (* representing for the terminator code).

The *RPLP2* gene was overexpressed in *E. coli*, and the results indicated that the fusion of *RPLP2* with the N-terminally His-tagged form gave rise to the accumulation of an expected 17.5 kDa polypeptide that formed inclusion bodies (Figure 4). Apparently, the recombinant protein was expressed after half an hour of induction and

Pd-p	MRYVASYLLAALGGNASP SAKDIKKILDSVGIEADDDRLNKVISELNGKN	50
Ho-p	MRYVASYLLAALGGNSSP SAKDIKKILDSVGIEADDDRLNKVISELNGKN	50
Mu-p	MRYVASYLLAALGGNSSP SAKDIKKILDSVGIEADDDRLNKVISELNGKN	50
Bo-p	MRYVASYLLAALGGNSSP SAKD IKKILD SVGIEADDDRLNKVI SELHGKN	50
Ra-p	MRYVASYLLAALGGNSNP SAKD IKKILD SVGIEADDERLNKVI SELNGKN	50

Pd-p	IEDVIAQGIGKLASVPA GGAVTV SAAPGS AAPAA GAAPAA AEEKKDEKKE	100
Ho-p	IEDVIAQGIGKLASVPA GGAVAV SAAPGS AAPAA GSAPAA AEEKKDEKKE	100
Mu-p	IEDVIAQGVGKLASVPA GGAVAV SAAPGS AAPAA GSAPAA AEEKKDEKKE	100
Bo-p	IEDVIAQGIGKLASVPA GGAVAV SAAP GSAAP AA GSAP AA AEEKKEEKKE	100
Ra-p	IEDVIAQGVGKLASVPA GGAVAV SAAP GSAAP AA GSAP AA AEEKKDEKKE	100

Pd-p	ESEESDDDMGFGLFD	115
Ho-p	ESEESDDDMGFGLFD	115
Mu-p	ESEESDDDMGFGLFD	115
Bo-p	ESEESDDDMGFGLFD	115
Ra-p	ESEESDDDMGFGLFD	115

Figure 3. Comparison of the RPLP2 amino acid sequences among the different species (N-myristoylation site: protein kinase C phosphorylation site: asein kinase II phosphorylation site:_).

Table 1. Comparison of secondary structure of RPLP2 protein among 5 mammal species.

Species	Amino acid sites					
	11	32-36	46	51	149	81-86
A. melanoleuca	H	C	C	C	C	H
H. sapiens	C	C	C	C	C	C
B. taurus	C	C	H	H	H	C
M. musculus	C	C	C	C	C	C
R. norvegicus	H	H	C	C	C	C

Note: H: Alpha helix; C: Random coil.

the after 2 h reached the highest level. These results suggested that the protein is active and it is just the protein encoded by the *RPLP2* from *Ailuropoda melanoleuca*. The expression product obtained could be used to purify the protein and to study its function further.

DISCUSSION

Recently, the studies on ribosomal protein have made much progress in Giant Panda (Hou et al., 2009a, b), however, there is no reports on RPLP2 protein in this species. In this study, we cloned genomic sequence and cDNA encoding RPLP2 from Giant Panda. The genomic sequence of *rpLp2* is 1838 bp in size. Compared with some mammals including *H. sapiens, B. taurus, M. musculus* and *R. norvegicus,* there are four exons and three introns. Further study indicated that the genomic, the introns, the 5'-untranslated sequence and the 3'-untranslated sequence are different in length (Table 2). The variations in lengths of the introns determine the lengths of the *rpLP2* gene.

Physical and chemical analysis showed that the molecular weight of the putative protein among the five mammalians is very close and that the theoretical pI is exactly identical (Table 3). Secondary structure analysis showed that although the amino acid sequence of the 11, 32 to 36, 46, 51, 149, 81 to 86 site in five kinds of mammals have different structure, this does not cause changes in their functional genes, subsequently the corresponding functional sites has not changed (Table 1).

From the alignment analysis of the cDNA sequence of *rpLP2* gene and the deduced amino acid sequence between Giant Panda and other mammals reported

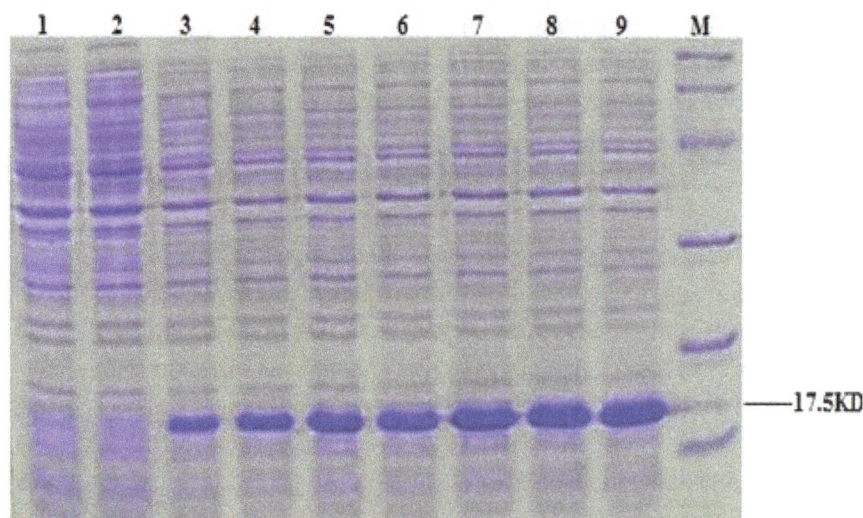

Figure 4. Overexpress of RPLP2 protein in recombinant *E. coli* BL21. Lane 1: control; Lines 2 to 9: overexpressing for 0, 0.5, 1, 1.5, 2, 2.5, 3 and 4 h; Line M: marker; Arrow: indicating recombinant protein.

Table 2. Comparison of *RPLP2* genomic sequence among 5 mammalian species.

Species	Size bp	No. of exons	Join sites in the CDS	Accession No.
A. lanoleucame	1838	4	1..123, 598..646, 1578..1676, 1762..1838	HQ_318037
H. sapiens	2941	4	300..422, 1662..1710, 2600..2698, 2825..2901	NC_000011
B. taurus	2139	4	258..380, 1195..1243, 1822..1920, 2026..2102	AC_000186
M. musculus	3693	4	260..382, 1077..1125, 3355..3453, 3583..3659	NC_000073
R. norvegicus	2272	4	247..369, 1047..1095, 1942..2040, 2160..2236	NC_005100

Table 3. Comparison of RPLP2 in *A. melanoleuca* with other 4 mammals.

Items	Species			
	H. sapiens	*B. taurus*	*R. norvegicus*	*M. musculus*
Cds similarity (%)	90.8	87.64	88.79	87.64
Aa Similarity (%)	97.39	97.39	97.39	97.39
Molecular weight (KD)	11.66	11.70	11.69	11.65
PI	4.14	4.23	4.15	4.14

including *H. sapiens, B. taurus, M. musculus* and *R. norvegicus,* it was found that Giant Panda shares high homology in nucleotide sequence with those mammals. Among them, the Giant Panda shares the highest homology in nucleotide sequence with *H. sapiens*, and shares the same high homology in amino acid sequences with other 4 mammals (Table 3). Comparison of RPLP2 genetic coding sequence of Giant Panda with these species, some variable sites exist among them, parts of which are degeneration sites and others are single variable sites. Further analysis indicated that those variable sites are caused by transformation or transition

of bases, which do not result to changes in the amino acid sequences encoded. So, protein functional sites in RPLP2 protein of the five mammalians have the same positions and numbers. The cDNA and the genomic sequence of *RPLP2* were cloned successfully from the Giant Panda, and the cDNA of the *RPLP2* gene was also overexpressed in *E. coli* BL21 strains, which confirms the gene cloned was the *RPLP2* gene from giant panda. The data will enrich and supplement the information about *RPLP2.* In addition, it will contribute to the protection for gene resources and the discussion of the genetic polymorphism. Also, it will provide a reliable basis for the study on the evolution of species.

ACKNOWLEDGEMENTS

This research was supported by the Key Chinese National Natural Science Foundation (30470261), Key Scientific Research Foundation of Educational Committee of Sichuan Province (07ZA120), Key Discipline Construction Project in Sichuan Province (SZD0420), Key Discipline of Zoology Construction Project in Sichuan Province (404001), Application Foundation Project of Sichuan Province (2009JY0061), Youth Fund Project of Educational Committee of Sichuan Province (09ZB088), Foundation Project of Educational Committee of Sichuan Province (10ZC120) and Application Foundation Project of Sichuan Province (2011JY0135).

REFERENCES

Hagiya A, Takao N, Yasushi M, Jun O, Yukiko T, Tomomi S, Takaomi N, Akira H, Toshio U (2005). A mode of assembly of P0, P1 and P2 Proteins at the GTPase-Associated Center in Animal Ribosomal: *In vitro* Analyses with P0 Truncation Mutants. J. Biol. Chem., 47: 39193-39199.

Krokowski D, Aleksandra B, Dariusz A, Anders L, Marek T, Nikodem G (2006). Yeast Ribosomal P0 Protein Has Two Separate Binding Sites for P1/P2 Proteins. Mole. Microbiol., 60: 386-400.

Qiu D, Pilar P, Alberto GM, David C, Miguel R, Juan PGB (2006). Different roles of P1 and P2 *Saccharomyces cerevisiae* Ribosomal Stalk Proteins Revealed By Cross-Lingking. Mole. Microbiol., 62: 1191-202.

Du YJ, Luo XY, Hao YZ, Zhang T, Hou WR (2007). Cloning and Overexpression of Acidic Ribosomal Phosphoprotein P1 Gene (RPLP1) from the Giant Panda. Int. J. Biol. Sci., 3: 428-433.

Du YJ, Hou WR, Peng ZS, Zhou CQ (2008). cDNA Cloning and Sequences Analysis of Acidic Ribosomal Phosphoprotein P1 (RPLP1) from Giant Panda. Acta Theriol. Sin., 28: 75-80.

Hou WR, Chen Y, Peng ZS, Wu X, Tang ZX (2007). cDNA cloning and sequences analysis of ubiquinol-cytochrome c reductase complex ubiquinone-binding protein (QP-C) from Giant Panda, Acta Theriol. Sin., 27: 190-194.

Hou WR, Du YJ, Chen Y, Wu X, Peng ZS, Yang J, Zhou CQ (2007). Nucleotide Sequence of cDNA Encoding the Mitochondrial Precursor Protein of the ATPase Inhibitor from the Giant Panda (*Ailuropoda melanoleuca*). DNA Cell Biol., 26: 799-802.

Hou WR, Luo XY, Du YJ, Chen Y, Wu X, Peng ZS, Yang J, Zhou CQ (2008). cDNA Cloning and Sequences analysis of RPS15 from the Giant Panda. Recent Patent on DNA Sequence, 2: 16-19.

Hou WR, Hou YL, Hao YZ, Du YJ, Zhang T, Peng ZS (2009). Sequence analysis and over-expression of ribosomal protein S28 gene (RPS28) from the Giant Panda. Afr. J. Biotechnol., 8(11): 2454-2459.

Hou YL, Du YJ, Hou WR, Zhou CQ, Hao YZ, Zhang T (2009). Cloning and sequence analysis of translocase of inner mitochondrial membrane 10 homolog (yeast) gene (*TIMM10*) from the Giant Panda, J. Cell Anim. Biol., 3: 9-14.

Hou YL, Hou WR, Ren ZL, Hao YZ, Zhang T (2009). cDNA Cloning and Overexpression of Ribosomal Protein S19 Gene (RPS19) from the Giant Panda. DNA Cell Biol., 28: 41-47.

Archibald JM, Evelyn MT, Patrick JK (2003). Novel Ubiquitin Fusion Proteins: Ribosomal Protein P1 and Actin. J. Mol. Biol., 328: 771-778.

Hwang KC, Xiang SC, Se PP, Mi RS, Sae YP, Eun YK, Nam HK (2004). Identification of Differentially Regulated Genes in Bovine Blastocysts Using an Annealing Control Primer System. Mole. Reprod. Dev., 69: 43-51.

Liao MJ, Zhu MY, Zhang ZH, Zhang AJ (2003). Cloning and sequence analysis of FSH and LH in the Giant Panda (*Ailuropoda melanoleuca*), Anim. Reprod. Sci., 77: 107-116.

Rodriguez-Gbriel MA, Bou G, Briones E, Zambrano R, Remacha M, Ballesta JPG (1999). Structure and Function of the Stalk, a Putative Regulatory Element of the Yeast Ribosome. Role of Stalk Protein Phosphorylation. Folia Microbiol., 44: 153-163.

Yoshihama M, Tamayo U, Shuichi A, Kazuhiko K, Seishi K, Sayomi H, Noriko M, Shinsei M, Tatsuo T, Nobuyoshi S, Naoya K (2002). The Human Ribosomal Protein Genes: Sequencing And Comparative Analysis of 73 Genes. Genome Res., 12: 379-390.

Remacha M, Jimenez-Diaz A, Santos C, Briones E, Zambrano R, Rodriguez Gabriel MA, Guarinos E, Ballesta JP (1995). Proteins P1, P2, and P0, Components of the Eukaryotic Ribosome Stalk. New Structural and Functional Aspects. Biochem. Cell Biol., 73: 959-968.

Sun B, Hou YL, Hou WR, Su XL, Li J, Wu GF, Song Y (2011). cDNA, Genomic sequence cloning and over-expression of ribosomal protein L15 gene (*RPL15*) from the Giant Panda. Afr. J. Agric. Res., 6(9): 2108-2114.

Wu ZA, Liu WX, Murphy C, Gall J (1990). Satellite DNA sequence from genomic DNA of the Giant Panda. Nucl. Acids Res., 18: 1054.

Zhang T, Hou WR, Hou YL (2009). Genomic sequence cloning and over-expression of ribosomal protein s20 gene (RPS20) from the Giant Panda (*Ailuropoda melanoleuca*). Afr. J. Biotechnol., 8: 5627-5632.

13

Variations in the epithelial cords of the ovaries of a microchiropteran bat, *Hipposideros speoris* (Schneider) during reproductive cycle: An enzymic approach

M. S. Sastry[1] and S. B. Pillai [1]

[1]Department of Zoology, Rashtrasant Tukdoji Maharaj Nagpur University, Nagpur University Campus, Nagpur-440033, India.

The ovaries of *Hipposideros speoris* were studied histologically and histochemically for the enzymes, 3β-hydroxysteriod dehydrogenase (3β-HSDH), Succinic dehydrogenase (SDH) and lipid from July 2005 to 2006. The interstitial cells or so called "epithelial cords" showed variations in their distribution, morphology, enzymic and their association with other ovarian structures. These cords appear to be formed in the ovarian cortex by the transformation of granulosa of the primordial follicles and small preantral follicles whose ova regress and disappear. Mostly these cords were conspicuous, hypertrophied, abundant and in clusters or in zones occupying a major portion of the cortex during 4 to 5 months of gestation and also during lactation. Both histological and histochemical studies revealed their significance as steroidogenic cells. The frequency with which these structures were observed during pregnancy made it obligatory to conclude that they have a certain significant role in ovarian physiology in overtaking the function of corpus luteum after its regression.

Key words: Chiroptera, *Hipposideros*, ovary, epithelial cords.

INTRODUCTION

Interstitial gland cells constitute an important ovarian component with steroidogenic function. The ovaries of *Hipposideros speoris* showed occurrence of three types of interstitial gland cells (Igc) – the thecal type Interstitial cells, stromal type Interstitial cells and epithelial cords (EC), all showing cyclical alterations histologically and histochemically with the reproductive cycle. These EC appear to be formed in the ovarian cortex either by the invagination of the germinal epithelium or by the transformation of granulosa cells of primordial follicles and small preantral follicles whose ova regress and disappear (Pillai, 2004). The existences of epithelial cords were documented among a variety of mammalian species (MacLeod, 1880; Matthews, 1935; Barker, 1951; Dawson and McCabe, 1951; Rennels, 1951; Brambell, 1956; Guraya

and Greenwald, 1964a, 1964b, 1965; Guraya, 1968; Mori and Matsumoto, 1970; Motta, 1974; Guraya, 1985, 2000) including a single chiropteran species, *Myotis grisescens* (Guraya, 1967c). The present study gives an account of EC during the complete reproductive cycle. Such observations may help in understanding the functional significance of EC in the cycling ovaries.

MATERIALS AND METHODS

Specimens of *H. speoris* were collected twice every calender month throughout the year with the help of a mist net from the natural population inhabiting abandoned mines in Khapa, Nagpur, Maharashtra. For histological studies, the ovaries were fixed quickly in Bouin's fixative, dehydrated in ethanol and embedded in paraffin

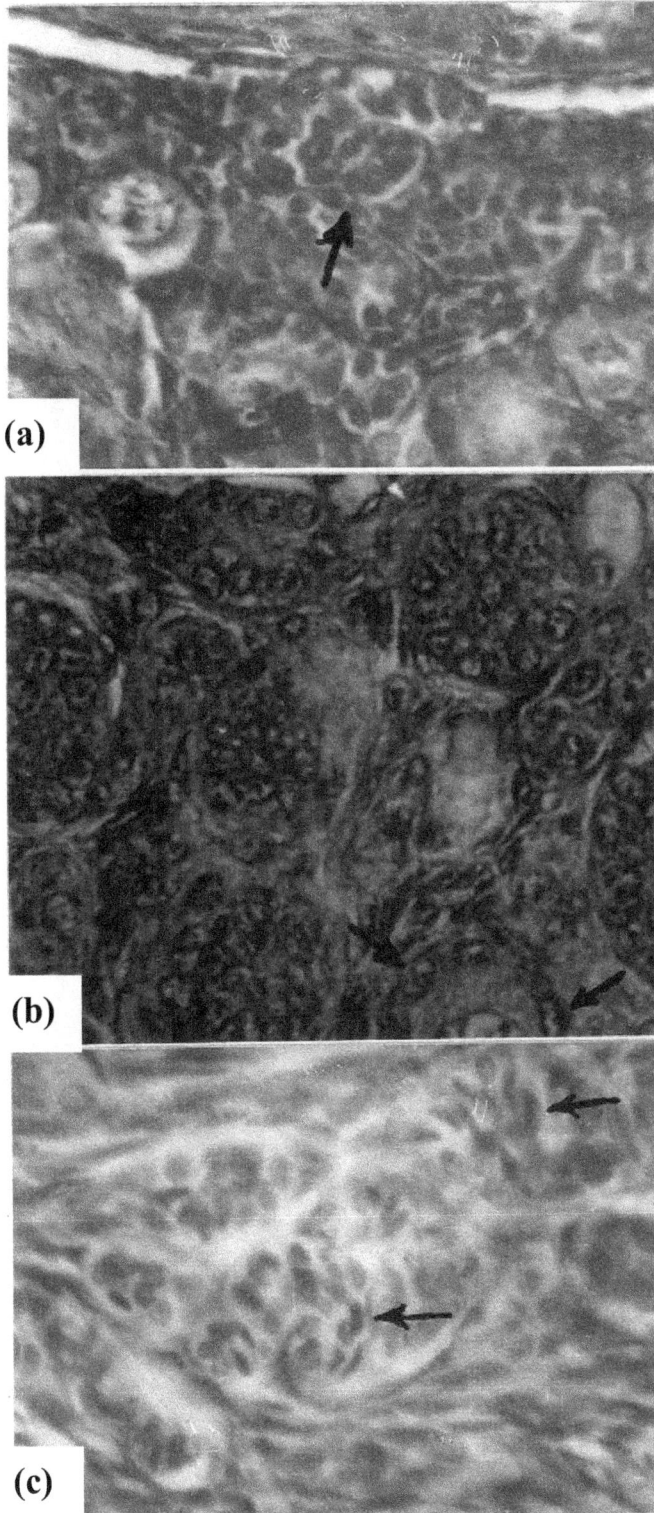

Figure 1. (a) Magnified view of the ovarian cortex clearly demonstrating the formation of epithelial cords by sinking of germinal epithelial nodules (arrow). **(b)** Cortical region magnified to show formation of epithelial cords from the degenerating primordial follicles by the hypertrophy of preganulosa cells (arrow). **(c)** Magnified view of the deep cortical part showing the epithelial cord formation from the remnants of atretic small primary follicles (arrow) X 250.

wax. The sections cut at 5 µm were stained with haematoxylin and eosin.

For histochemical studies, calcium-formal and buffered formalin fixed and unfixed frozen tissues were cut on freezing microtome at -20°C and stored at -25°C until stained (Pearse, 1972; Lillie and Fullmer, 1976).

Lipids appeared black or bluish black when cut frozen sections were washed in water for 2 to 5 min to remove formaldehyde, dehydrated for 3 to 5 min in pure propylene glycol by moving sections at intervals, transferred to the dye solution (0.7 g of Sudan black B dissolved in 100 ml propylene glycol at 100 to 110°C) for 3 to 7min with occasional agitation. Then differentiated in 85% propylene glycol for 2 to 3 min and washed in glycerol jelly.

For the determination of 3β-HSDH fresh frozen cryostat sections were incubated aerobically at 37°C in incubation medium containing dimethyl formamide (DMF); β-Nicotinamide Adenine Dinucleotide (β-NAD); Nitro blue tetrazolium salt (NBT) and pregnenolone for 30 to 45 min Then post fixation of sections were done by 10% N-formalin for 10 min, washed and rinsed in DMF. Again washed in water and mounted in glycerol jelly.

Acetone fixed cryostat sections were employed for SDH localization. Sections were incubated in a medium consisting of Nitroblue – tetrazolium salt in 0.1 M phosphate buffer (pH = 7.6), and sodium succinate at 37°C for 10 to 30 min washed, dehydrated, in series of alcohols, cleared and mounted in permount.

RESULTS

H. speoris is a monovular and monotocous bat, breeding once in a year from December to May and the gestation lasts for 135 ± 5 days starting from second week of December upto last week of April.

The EC were encountered throughout the reproductive cycle but a variation in their distribution, morphology, enzymity and their association with other ovarian structures were observed. They were found to have originated either by the invagination of the germinal epithelium or from the hypertrophy of persistent granulosa of atretic primordial and small preantral follicles (Figure 1 a, b and c). Each cord consisted of a single layer of either spherical or ovoid epithelial cells with vesicular nuclei lined by basement membrane.

During anoestrous, proestrous and estrus (July to November) the EC were mostly individual, few in number, small in size and irregularly distributed among the primordial follicles (Figure 2a, b, c). The cords showed less accumulation of lipid droplets and low to negligible enzyme activity.

A sudden spurt in development was observed at the approach of pregnancy. The early gestation was thus marked by EC scattered in the cortex either in groups of 2 to 3 or singly. Each cord consisted of few small cells with vesicular nuclei (Figure 2e). Early pregnancy was marked by an extrovert functional corpus luteum which was fully functional. The cords were sudanophilic with fine lipid droplets aggregated along the cells borders (Figure 2f), similarly, the cords exhibited moderate 3β-HSDH activity concentrated particularly along the borders of the cells (Figure 2g). The histochemical distribution of SDH was low to moderate (Figure 2h).

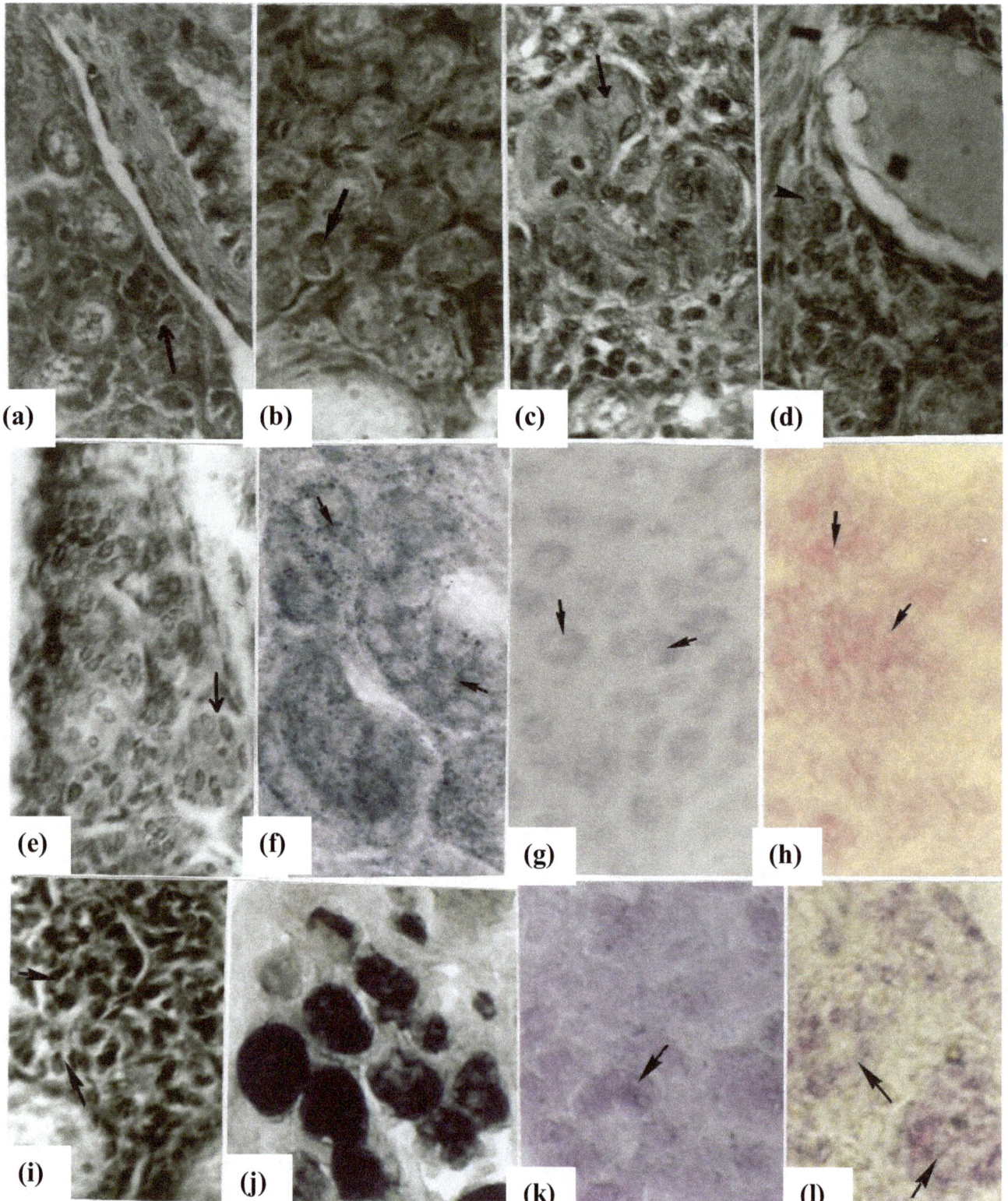

Figure 2. (a) Epithelial cords (arrow) distributed individually between primordial follicles at anoestrous. **(b)** Reduced but well-defined epithelial cords (arrow) at proestrus. **(c)** Arrow marks the regressed epithelial cords, with inactive cells during estrus phase **(d)** Epithelial cords during mid pregnancy shifted towards medullary region in the vicinity of blood capillary (arrow head). **(e)** Epithelial cords during early pregnancy scattered in cortical region either in groups of 2 - 3 or singly. Note the smaller size of the cords with few cells in each cord (arrow). **(f)** Aggregation of epithelial cords during early pregnancy demonstrating diffuse lipid droplets along the cell outlines (arrow). **(g)** Epithelial cords at early pregnancy exhibiting moderate 3β-HSDH activities (arrow). **(h)** Epithelial cords (arrow) from early pregnancy illustrating moderate SDH activity. **(i)** A cluster of highly hypertrophied and active epithelial cords (arrow) consisting of 20 – 30 cells X 250. **(j)** Well developed epithelial cords intensely stained with Sudan black B **(k)**   moderate to strong 3β-HSDH activities in the epithelial cords along cell borders (arrow) during early pregnancy. **(l)** Epithelial cords (arrow) during early pregnancy showing moderate SDH activity X 250.

At neural groove stage of embryo development (early February) the newly formed placenta and the corpus leuteum both were functional and the EC were extensive, highly hypertrophied large in size, coiled, clustered, each cord was demarcated by separate basal lamina and was lined by a single layer of 20 to 30 cells (Figure 2i). More often the cords get shifted towards the cortico-medullary region and mostly found in the vicinity of capillary loops. The cords displayed accumulation of diffuse sudanophilic lipid droplets (Figure 2j) and demonstrated intense steroidogenic activity (Figure 2k) but SDH activity was moderate (Figure 2l).

The EC were at their peak of development, more or less occupying the complete peripheral cortex, were distributed into zones, each showing presence of highly hypertrophied 30 to 40 cells at the limb-bud stage of embryonic development during March (Figure 3a) when the placenta was functional but the corpus luteum was regressed. Many of them were in medullary region in close association with the capillary loops (Figure 2d). The cords histochemically exhibited rich sudanophilic lipid droplets (Figure 3b), marked 3β-HSDH and SDH activities (Figure 3c and d).

During the near term period, that is late April-Early May, the EC were restricted only to some peripheral cortical portions. There was a decline in their number and appeared inactive. Even there was a reduction in the size of clusters; now only 2 to 3 cords formed a cluster, which was previously observed to be 10 to 12. The cells of each cord were also reduced in size and numbers when the placenta was well established (Figure 3e) but the corpus luteum was completely absorbed. The cords revealed large clumps of lipids of coarse nature (Figure 3f), faint 3β-HSDH activity (Figure 3g) and SDH activities (Figure 3h).

However, during lactation (May), there was a pick up in the activities of EC showing an increase in cord number and size with 15 to 30 hypertrophied vesicular cells in each cord (Figure 3i) but some ovaries showed contrast activities of epithelial cords, as there is an asynchrony in the reproductive cycle from the same period. Histochemically the cords demonstrated abundant diffuse sudanophilic lipid droplets (Figure 3j), enhanced activities of 3βHSDH (Figure 3k) and marked activity of SDH (Figure 3l).

DISCUSSION

The origin of these structures remains controversial. They were believed to be originating either by the invagination of the germinal epithelium (Simpson and Van Wagenen, 1953; Guraya and Greenwald, 1964a, 1964b; Motta, 1974) or from the hypertrophy of granulosa cells of the primordial follicles and small preantral follicles, whose ova were lost by atresia (Guraya, 1967a, 1968a, b, d). In H. speoris the origin of epithelial cords were observed to be of dual nature that is, from the germinal epithelium by its in growth and from the persisting granulosa of atretic primordial follicles and small preantral follicles.

In the present study, cyclic changes in the epithelial cords are related with the reproductive cycle of H. speoris. Thus, during early pregnancy, the EC were found scattered in the cortex either in groups of 2 to 3 or singly. But during mid-pregnancy, there was a sudden burst in the histological and histochemical activities. These activities were more enhanced during advanced pregnancy as evident by the occurrence of densely populated, highly hypertrophied zones in the cortex, may be equivalent structures to - accessory corpora lutea, supporting the chorio-allantoic placenta to sustain the development of the embryo. Also during the lactation, the cords were abundant and histochemically functional, might be for the synthesis of hormones, progesterone and estrogen necessary for the synthesis and secretion of milk. The EC during proestrus and estrus were mostly individual and irregularly distributed, this insignificance in their development might be due to the development of the thecal lgc only during proestrus and estrus, thus, subsiding the pre-existing progesterone secreting EC which were important in the implantation of the blastocyst, its further development and the pro-gestational changes undergone by the uterus (Pillai, 2004).

Lipids are the precursors for steroid biosynthesis and its accumulation in the tissues suggests hormone storage and the amount decreases when hormone gets released (Guraya, 1985). Similarly, the enzyme 3β-HSDH is associated with the microsomes derived from extensively developed granular endoplasmic reticulum, whose function in the cell is conversion of pregnenolone to progesterone (Kovarik and Velardo, 1979). Thus, highly sudanophilic and 3β-HSDH positive cords suggest their role in steroid biosynthesis as earlier discussed (Baker, 1951; Guraya and Greenwald 1964a; Guraya, 1985, 2000). As the EC shift towards the medullary parts develop the histochemical features specific to steroid secreting cells, which consists of the presence of diffusely distributed sudanophilic lipid droplets (Guraya, 2000; Guraya and Motta, 1980). The shifting of EC to the medullary region and association with the blood capillaries also suggested their role in the metabolism of steroidal compounds particularly progestins (Savard et al., 1965; Guraya, 1985).

Though the activity of SDH is used as a criterion of luteal function, its activity is correlated with presumed sites of hormone production and places of cellular proliferation (Meyer et al., 1945, 1947). The EC exhibited conspicuous SDH reactivity being highest during mid-pregnancy and lactation as observed in rabbit (Foraker et al., 1955).

The fate of the cord could not be determined, however, from the foregoing study the epithelial cords were observed to be degenerating and finally reverting back to the stromal tissue.

Conclusion

The ovary of H. speoris shows extensive development of

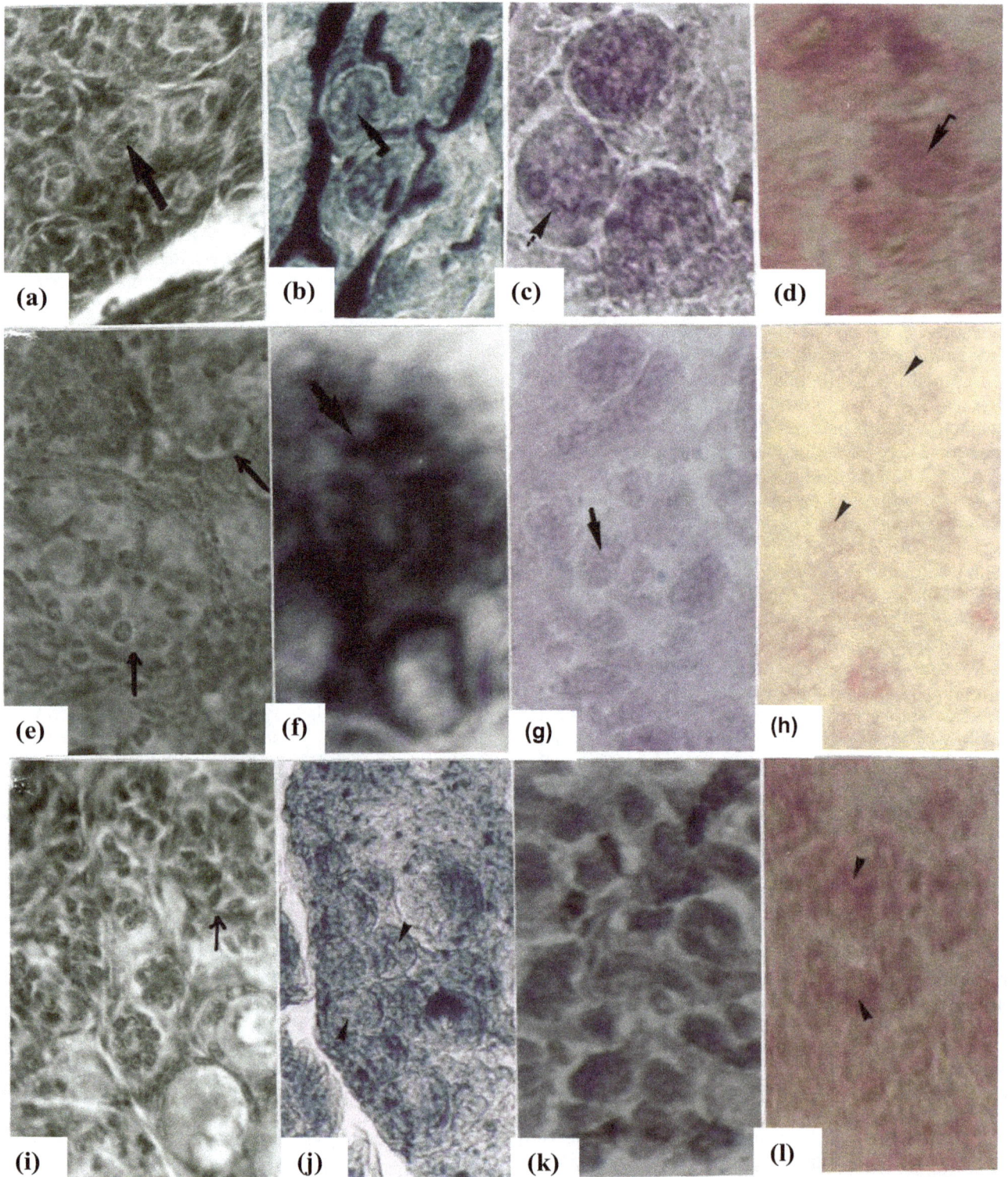

Figure 3. (a) Highly hypertrophied, closely clustered cortical zones of epithelial cords, each with 20 – 30 hypertrophied cells in each cord (arrow). **(b)** Dense accumulation of sudonophilic lipids in the cords (arrow). **(c)** An abrupt shoot up in the 3β-HSDH reactivity in highly developed epithelial cords (arrow). **(d)** Closely packed intense SDH granules giving a diffused pattern of distribution in the epithelial cord (arrow). **(e)** The reduction in the size of cluster with degenerating cells (arrow). **(f)** Reduced number of epithelial cords revealing large clumps of lipid droplets of coarse nature (arrow). **(g)** Epithelial cords (arrow), exhibiting very faint 3β-HSDH reaction. **(h)** Faint SDH reactivity in the ill-developed epithelial cords (arrow). **(i)** An increase in size and number of epithelial cords with vesicular cells (arrow). **(j)** Epithelial cords in the cortical region showing abundant diffuse sudanophilic lipid droplets. **(k)** On enhanced 3β-HSDH reactivity in well-developed epithelial cords. **(l)** Few epithelial cords (arrow) demonstrating high to moderate SDH activity X 250.

epithelial cords particularly during mid and advanced pregnancy and lactational. These EC present wide variations in morphology and histochemistry throughout the reproductive cycle. Both morphological and the histochemical studies revealed their significance as steroidogenic cells in *H. speoris*. The frequency with which these structures were observed during late stages of pregnancy and lactation made it obligatory to conclude that they have a certain significant role in ovarian physiology.

REFERENCES

Barker WL (1951). A cytological study of lipids in sow's ovaries during the estrous cycle. J. Endocrinol. 48: 772-785.

Brambell FWR (1956). Ovarian changes. In *Marshall's Physiology of Reproduction*. Ed. A. S. Parkes Vol-I Longmans Green and Co. London. pp. 397-542.

Dawson AB, McCABE M (1951). The interstitial tissue of the ovary in infantile and juvenile rats. J. Anat. 88: 543-571.

Foraker AG, Denham SW, Mitchell DD (1955). Succinic dehydrogenase and endogenous reductase activity in the rabbit ovary in pregnancy. J. Obstet. Gynaecol. Br. Emp. 62: 447-451.

Guraya SS (1967a). Cytochemical observations concerning the formation, release and transport of lipid secretary products in the interstitial (thecal) cells of the rabbit ovary. *Zeitschrift fur Zellforschung* 83: 187-195.

Guraya SS (1967c). Cytochemical study of interstitial cells in the bat ovary. Nature (London). 214: 614-616.

Guraya SS (1968). Histophysiology and histochemistry of interstitial gland tissue in the ovaries of non-pregnant marmosets. Acta Anatomica (Basal) 70: 623-640.

Guraya SS (1968a). Histochemical study of interstitial cells in the cattle ovary. Acta Anatomica 70: 447-458.

Guraya SS (1968b). Comparative observations on the origin and function of epithelial cords in the mammalian ovary. VI *Congress for International Reproduction & Animals Inseminated Artificially*, Paris. 1:141-143.

Guraya SS (1968d). A histochemical study of preovulatory and postovulatory follicles in the rabbit ovary. J. Reprod. Fertil. 15 : 381-387.

Guraya SS (1985). Biology of ovarian follicles in mammals. Springer—Verlag (Heidelberg- Berlin, New York).

Guraya SS (2000). Comparative cellular and Molecular biology of ovary in mammal: Fundamental & applied aspects. Oxford and IBH publishing Co. Pvt. Ltd. New Delhi.

Guraya SS, Greenwald GS (1964a). Histochemical studies on the interstitial gland in the rabbit ovary. Am. J. Anat. 114: 495-519.

Guraya SS, Greenwald GS (1964b). A comparative histochemical study of interstitial tissue and follicular atresia in mammalian ovary. Anat. Rec. 149 : 411-434.

Guraya SS, Greenwald GS (1965). A histochemical study of the hamster ovary. Am. J. Anat. 116: 257-268.

Guraya SS, Motta PM (1980). Interstitial cells and related structure. In : Biology of the ovary. (Eds.) P.M. Motta and E.S.E. Hafez, Martinus Nijhoff, London. pp. 66-85.

Kovarik FA, Velardo JT (1979). Histochemical study of ovarian hydroxysteroid dehydrogenase activity during normal pseudopregnancy in the rat. Anat. Rec. 194: 273-282.

Lillie RD, Fullmer MH (1976). *Histopathologic technique and practical histochemistry*. IV Edition, J. D. Jeffers & Anne, T. Vinnicombe (Eds.) McGraw-Hill Inc. U.S.A. p. 568.

Macleod MJ (1880). Contribution a l'etude de la structure de l'ovaire des Mammiferes. Arch. Biol. 1: 241-278.

Matthews LH (1935). The oestrous cycle and intersexuality in the female mole (*Talpa europaea*). Proc. Zool. Soc. London 39: 347-383.

Meyer RK, McShan WH, Erway WF (1945). Succinic dehydrogenase activity of ovarian and luteal tissue. Endocrinology 37: 431-436.

Meyer RK, Soukup SW, McShah WH, Biddulp C (1947). Succinic dehydrogenase in rat ovarian tissues during pregnancy and lactation. Endocrinology 41: 35-44.

Mori H, Matsumoto K (1970). On the histogenesis of the ovarian interstitial gland in rabbits. I Primary interstitial gland. Am. J. Anat. 129 : 289-306.

Motta P (1974). The fine structure of the ovarian cortical crypts and cords in mature rabbits. Acta Anatomica 90: 36-64.

Pearse AGE (1972). Histochemistry theoretical and applied. 3rd Edition (Vol. 2) Churchill Livingston.

Pillai SB (2004). Histo-Enzymological Changes In The Ovary Of A Microchiropteran Bat *Hipposideros Speoris* (Schneider) During Reproductive Cycle. Ph.D. Thesis, Nagpur University, Nagpur.

Rennels EG (1951). Influence of hormones on the histochemistry of ovarian interstitial tissue in the immature rat. Am. J. Anat. 88: 63-108.

Savard K, Marsh JM, Rice F (1965). Gonadotropins and ovarian steroidogenesis. Recent Prog. Horm. Res. 21:285-365.

Simpson ME, Van Wagenen G (1953). Response of the ovary of the monkey (*Macaca mulatta*) to the administration of the pituitary follicle stimulating hormone (FSH). Anat. Rec. 115: 570.

14

Influence of reproductive cycle, sex, age and season on haematologic parameters in domestic animals: A review

Yaqub, L. S., Kawu, M. U. and Ayo, J. O.

Department of Physiology, Faculty of Veterinary Medicine, Ahmadu Bello University, Zaria, Kaduna State, Nigeria.

Haematologic parameters play a pivotal role in clinical diagnosis, the evaluation of patient before surgical intervention and in monitoring responses to therapy. The paper reviews haematologic parameters as influenced by exogenous and endogenous factors, including reproductive cycle, sex, age and season, with emphasis on domestic animals reared in the tropics. It is concluded that reproductive cycle, sex, age and season modulate haematologic parameters, and that they should be considered in order to ensure accurate interpretation of the parameters in domestic animals.

Key words: Haematology, reproductive cycle, age, sex and season.

INTRODUCTION

From time immemorial blood has been regarded by humans as the essence of life, the seat of the soul and the progenitor of psychic and physical strength (Ajibola and Ogunsanmi, 2004). Haematological and serum biochemical profiles provide reliable information on the health status of animals (Kral and Suchy, 2000; Cetin et al., 2009). They also reflect the responsiveness of an animal to its internal and external environments (Esonu et al., 2001). Haematological tests have been widely used for the diagnosis of various livestock diseases (Tibbo et al., 2004; Cetin et al., 2009). The information obtained from blood parameters substantiates physical examination and, coupled with medical history, provide excellent basis for diagnosis of diseases (Tibbo et al., 2004). It is also useful in evaluating patients before commencing any surgical intervention and selecting appropriate treatment. For instance, haematocrit or packed cell volume (PCV), haemoglobin (Hb), total protein (TP), leucocyte count and whole blood coagulation time are important indices of animal health and production (Oladele et al., 2005). PCV is a reliable indicator of the value of haemoglobin and circulating erythrocytes (RBCs), while changes in plasma globulins reflect the severity of a disease in birds and, thus, serve as the basis for prognosis (Oladele et al., 2005). It also helps in distinguishing the normal state from the state of stress which can be nutritional, environmental or physical (Aderemi, 2004).

Several factors affect cellular and plasma haemodynamics. They include, age (Egbe-Nwiyi et al., 2000; Olayemi and Nottidge, 2007; Devi and Kumar, 2012), sex (Gabriel et al., 2004; Cetin et al., 2009), breed (Tibbo et al., 2004; Tibbo et al., 2008a; Tibbo et al., 2008b), season (Mira and Maria, 1994; Oladele et al., 2005), pregnancy (Ozegbe, 2001; Kim et al., 2002; Farooq et al., 2011; Okonkwo et al., 2011a) and nutritional status (Ekenyem and Madubruke, 2000; Iyayi, 2001). Other factors that affect haematological parameters include lactation (Harewood et al., 2000), egg laying (Oyewale and Fajimi, 1988), blood volume (Probst et al., 2006), stage of oestrous cycle (Alavi-Shoustari et al., 2006; Chaudhari and Mshelia, 2006), biological rhythms (Hauss, 1994; Greppi et al., 1996; Azeez et al., 2009) and altitude (Wickler and Aderson, 2000). The aim of the present paper is to briefly review the current state of existing knowledge on the influence of reproductive cycle, sex, age and season on haematologic indices, with emphasis on domestic animals reared under tropical conditions.

SEX DIFFERENCE IN HAEMATOLOGIC PARAMETERS

Erythrocytic parameters

Sex has been found to influence haematological values in many animal species, and values in females are almost always lower than in males. Olayemi et al. (2006), obser-ved no significant sex differences in PCV, RBC count and erythrocytic indices in the fruit bat (*Eidoln helvum*). The lack of sexual dimorphism in the RBC values observed was attributed to the fact that the bats were bled outside their breeding season, when the influence of hormone was minimal on the blood values. Higher values in males than females in parameters relating to RBCs were repor-ted in pheasant birds (Hauptmanova et al, 2006), geese (Lazar et al., 1991), Japanese quails (Mihailov et al., 1999), budgerigars (Itoh, 1992), chickens (Oladele et al., 2000). and guinea fowls (Oladele et al., 2005; Obinna et al., 2011). In the mallard duck, PCV and Hb were reported to be higher in the male than female, with values of 41.5 % vs 39.0 %, and13.8 g% vs 13.00 g% for packed cell volume and haemo-globin, respectively (Oladele et al., 2007). Oladele *and* Audu, (2010) reported insignificant lower PCV for female than male geese (*Anser anser*) in Zaria. Other studies by Oyewale and Ajibade (1990) and Awotwi and Boohene (1992) also showed that male birds have higher PCV and Hb than their female counterparts.

The rise in blood parameters in males in comparison with females is often attributed to the effect of androgens, which stimulate erythropoiesis and, thus, cause increase in the number of circulating RBCs, PCV and Hb concen-tration (Villiers and Dunn, 1998). Higher PCV, Hb and RBC values were also observed in the male Agora rabbit in comparison with the female (Cetin et al., 2009). Similar higher PCV, Hb, MCH and MCHC were reported in the male than female (Chineke et al., 2006). It was also reported that haemoglobin was higher in female New Zealand rabbits (Poljicak-Milas et al., 2009). However, a few other workers reported similar RBC values in male and female Nigerian laughing doves (Olayemi et al., 2006), pigeons, peafowls (Oyewale, 1994) and ducks (Olayemi et al., 2002). In Red Sokoto goats, the males have been shown to have a higher PCV than the females (Tambuwal et al., 2002; Okonkwo, 2011b). On the other hand, in West African Dwarf and Sahel goats, PCV values were reported to be similar for both sexes (Daramola et al., 2005; Okonkwo, 2011b). Adamu et al. (2010) did not

obtain a significant effect of sex on PCV and total white blood cell count, but recorded significantly higher plasma fibrinogen in the female than male Polo horses.

Leukocytic parameters

Circulating total leucocyte count represents the outcome of the dynamic production of bone marrow, the release of the cells to the peripheral circulation and the storage in different organs or pools. Sex differences in immune function are well established in vertebrates (Schuurs and Verheul, 1990; Kaushalendra, 2012). Male generally ex-hibit lower immune response than female and under pathogenic conditions (Schuurs and Verheul, 1990; Zuk and McKean, 1996). Male goats have higher lymphocyte count as compared to females, whereas the females have a higher neutrophil count as compared to the males (Tambuwal et al., 2002; Daramola et al., 2005). Similarly, in the Red Sokoto goat, higher leucocyte count had been reported in females than in males (Tambuwal et al., 2002).

No significant sex difference in total leucocyte count was observed in African Fruit bats (Olayemi et al., 2006), African White-bellied pangolins and guinea fowls (*Numida meleagris pallas*) (Oyewale et al., 1997). However, a sig-nificantly higher total leukocyte in male than female African Giant rats was observed (Oyewale et al., 1998). In rabbits, total leucocyte was significantly higher in females than males (Chineke et al., 2006). Total leucocyte are higher in females than in stallion, as reported in Spanish purebred horses (Hernandez et al., 2008), while other study failed to find significant differences between sexes (Lacerd et al., 2006).

Plasma proteins

Plasma proteins are the key components of plasma and they play crucial role in maintaining homeostasis. Plasma proteins consist of albumin, globulin and fibrinogen (Okonkwo et al., 2011b). These proteins have multiple functions; albumin is the most abundant and osmotically active plasma protein, and it is an important carrier of many substances in the peripheral circulation (Alberghina et al., 2010). Globulins are classified on the basis of their electrophoretic mobility as alpha-, beta- and gamma- glo-bulins. While fibrinogen is important in blood clot forma-tion (Harper et al., 1977), thereby preventing loss of blood from ruptured blood vessel.

The effect of sex on plasma proteins has been shown to vary in birds, depending on the breed of the birds. Significantly higher total protein level had been reported in the females than in male guinea fowls (Oladele et al., 2005) and chickens (Oladele et al., 2000). However, no significant sex variation in total protein was observed in local ducks (Oladele et al., 2001a) and pigeons (Oladele et al., 2001b). In West African Dwarf goats, there were no significant sex effects on albumin and globulin, but the male had significant higher fibrinogen than the female. This is in contrast to the finding of Adamu et al. (2010), who documented higher fibrinogen in the female of Polo horses.

EFFECT OF AGE ON HAEMATOLOGIC PARAMETERS

Erythrocytic parameters

The influence of age on the blood parameters of animals has been determined in several species of mammals and birds in Nigeria, such as the New Zealand rabbits (Olayemi et al., 2007), local dogs (Awah and Nottidge, 1998), cats (Nottidge et al., 1999), African Giant rats (Nssien et al., 2002), goats (Egbe-Nwiyi et al., 2000; Addass et al., 2010b), local ducks (Olayemi et al., 2003) and exotic ducks (Oyewale and Ajibade, 1990; Hatipoglu and Bagci, 1996). Mean PCV, Hb and RBC indices were similar in young and adult New Zealand rabbits (Olayemi et al., 2007). A similar observation was made in Nigerian local cats (Nottidge et al., 1999). However, dogs that were more than three months' old were found to have lower PCV values than adult dogs (Oduye, 1978). Higher PCV values were observed in old than young goats (Addass et al., 2010b). Similarly, in sheep, PCV showed a gradual increase with age (Addass et al., 2010a), with lowest values occurring within the first three months of life (Egbe-Nwiyi et al., 2000). Furthermore, also documented an increase in PCV with advancing age in cattle. This trend is also true for donkeys with increasing RBC, PCV, Hb and erythrocytic indices with advancing age (Terkawi et al., 2002; Etana et al., 2011). This finding is also corroborated by the observation in buffalo (Patil et al., 1992; Jabbar et al., 2012). Jabbar et al. (2012) concluded that higher erythrocyte count was responsible for increased PCV value in growing buffalo heifers as compared to adult heifers; apparently due to high basal metabolic rate, leading to increased rate of erythropoiesis and hence increase in erythrocyte count. Contrary to the trend of erythrocyte count and PCV, haemoglobin concentration tends to be higher at birth when compared with the value in adults (Patil et al., 1992; Jabbar et al., 2012).

Leucocyte count

Total leucocyte count and lymphocyte counts (LC) are lower, but higher heterophil and eosinophil counts in adult than in young Hawaiian dark-rumped petrels (*Pterodoma phaepygia*) have been reported (Work, 1996). In addition, Addass et al. (2010a) reported a significant age effect on LC in Nigerian indigenous sheep. Similarly, decreased total leucocyte count and LC and increased NC and eosinophil count with age were obtained in Ethiopian indigenous goats (Tibbo et al., 2004). However, this is in contrast to the finding in indigenous goats in Nigeria, in which age had no significant influence on TLC in four

breeds (West African dwarf, Red Sokoto, Kano Brown and Borno White goats) of goats (Addass et al., 2010b). This is in consonant with the finding in New Zealand rabbit in Nigeria (Olayemi and Nottidge, 2007). Furthermore, lower TLC and lymphocyte count in older pregnant Andalusian Carthusian strain were documented (Satue et al., 2009).

SEASONAL VARIATION IN HAEMATOLOGIC PARAMETERS

Erythrocytic parameters

Generally, the haematological profile is an important indicator of the physiological changes in animals (Jain, 1993; Kumar and Pachaura, 2000). Seasonal changes in the thermal environment influence the physiological responses of animals. Changes in haematological parameters such as total RBC count (Koubkova et al., 2002), PCV (El-Nouty et al., 1990) and RBC indices of mean corpuscular volume (MCV), mean corpuscular haemoglobin (MCH), mean corpuscular haemoglobin concentration (MCHC) are of value in determining the adaptation of animals to the environment. Haemoglobin concentration (Kumar and Pachaura, 2000) and TLC are also indicative of adaptation to adverse environmental conditions. Indeed, haematological values are used to asses stress and welfare in animals (Anderson et al., 1999), especially the neutrophil/lymphocyte ratio (Stanger et al., 2005; Minka and Ayo, 2007).

An increase in body temperature of goats is usually associated with a rise in water intake and depression of food intake (Quartermain and Broadbent, 1997). Thermal stress causes the rostral cooling centre of the hypothalamus to stimulate the medial satiety centre, which inhibits the appetite centre, resulting in reduced feed intake (Albright and Allison, 1972). Under subtropical conditions, the water consumption of goats was greater in summer than winter and spring (Hadjipanayioton, 1995). Such nutritional changes influence the composition of blood in goats (Abdelatif et al., 2009). Furthermore, at high ambient temperature, peripheral vasodilatation and redistribution of cardiac output are associated with expansion of blood volume and haemodilution (Olson et al., 1995).

Aengwanich et al. (2009) reported no significant effect of season on haematological values of crossbred beef cattle at slaughterhouse in northern part of Thailand. However, Tibbo et al. (2004) showed that RBC, PCV and Hb values decreased more during the rainy than any other season. They attributed the decrease to a possible increase in parasite challenge and/or increased water intake through the lush grasses that were available for grazing in that season. Similarly, MCV, MCH and MCHC were higher in summer, while PCV was lower during winter (Kumar and Pachura, 2000). Abdelatif et al. (2009) reported the highest RBC count, PCV and Hb concentration during wet summer and lowest during dry summer; while MCV and MCH were significantly higher during winter than in either wet or dry summer in Nubian goats. The same trend was observed in RBC parameters of goats and in Angora rabbits (Cetin et al., 2009). However, the highest PCV and Hb were recorded during the rainy season in pigeons in Nigeria (Oladele et al., 2001).

Leucocyte count and biochemical parameters

Total leucocyte, lymphocyte and heterophil counts have been reported to be higher in Nigerian local ducks during the dry than wet season, but monocyte and eosinophil counts were not affected by season (Olayemi and Arowolo, 2009). Cetin et al. (2009) also demonstrated a decrease in leucocyte and lymphocyte ratio during the month of July and October in Agora rabbits. Abdelatif et al. (2009) did not observe any significant effect of season on TLC, but the monocyte ratio was significantly higher during the wet and dry summer, as compared to that of the winter.

REPRODUCTIVE CYCLE VARIATION IN HAEMATOLOGIC PARAMETERS

Variations in haematological parameters during the reproductive cycle

Haematological parameters vary with normal physiological and pathological status (Bobade et al., 1985). Various workers have reported changes in haematological parameters during the different phases of the oestrous cycle (Harewood et al., 2000; Alavi-Shoushtari et al., 2006; Chaudhari and Mshelia, 2006). Significantly higher erythrocyte and leucocyte counts have been reported in oestrus in comparison with the dioestrus phase in cattle (Soliman and Zaki, 1963; Hussain and Daniel, 1991). The finding in RSG revealed no significant fluctuation of blood cellular component during the oestrous cycle (Yaqub et al., 2011a). Ijaz et al. (2003) reported an increase in Hb concentration, erythrocyte sedimentation rate, MCH, MCHC in cyclic as compared to non-cyclic cows. In bitches, Chudhari and Mshelia (2006) observed the highest RBC values at pro-oestrus and the lowest during pregnancy. WBC, PCV and Hb values showed increasing pattern from anoestrus to pro-oestrus and decreasing pattern with transition from pro-oestrus to estrus. Also in the study, the lowest TLC was recorded during pregnancy, while the highest was obtained during dioestrus. Harewood et al. (2000) observed a decrease in Hb, PCV and RBC, but an increase in MCV during pregnancy. By contrast, significant higher PCV was recorded in pregnant West African Dwarf ewes than either the lactating or dry ewes (Durotoye and Oyewale, 2000). Tewes et al. (2007) demonstrated significant cyclic changes in TLC, blood pH and total protein, but not in RBC and its indices in sows.

In Baladi does, the blood cellular components decreased during the last four weeks of pregnancy, but leuco-cyte increased on the day of parturition (Azab and Abdel-Maksad, 1999). The erythrocytic indices of MCH and MCV increased during the last three weeks of pregnancy.

Effect of reproductive cycle phase on plasma protein concentrations

Some conditions such as dehydration, external haemorrhage, inflammatory disorders, stress, pregnancy, lactation (Thomas, 2000) and stage of oestrous cycle have been reported to affect plasma protein concentration (Alavi-Shoushtari et al., 2006). In cows, serum total protein was reported to be lower during the oestrus than other phases, and this was attributed to a reduction in serum α_1, γ_1 and γ_2 globulin during oestrus (Alavi-Shoushtari et al., 2006). Similar low serum concentration of total protein was documented during oestral phase of oestrous cycle in Red Sokoto goats (Yaqub et al., 2011b). However, Khan et al. (2010) demonstrated significantly higher level of plasma globulin in normally cycling cows on day 0 of the cycle (3.82 ± 0.01 g/dl) in comparison with day 20 (3.58 ± 0.11 g/dl).

Repeat-breeder cows had significantly lower plasma proteins as compared to normal cycling cows, irrespective of the days of the cycle. In repeat breeding cows, highest and lowest concentrations of plasma proteins were recorded on day 5 and 20 of the cycle, respectively (Khan et al., 2010). In repeat breeding cows, lowest level of albumin was observed on day 15 of the cycle, and highest concentration on day 20; while the lowest and highest levels of globulin were recorded on day 20 and 5 of the cycle, respectively (Khan et al., 2010). Similar cyclic variation in plasma protein in repeat-breeders in comparison with normal cycling animals has been documented by many workers (El-Belely, 1993; Burle et al., 1995; Jani et al., 1995). However, Gandotra et al. (1993) and Ramakrishima (1996) observed no significant variation in protein levels between normal cycling and repeat breeding cows. High incidence of repeat breeding and anoestrous in cows has been attributed to a decrease in circulation of cholesterol (Kumar and Sharma, 1993), glucose (Jani et al., 1995), total protein, albumin and globulin (Joe Arosh et al., 1998).

There were no differences in plasma total protein and albumin between pregnancy and early lactation mares (Milinkovic-Tur et al., 2005). This finding is in congruent with the result obtained in Sahel goats during pregnancy (Waziri et al., 2010). In cows, there were differences in total protein, albumin and globulin fractions between pregnant and non-pregnant cows (Zvorc et al., 2000). In addition, many of the globulin fractions decreased during the last month of gestation.

CONCLUSIONS

Haematologic parameters of domestic animals are significantly influenced by reproductive cycle, sex, age and season. These factors should be considered when interpreting the parameters in order to ensure accuracy.

ACKNOWLEDGEMENTS

The authors acknowledge the tireless efforts of numerous researchers, who have helped in improving our current understanding of various exogenous and endogenous factors influencing haematologic parameters in domestic animals.

RFERENCES

Abdelatif AM, Ibrahim YM, Hassan MY (2009). Seasonal variation in erythrocytic and leukocytic indices and serum proteins of female Nubian goats. Middle-East. J. Sci. Res. 4(3): 168 – 174.

Adamu S, Danladi BN, Mishelia WP, Esievo KAN (2010). Reference values and usefulness of determination of plasma fibrinogen level in polo horses. Zariya Vet. 7(1): 11-17.

Addass PA, Midau A, Babale DM (2010b). Haemato-biochemical findings of Indigenenous goats in Mubi, Adamawa State, Nigeria. J. Agric. Soc. Sci. 6(1): 14 – 16.

Addass PA, Perez KA, Midau A, Lawan, AU, Tizhe MA (2010a). Haemato-Biochemical findings of indigenous sheep breeds in Mubi Adamawa State, Nigeria. Global Vet. 4(2): 164-167.

Aderemi FA (2004). Effect of replacement of wheat bran with cassava root sieviate supplemented or unsupplemented with enzyme on the haematology and serum biochemistry of pullet chicks. Trop. J. Anim. Sci. 7: 147-153.

Aengwanich W, Chantritratikul A, Pamok S (2009). Effect of seasonal variations on haematological values and health monitor of crossbreed cattle at slaughterhouse in Northern eastern part of Thailand. American-Eurasian J. Agric. Environ. Sci. 5 (5): 644-648.

Ajibola OO, Ogunsanmi VT (2004). Comparative studies on erythrocyte calcium, potassium, haemoglobin concentration, osmotic resistance and sedimentation rates in grey duiker (Sylvicapra grimma), sheep and goats experimentally infected with Trypanososma congolense. Vet. Arhiv. 74(3): 201 - 216.

Alavi-Shoushtari SM, Asri-Rezai S, Abshenas J (2006). A study of the uterine protein variations during the estrous cycle in the cow: a comparison with serum proteins. Anim. Rep. Sci. 96 (1 - 2): 10 – 20.

Alberghina D, Casella S, Vazzana I, Ferrantelli V, Giannetto C, Piccione G (2010). Analysis of serum proteins in clinically healthy goats (Capra hircus) using agarose gel electrophoresis. Vet. Clin. Path. 317-321.

Albright JL, Allison CW (1972). Effects of varying the environment upon performance of dairy cattle. J. Anim. Sci. 32: 57- 66.

Anderson BH, Watson DI, Colditz GI (1999). The effect of dexamethasone on some immunological parameters in cattle. Vet. Res. Comm. 23: 399-413.

Awah JN, Nottidge HO (1998). Serum biochemical parameters in clinically healthy dogs in Ibadan. Trop. Vet. 16: 123-129.

Awotwi EK, Boohene YG (1992). Haematological studies on some poultry species in Ghana. Bull. Anim. Health Prod. Afr. 40: 65 – 71.

Azab ME, Abdel-Maksad HA (1999). Changes in some haematological and biochemical parameters during prepartum and postpartum periods in female Baladi goats. Small Rum. Res. 34(1): 77-85.

Azeez OI, Oyagbemi AA, Oyewale JO (2009). Diurnal fluctuation in haematological parameters of the domestic fowls in the hot humid tropics. Int. J. Poult. Sci. 8(3): 247-251.

Bobade PA, Oduye OO, Helen O, Agbona O (1985). Haemogram of clinically normal dogs with particular reference to local (Nigerian) and German shepherd dogs. Nig. Vet. J. 14: 7 - 11.

Burle PM, Mangle NS, Kothekhan MD, Lalorey DR (1995). Blood biochemical profile on postpartum reproduction and energy balance in dairy cattle. J. Dairy Sci. 73: 2342-2349.

Cetin N, Bekyurek T, Cetin E (2009). Effect of sex, pregnancy and season on some haematological and biochemical blood values in

Angora rabbits. Scand. J. Lab. Anim. Sci. 36 (2): 155-162.

Chaudhari SUR, Mshelia GD (2006). Evaluation of the haematological values of bitches in Northern Nigeria for the staging of pregnancy. Pak. J. Biol. Sci. 9 (2): 310 – 312.

Chineke CA, Ologun AG, Ikeobi CON. (2006). Haematological parameters in rabbit breeds and crosses in humid tropics. Pak. J. Biol. Sci. 9 (11): 2102 – 2106.

Daramola JO, Adeloye AA, Fatoba TA, Soladoye AO (2005). Haematological and biochemical parameters of West African Dwarf goats. Livestock Research for Rural Development, 17, Art. 95. Retrieved January 19, 2011, from http://www.lrrd.org/lrrd/lrrd17/8/dara17095. htm. 3:05 pm.

Devi R, Kumar MP (2012). Effect of ageing and sex on the ceruloplasmin (Cp) and the plasma protein levels. J. Clini. Diagn. Res. 6(4): 577-580.

Durotoye LA, Oyewale JO (2000). Blood and plasma volume in normal West African dwarf sheep. Afr. J. Biomed. Res. 3: 135-137.

Egbe-Nwiyi TN, Nwosu SC, Salami HA (2000). Haematological values of apparently healthy sheep and goats as influenced by age and sex in arid zone of Nigeria. Afr. J. Biomed. Res. 3: 109-115.

Ekenyem BU, Madubruke FN (2000). Haematological and serum biochemistry of grower pigs fed varying levels of Ipomoea asarifolia leaf meal. Pak. J. Nut. 6(6): 603-606.

El-Belely NS (1993). Progesterone, estrogen and selected biochemical constituents in plasma and uterine flushings of normal and repeat breeder buffalo cows. J. Agric. Sci. 120: 241-250.

El-Nouty FD, Al-Haidary AA, Salah MS (1990). Seasonal variation in hematological values of high and average yielding Holstein cattle in semi-arid environment. J. King Saud. Univ. 2(2): 173-182.

Esonu BO, Emenalom OO, Udedebie U, Herbert DF, Ekpori IC, Iheukwuemere FC (2001). Performance and blood chemistry of weaner pigs fed raw mucuna bean (velvet bean) meal. Trop. Anim. Prod. Invest. 4: 49 - 54.

Etana KM, Jenbere TS, Bojia E, Negusie H (2011). Determination of reference haematological and serum biochemical values for working donkeys of Ethiopia. Vet. Res. 4(3): 90-94.

Farooq H, Samad HA, Sajjad S (2011). Normal reference Haematological values of one-humped camels (Camelus Dromedarius) kept in Cholistan desert. J. Anim. Plant Sci. 21(2): 157-160.

Gabriel UU, Ezeri GNO, Opabunmi OO (2004). Influence of sex, source, health status and acclimation on the haematology of Clarias gariepinus. Afr. J. Biotech. 3(9): 460-467.

Gandotra VK, Chaudhary RK, Sharma RD (1993). Serum biochemical constituents in normal and repeat breeding cows and buffaloes. Indian Vet. J. 70: 84-85.

Greppi GF, Casini L, Gatta D, Orlandi M, Pasquini M (1996). Daily fluctuations of haematology and blood biochemistry in horses fed varying levels of proteins. Equine Vet. J. 28(5): 350-353.

Hadjipanayioton M (1995). Fractional outflow of soybean meal from the rumen, water intake and ruminal fermentation pattern in sheep and goats at different seasons and age group. Small Rum. Res. 17: 137-143.

Harewood WJ, Gillin A, Hennessys A, Armistead J, Horvath JS, Tiller DJ (2000). The effects of the menstrual cycle, pregnancy and early lactation on haematology and plasma biochemistry in the baboon (Papio hamadryas). J. Med. Primatol. 29: 415 – 420.

Harper HA, Rodwell VW, Mayer PA (1977). Review of Physiological Chemistry, 6th (eds). Lange Medical Publication, California.

Hatipoglu S, Bagci C (1996). Some haematological values from Pekin ducks. Berl. Munch. Tierarztl Wschr. 109: 172-176.

Hauptmanova K, Maly M, Literak I (2006). Changes of haematological parameters in common pheasants throughout the year. Vet. Med. 51(1): 29 - 34.

Hauss E (1994). Chronobiology of circulating blood cells and platelets. In: Biological Rhythms in clinical and laboratory Medicine. Touitou, Y. 8 Hauss, E. (eds)., Springer-Verlag. pp. 504-526

Hernandez AM, Satue K, Lorente C, Garces C, Oçonnor JE (2008). The influence of age and gender on haematologic parameters in Spanish Horses. Proceeding of Veterinary European equine meeting – XIV SIVE congress, Venice (Italy).

Hussain AM, Daniel RCW (1991). Studies on some aspects of neutrophil functions and uterine defences in cows during the oestrous cycle. Rep. Dom. Anim. 26: 290 - 296.

Ijaz A, Gohar NA, Ahmad M (2003). Haematological profile in cyclic and non-cyclic and endometric cross-breed cattle. Int. J. Agric. Biol. 5(3): 332-334.

Itoh N (1992). Some haematologic values in budgerigar. J. Rakuno Gakuen Univ. 17: 61-64.

Iyayi EA (2001). Cassava leaves as supplements for feeding weaner swine. Trop. Anim. Prod. Invest. 4: 141-150.

Jabbar L, Cheena A, Riffat S (2012). Effect of different dietary energy levels, season and age on haematological indices and serum electrolytes in growing buffalo heifers. J. Anim. Plant Sci. 22(3 Suppl): 279-283.

Jain NC (1993). Essentials of Veterinary Haematology (1st Edition), Lea and Febiger, Philadelphia. pp. 1-18.

Jani RG, Prajapati BR, Dave MR (1995). Hematological and biochemical changes in normal fertile and infertile Surti buffaloe heifers. Indian J. Anim. Rep. 5: 14-22.

Joe-Arosh K, Devanathan AD, Rajasundaram TG, Rajasekaran J (1998). Blood biochemical profile in normal cyclical and anoestrous cows. Indian J. Anim. Sci. 68: 1154-1156.

Kaushalendra CH (2012). Correlation between peripheral melatonin and general immune status of domestic goat, Capra hircus: A seasonal and sex dependent variation. Small Rum. Res. 107: 147-156.

Khan S, Thangawel A, Selvasubramaniyan S (2010). Biochemical profile in repeat breeding cows. Tamilnadu J. Vet. Anim. Sci. 6(2): 75-80.

Kim JC, Yun HI, Chas SW, Kim KH, Koh WS (2002). Haematological changes during the normal pregnancy in New Zealand White rabbis. Comp. Clin. Pathol. 11: 98-106.

Koubkova M, Knizkova I, Kunc P, Hartlova H, Flusser J, Dolezal O (2002). Influence of high environmental temperatures and evaporative cooling on some physiological, haematological and biochemical parameters in high-yielding dairy cows. Czech J. Anim. Sci. 47: 309-318.

Kral I, Suchy P (2000). Haematological studies in adolescent breeding cocks. Acta. Vet. Brno. 69: 189 - 194.

Kumar B, Pauchaura SP (2000). Haematological profile of crossbred dairy cattle to monitor herd health status at medium elevation in central Himalayas. Res. Vet. Sci. 69: 141-145.

Kumar S, Sharma P (1993). Hematological changes during fertile and nonfertile estrous in rural buffaloes. Buffaloe J. 9: 69-73.

Lacerd L, Compos R, Sperb M, Soares E, Barbosa E, Rerreira R, Santos V, Gonzalez FH (2006). Haematologic and biochemical parameters in three high performance horse breeds from southern Brazil. Archiv. Vet. Sci. 11(2): 40-44.

Lazar V, Pravda D, Stavkova J (1991). Analysis of the sources of variability of the haematological characteristics of geese (in Czech). Zivocisma Vyroba, 36 : 517-523.

Mihailov R, Lasheva V, Lashev L (1999). Some haematological values in Japanese quails. Bulg. J. Vet. Med. 2: 137 – 139.

Milinkovic-Tur S, Peric V, Stojevic Z, Zdelar-Tuk M, Pirljin J (2005). Concentrations of total proteins and albumins, and AST, ALT and GGT activities in the blood plasma of mares during pregnancy and early lactation. Vet. Arhiv. 75(3): 195-202.

Minka NS, Ayo JO (2007). Physiological responses of transported goats treated with ascorbic acid during hot dry season. Anim. Sci. J. 78(2): 164-172.

Mira A, Maria ML (1994). Seasonal effects on the haematology and blood plasma proteins of two species of mice Mus musculus domesticus and M. spretus from Portugal. Hystririx 5: 63-72.

Nottidge HO, Taiwo VO, Ogunsanmi AO (1999). Haematological and serum biochemical studies of cats in Nigeria. Trop. Vet. 17: 9-16.

Nssien MAS, Olayemi FO, Onwuka SK, Olusola A (2002). Comparison of some plasma biochemical parameters in two generations of African giant rat (Cricetomys gambianus, Waterhouse). Afr. J. Biomed. Res. 5: 63-67.

Obinna, OVM, Emmanuel OU, Princewill OI, Helen O, Christopher E (2011). Effect of sex and systems of production on the haemato-logical and serum biochemical characters of helmeted guinea fowls (Numida meleagris pallas) in South Eastern Nigeria. Int. J. Biosci. 1(3): 51-56.

Oduye OO (1978). Haematological studies on clinically normal dogs in

Nigeria. Zentralblalt Fur Veterina Medizine 25: 548-555.

Okonkwo JC, Okonkwo IF, Ebyh GU (2011a). Effect of breed, sex and source within breed on the haematological parameters of the Nigerian goats. Online J. Anim. Feed Res. 1(1): 8-13.

Okonkwo JC, Omeje JS, Okonkwo IF (2011b). Effect of source and sex on blood protein fractions of West African dwarf goats (WADG). Res. Opin. Anim. Vet. Sci. 1(3): 158-161.

Oladele SB, Audu SB (2010). Packed cell volume, haemoglobin, total protein and whole blood coagulation time of the geese (Anser anser) reared in Zaria, Northern Nigeria. Zariya Vet. 7(1): 1 – 10.

Oladele SB, Ayo JO, Esievo KAN, Ogundipe SO (2000). Effect of season and sex on packed cell volume, haemoglobin and total protein of indigenous chickens in Zaria, Nigeria. J. Med. Allied Sci. 3: 173-177.

Oladele SB, Ayo JO, Esievo KAN, Ogundipe SO (2001a). Seasonal and sex variations in packed cell volume, haemoglobin and total protein of indigenous duck in Zaria, Nigeria. J. Trop. Biosci. 1(1): 84-88.

Oladele SB, Ayo JO, Ogundipe SO, Esievo KAN (2005). Seasonal and sex variations in packed cell volume, haemoglobin and total protein of the guinea fowl (Numida meleagris) in Zaria, Northern Guinea Savannah zone of Nigeria. J. Trop. Biosci. 5 (2): 67- 71.

Oladele SB, Isa IH, Sambo SJ (2007). Haematocrit, haemoglobin, total protein and whole blood coagulation time of the Mallard duck, Niger Vet. J. 28 (1): 14 – 20.

Oladele SB, Ogundipe S, Ayo JO, Esievo KAN. (2001b). Effects of season and sex on packed cell volume, haemoglobin and total protein of indigenous pigeons in Zaria, Northern Nigeria. Vet. Arhiv. 71(5): 277-286.

Olayemi F, Oyewale J, Rahman S, Omolewa O (2003). Comparative assessment of the white blood cell values, plasma volume and blood volume in the young and adult Nigerian duck (Anas platyrhynchos). Vet. Arhiv. 73(5): 27 – 276.

Olayemi FO, Arowolo ROA (2009). Seasonal variations in the haema-tological values of the Nigerian Duck (Anas platyrhynchos). Int. J. Poult. Sci. 8(8): 813-815.

Olayemi FO, Nottidge HO (2007). Effect of age on the blood profiles of the New Zealand rabbit in Nigeria. Afr. J. Biomed. Res. 10: 73 – 76.

Olayemi FO, Ojo EA, Fagbohun OA (2006). Haematological and plasma biochemical parameters of the Nigerian Laughing dove (Anas platyrhynchos), Vet. Arhiv, 76 (2); 145 – 151.

Olayemi FO, Oyewale JO, Omolewa OF (2002). Plasma chemistry values in the young and adult Nigerian duck (Anas platyrhynchos). Israel. J. Vet. Med. 57: 1 - 5.

Olson K, Joaster-Hermelin M, Hossaini-Hilali J, Hydbrig E, Dahlborn K(1995). Heat stress causes excessive drinking in feed and food-deprived pregnant goats. Comp. Bioch. Physiol. (A), 110 (4): 309-317.

Oyewale JO (1994). Further studies on osmotic resistance of nucleated erythrocyte: observation with pigeon, peafowl, lizard and toad erythrocyte during changes in temperature and pH. J. Vet. Med. 41: 62 – 71.

Oyewale JO, Ajibade HA (1990). Osmotic fragility of erythrocytes of the white Pekin duck. Vet. Arhiv, 60: 91-100.

Oyewale JO, Fajimi JL (1988). The effect of egg lying on haematological and plasma biochemistry of guinea hen. Bull. Anim. Health Prod. Afr. 36: 229 - 232.

Oyewale JO, Okewumi TO, Olayemi FO (1997). Haematological changes in West African Dwarf goats following haemorrhage. J. Vet. Med. 44: 619-624.

Oyewale JO, Olayemi FO, Oke OA (1998). Haematology of the wild adult African giant rat (Cricetomys gambianus, Water-house). Vet. Arhiv. 68: 91-99.

Ozegbe PC (2001). Influence of pregnancy on some erythrocyte biochemical profile in the rabbits. Afr. J. Biomed. Res. 4: 135-137.

Patil MD, Talvelker BA, Joshi VG, Deshmukh BT (1992). Haematologic studies during the oestrous cycle in Murrah buffalo heifers. Indian Vet. J. 69: 894-897.

Poljicak-Milas N, Kardum-Skelin, I, Vudan M, Marenjak TS, Ballarin-Perharic A, Milas Z (2009). Blood cell count analyses and erythrocyte morphometry in New Zealand white Rabbits. Vet. Arhiv. 79(6): 561-571.

Probst RJ, Jenny ML, Bird DN, Pole GL, Aileen KS, John RC (2006).

Gender differences in the blood volume of conscious Sprague-Dawley rats. J. Am. Assoc. Lab. Anim. Sci. 45(2): 49-52.

Quartermain AR, Broadbent MP (1997). Some patterns of response to climate by Zambian goats. East Afr. Agric. Forest J. 40: 115-124.

Ramakrishima KV (1996). Microbial and biochemical profile in repeat breeder cows. Indian J. Anim. Rep. 17: 30-32.

Satue K, Blanco O, Munoz A (2009). Age-related differences in the haematological profile of Andalusian broodmares of Carthusian strain. Vet. Med. 54: 175-182.

Schuurs AH, Verheul HA (1990). Effects of gender and sex steroids on the immune response. J. Steroid Biochem. 35(2):157-172.

Soliman MK, Zaki K (1963). Blood picture of Friesian cows during the oestrous cycle and pregnancy. J. Arab Vet. Med. Assoc. 4: 343 - 54.

Stanger KJ, Ketheesan AJ, Parker CJ (2005). The effect of transportation on the immune status of Bos indicus steers. J. Anim. Sci. 83: 2632-2636.

Tambuwal FM, Agale BM, Bangana A (2002). Haematological and biochemical values of apparently healthy Red Sokoto goats. Proceeding of the 27th Annual Conference of the Nigerian Society for Animal Production, Federal University of Technology, Akure, Nigeria. pp. 50-53.

Terkawi AD, Tabba D, Al-Omari A (2002). Estimation of normal haematology values of local donkeys in Syria. Proceedings of the 4th International Colloquium on working Equines, April 20-25, Spana-Alabaata University, Hama, Syria. pp. 115-118.

Tewes H, Steinbach J, Smidt D (2007). Investigations on the blood composition of sows during the reproductive cycle. Rep. Domest. Anim. 12(3): 117 - 124.

Thomas JS (2000). Overview of plasma protein. In: Feldman, B. F., Zinkl, J. G. and Jain, N. C. (Editors), Schalm's Veterinary Haematology, Fifth Edition, Lippincott Williams and Winlkins, Philadelphia. pp. 891-909.

Tibbo M, Jibril T, Woldemeskel M, Dawo F, Aragaw K, Rege JEO (2004). Factors affecting haematological profiles in three Ethiopian indigenous goat breeds. Intern. J. Appl. Res. Vet. Med. 2(4): 297 – 309.

Tibbo M, Jibril Y, Woldemeskel M, Dawo F, Argaw K, Rege JEO (2008a). Serum enzyme levels and influencing factors in three indigenous Ethiopian goats breeds. Trop. Animl. Health Prod. 40: 657-666.

Tibbo M, Woldemeskel M, Argaw K, Rege JEO (2008b). Serum enzyme levels and influencing factors in three indigenous Ethiopian sheep breeds. Comp. Clin. Pathol. 17 (3): 149-155.

Villiers E, Dunn JK (1998). Basic haematology. In: Davidson, M., Else, R. and Lumsden, J. (Editors). Manual of Small Animal Clinical Pathology, Shurdngton, Cheltenham, United Kingdom. pp. 33-60.

Waziri AM, Ribadu AY, Sivachelvan N (2010). Changes in the serum proteins, haematological and some srum biochemical profiles in the gestation period in the Sahel goats. Vet. Arhiv. 80(2): 215-224.

Wickler SJ, Aderson TP (2000). Haematological changes and athletic performance in horses in response to high altitude (3800 m). Am. J. Physiol. – Regul. Intergr. Comp. Physiol. 279: 1176-1181.

Work TM (1996). Weight, haematology and serum chemistry of seven species of free-ranging tropical pelagic seabird. J. Wildlife Dis. 13: 1051-1055.

Yaqub LS, Ayo JO, Rekwot PI, Onyeanusi BI, Kawu MU, Ambali SF, Unchendu C (2011a). Changes in haematologic parameters and erythrocyte osmotic fragility during oestrous cycle in Red Sokoto goats. Book of Abstract 48th Annual Congress of NVMA, Kwara, Kwara State government Banquet Hall, Ilorin, Nigeria.

Yaqub LS, Ayo, JO, Rekwot, PI, Oyeanusi BI, Kawu MU, Ambali SF, Shittu M, Abdullahi A (2011b). Changes in serum proteins and urea during the oestrous cycle in Red Sokoto goats. Adv. Appl. Sci. Res. 2(6): 197-205.

Zuk W, Mckean KA (1996). Sex difference in parasites infections: pattern and process. Int. J. Parasitol. 26: 1009-1024.

Zvorc Z, Matijatko V, Beer B, Forsek J, Bedrica L, Kucer N (2000). Blood serum proteinograms in pregnant and non-pregnant cows. Vet. Arhiv. 70(1): 21-30.

15

Reproductive biology of *Oreochromis niloticus* in Lake Beseka, Ethiopia

Lemma Abera Hirpo

Zwai Fishery Research Center, P.O. Box 229, Zwai, Ethiopia.

Reproductive biology of *Oreochromis niloticus* in Beseka was studied. Samples of *O. niloticus* were collected monthly during September 2010 to August 2011 using different centimeter mesh sizes of gillnets. The relationship between total length and total weight was curvilinear and sex ratio was different throughout the sampling periods. The 50% sexual maturity length (L_{50}) was estimated at 14 cm TL for females and 17 cm TL for males. Estimated fecundity was linearly related with total length and total weight of the fish. Absolute fecundity was estimated in number and range from 125 to 251 with a mean of 161 ± 2.5. The frequency of ripe gonads suggested that *O. niloticus* in Lake Beseka breeds throughout the year and intensive breeding coincided with the rainy seasons.

Key words: Breeding season, length-weight, *Oreochromis niloticus*, sex ratio, fecundity, Lake Beseka.

INTRODUCTION

A cheap source of protein is urgently required to support an ever increasing human population. Fishery resources definitely can offer one of the solutions to the problem of food shortage in a country like Ethiopia. Moreover, the Nile tilapia (*Oreochromis niloticus*) is the most preferred fish species in Ethiopia for human consumption and the demand has increased rapidly over the last few years.

Therefore, information on the breeding and fecundity of *O. niloticus* can provide basic knowledge for the proper management of the resource. However, such knowledge is not recently available for the species in the lake and this has hindered proper management of the fishery. Therefore, the major objective of the present study was to generate basic biological information that could help to ensure proper exploitation and management strategies on the Ethiopian fishery in general and *O. niloticus* in the lake in particular. The specific objectives were to assess breeding season and fecundity of *O. niloticus* in Lake Beseka in Ethiopia.

MATERIALS AND METHODS

Field sampling and measurement

Samples of *O. niloticus* were collected using gill nets monthly between September 2010 and August 2011 from different sites. The gear were set parallel to the shoreline in the afternoon (5:00 pm) and lifted in the following morning (7.00 am). Immediately after capture, total length (TL) and total weight (TW) of each specimen were measured to the nearest 0.1 cm and 0.1 g, respectively. The ripe ovaries were split longitudinally and turned inside out, to ensure the penetration of the preservative before they were stored in labeled jars. Finally, ripe ovaries were preserved in Gilson's fluid to estimate fecundity (Bagenal and Tesch, 1978). Preserved samples were then transported for further laboratory analysis.

Estimation of sex - ratio and length at maturity

The number of female and male *O. niloticus* caught was recorded for each sampling occasion. Sex-ratio (female : male) was then calculated for each month and total sample. The average length at

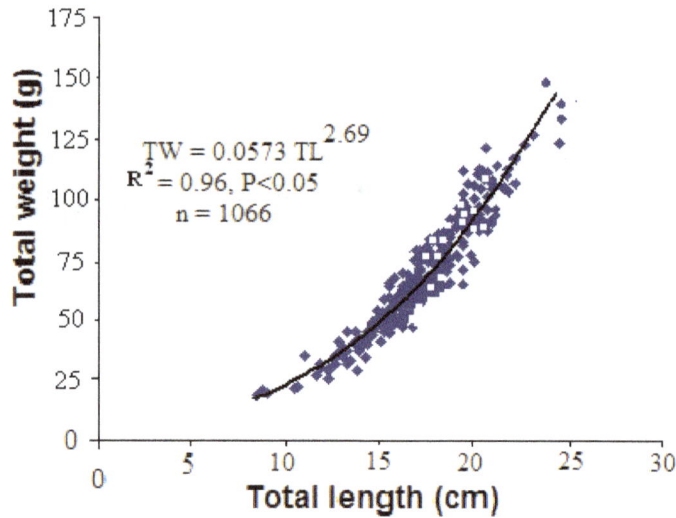

Figure 1. Length-weight relationship of *O.noloticus* in Lake Beseka.

Table 1. Number of females and males fish sampled in Lake Beseka.

Month	F	M	F:M
Sep. 2010	94	21	1:0.22
Oct	65	17	1:0.26
Nov.	57	22	1:0.39
Dec.	42	13	1:0.31
Jan. 1011	14	9	1:0.64
Feb.	17	11	1:0.06
Mar.	33	18	1:0.55
Apr.	45	22	1:0.49
May	66	29	1:0.44
Jun.	76	37	1:0.49
Jul.	120	44	1:0.37
Aug.	164	30	1:0.18
Total	793	273	1:0.34

first maturity (L_{50}) has been defined as the length at which 50% of the individuals in a given length classes reach maturity (Willoughby and Tweddel, 1978). Thus, after classifying data by length class, the percentages of male and female *O. niloticus* with mature gonads were plotted against length to estimate L_{50} (Tweddle and Turner, 1977).

Determination of breeding season and fecundity estimation

The breeding season of *O. niloticus* was determined from monthly frequency of fish with ripe gonads. The ovaries were split longitudinally and turned inside out, to ensure the penetration of the preservative before they were stored in labeled jars (Bagenal and Tesch, 1978). Finally, ripe ovaries were preserved in Gilson's fluid to estimate fecundity (Simpson, 1959). The fecundity of ripe gonads preserved in Gilson's fluid was estimated gravimetrically. To estimate fecundity, the preservative was replaced with water, and the eggs were washed repeatedly, and decanting the supernatant.

Estimated fecundity was then obtained by weighing the entire eggs, and two sub-samples were taken and counted, each of which were all similarly, dried. The eggs were visually counted and weighed using a sensitive balance (ACB plus-3000g). The total number (N) was computed using the following ratio:

$$N/n = W/w$$

Where, N = unknown total number of eggs; n = number counted in sub sample (1000); W = weight of all eggs (g); w = weight of the sub sample (g)

Least squares regression was then used to find the relationship between fecundity and total length, total weight and gonad weight (Admassu, 1994).

RESULTS

Length-weight relationship of the fish

The length-weight relationship of *O. niloticus* in Lake Beseka was curvilinear and statistically highly significant (P<0.05) (Figure 1). The equations separated by sex were as follows:

Males: $TW = 0.0124 \times TL^{2.61}$, $R^2 = 0.631$, n = 273
Females: $TW = 0.0141 \times TL^{2.73}$, $R^2 = 0.709$, n = 793

Therefore, an equation combined for both sexes was fitted and shown in Figure 1. The equation was for fish ranging in length from 8 to 25 cm, and in total weight from 18 to 149 g. The slope (b = 2.69) was close to the theoretical value of 3.

Sex ratio and Length at maturity

Sex ratio results are presented in Table 1. The ratio was different for all sampling months and total sample. Females numerically outnumbered males in all sampling periods.

The smallest sexually mature fish that was caught in this study was a female fish of 6 cm TL and a male fish of 7 cm TL. The 50% maturity length (L_{50}) was estimated to be 14 cm TL for females (Figure 2) and 17 cm TL for males (Figure 2). On the average, females appeared to attain sexual maturity at a relatively smaller size than males.

Breeding season

The frequency of temporal variation between ripe males and females was similar. The frequency was found to be high from September (2010), March to April and August (2011) (Figure 3). The lowest frequency of ripe fishes was recorded at time between October and February,

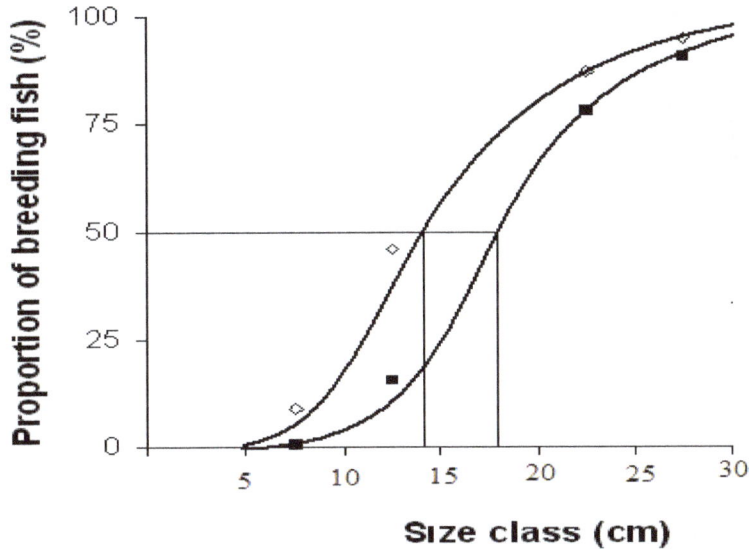

Figure 2. The proportion of size at L_{50} maturity of males (♦) and females (◊) in Lake Beseka.

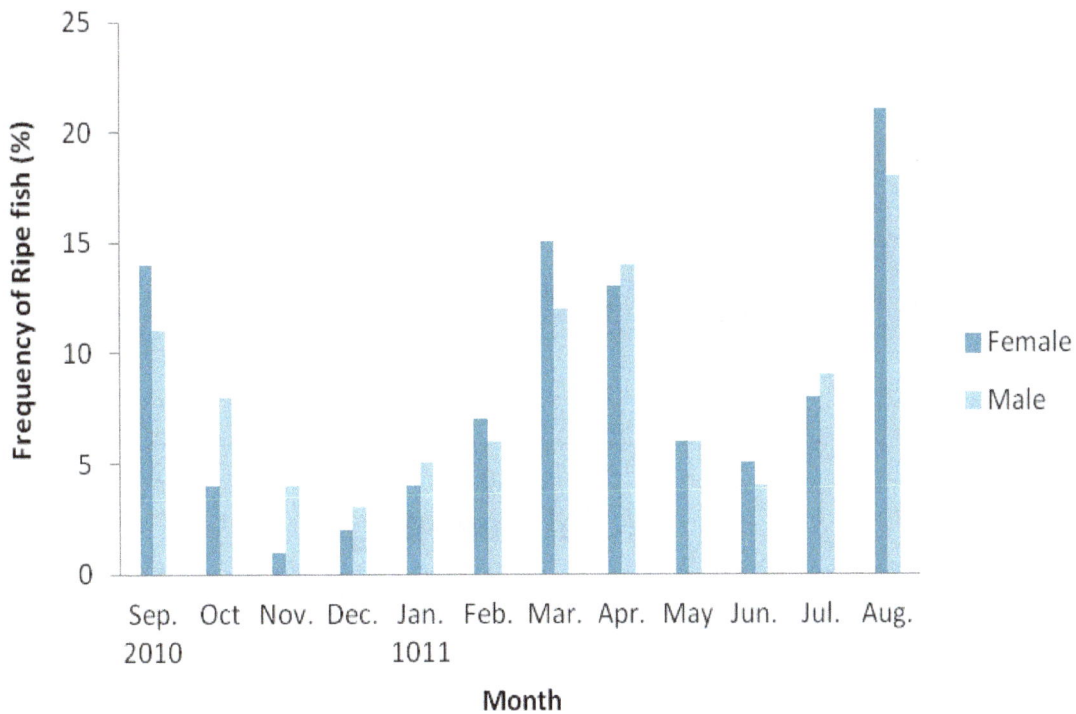

Figure 3. Temporal variation in frequency (%) of ripe female and male *O. noloticus* in Lake Beseka.

and May to July.

Fecundity estimation

A total of 37 ripe female were used for fecundity estimation. Their total length and total weight ranged from 12 to 25 cm and 26 to 149 g, respectively. The number of eggs per individual ranged from 125-351 with a mean of 261 ± 2.5. Fecundity was linearly related to total length and total weight (Figures 4 and 5).

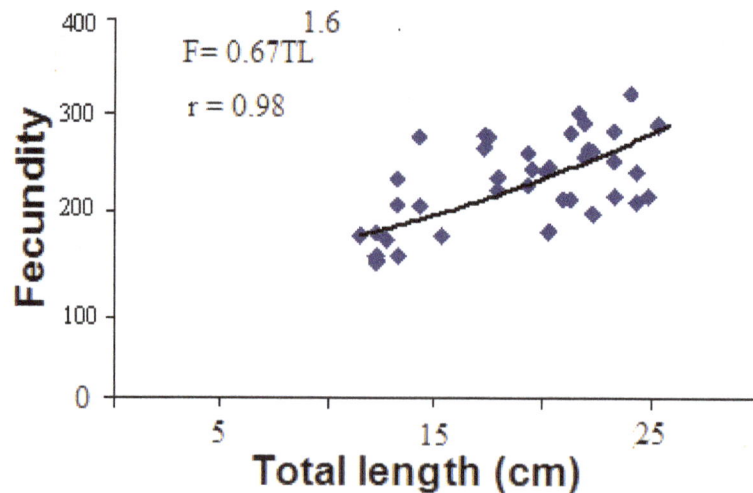

Figure 4. Relationship between fecundity and total length of *O. niloticus* in Lake Beseka.

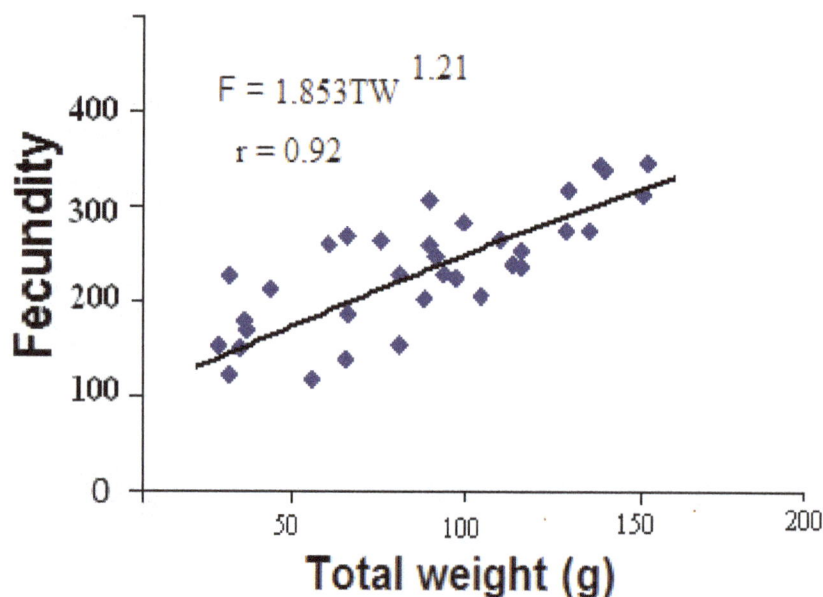

Figure 5. Relationship between fecundity and total weight of *O. niloticus* in Lake Beseka.

DISCUSSION

The largest fish caught in the present study was 25 cm (TL) which was smaller than Lake Babogaya (28 cm) (Lemma, 2012). There were a curvilinear relationship between total length and total weight of the fish in the lake (Figure 1). The value of b (2.69) was close to the theoretical value (b = 3), indicating isometric growth. These finding are in agreement with the principle of fish growth (Bagenel and Tesch, 1978). The study showed

that an unbalanced sex ratio existed for samples taken throughout the sampling period. The unbalanced sex ratio found in the present study is difficult to explain. Probably, it could be attributed to behavioral differences between the sexes, which might have made females more vulnerable and passive to fishing gears such as gill nets. The preponderance of females has been attributed to sexual segregation during spawning, activity differences, gear type and fishing site (Admassu, 1994). Hence, further study is required to see if the same factors could be responsible for sex ratio results for *O. niloticus* in the current study.

The size of 50% sexual maturity of the fish in this study was smaller than values estimated for the same species in Lake Awassa and Zwai. Length of maturity in many fish species depends on demographic conditions, and is determined by genes and the environment (Fryer and Iles, 1972; Lowe-McConnell, 1987). Generally, fish in poor condition mature at smaller size than those in good condition (Lowe-McConnell, 1958, 1959, 1987).

Intensive breeding activity of the fish in Lake Babogaya was coincident with the rainy season. Thus, rainfall and associated factors like temperature may act as cues for spawning by the fish so that offspring are produced at a time of better growth and survival. The role of rainfall in fish spawning is well documented (Fryer and Iles, 1972; Balarin and Hatton, 1979; Lowe-McConnell, 1982). Runoff, for instance, results in increased nutrient concentrations which in turn result in improved food quantity and quality (Jalabert and Zohar, 1982; Tadesse, 1988; Admassu, 1996). A correlation between rainfall and peak breeding activity has also been reported for different species (Dadebo, 1988) and other species (Tadesse, 1988; 1997; Admassu, 1994; 1996; Teferi, 1997) in Ethiopia, and elsewhere (Fryer and Iles, 1972; Jalabert and Zohar, 1982; Lowe-McConnell, 1982; Stewart, 1988) for the same species.

Fecundity in the current study was slightly lower than the same species in different water bodies of the country. This could be due to its lower body condition and growth as compared to the species in the other lakes. Fish in poor body condition are reported to have less fecundity than those in better condition (Lowe-McConnell, 1959). Even though this study has baseline information for the proper utilization of the resources further detailed studies are required on other biological aspects (growth, mortality, feeding habits, etc) of the fish, as well as on the limnology of the lake in general.

REFERENCES

Lemma AH (2012). Breeding seasons and condition factor of *Oreochromis niloticus* (Pisces: Cichlidae) in Leke Babogaya, Ethiopia. Int. J. Agric. Sci. 2 (3): 116-120

Admassu D (1994). Maturity, Fecundity, Brood size and sex ratio of Tilapia (*Oreochromis niloticus* L.) in Lake Awassa. SINET: Ethiop. J. Sci. 17(1): 53-96.

Admassu D (1996). The breeding season of tilapia, *Oreochromis niloticus* L. in Lake Awassa (Ethiopian rift valley). Hydrobiologia 337 :77-83.

Bagenal TB, Tesch FW (1978). Age and growth. In: Methods for assessment of fish production in Fresh waters, Bagenal, T.B. (ed.). Hand book No.3, Blackwell Scientific Publications, Oxford, England. pp.101-136.

Dadebo E (1988). Studies on the biology and commercial catch of *Clarias mossambicus* Peters (Pisces: Cariidae) in Lake Awassa, Ethiopia. M.Sc. Thesis, School of Graduate Studies, Addis Ababa University, Addis Ababa. pp.73

Fryer G, Iles TD (1972). The cichlid Fishes of the Great Lakes of Africa: Their Biology and Evolution. Oliver and Boyd, Edinburgh. pp. 6-72.

Jalabert B, Zohar Y (1982). Reproductive Physiology of cichlid fishes, with particular reference to Tilapia and Sarotherodon. In: Pullii and Low-McConnell, R.H. (eds). The Biology and culture of Tilapias. R.S.V. ICLARM Conference proceeding, Philippines, Manila. pp 129-140.

Lowe-McConnell RH (1958). Observations on the Biology of *Tilapia nilotica* Linne (Pisces: Cichlidae) in East African waters. Revue Zool. Bot. Afr. 57:129-170.

Lowe-McConnell RH (1959). Breeding behavior patterns and ecological differences between *Tilapia* species and their significance for evolution within *Tilapia* (Pisces: Cichlidae). Proc. Zool. Soc. Lond. 32: 1-30.

Lowe-McConnell RH (1982). Tilapias in fish communities. In: Pullin RSV. and Lowe- McConnell RH. (ed.). The biology and culture of tilapias. Proceedings of national conference, Manila, Philippines. pp 309-31.

Lowe-McConnell RH (1987). Ecological studies in tropical fish communities. Cambridge University Press. pp. 382.

Stewart KM (1988). Changes in condition and maturation of *Oreochromis niloticus* L. population of Ferguson's Gulf, Lake Turkana, Kenya. J. Fish Biol. 33: 181-188.

Tadesse Z (1988). Studies on some aspects of the biology of *Oreochromis niloticus* L. (Pisces. Cichlidae) in Lake Ziway, Ethiopia. M.Sc.thesis. School of Graduate Studies, Addis Ababa University, Addis Ababa. pp.78

Tadesse Z. (1997). Breeding season, Fecundity, Length-weight relationship and Condition factor of *Oreochromis niloticus* L. (Pisces: Cichlidae) in Lake Tana, Ethiopia. SINET: Ethiop. J. Sci. 20 (1):31-47.

Teferi Y. (1997). The condition factor, feeding and reproductive biology of *Oreochromis niloticus* Linn. (Pisces: Cichlidae) in Lake Chamo, Ethiopia. M.Sc. thesis. School of Graduate Studies, Addis Ababa University. Addis Ababa. pp.79.

Tweddle D, Turner JL. (1977). Age, growth and natural mortality rates of some cichlid fishes of Lake Malawi. J. Fish Biol. 10:385-398.

16

Assessment of viability, chromatin structure stability, mitochondrial function and motility of stallion fresh sperm by using objective methodologies

Giannoccaro Alessandra[1], Lacalandra Giovanni Michele[2], Filannino Angela[2], Pizzi Flavia[3], Nicassio Michele[2], Dell'Aquila Maria Elena[2] and Minervini Fiorenza[1]*

[1]Institute of Sciences of Food Production (ISPA), National Research Council (CNR), Via G. Amendola 122/O, 70125, Bari, Italy.
[2]Department of Animal Production, University of Bari, Strada Provinciale Casamassima km 3, 70010, Valenzano, Bari, Italy.
[3]Institute of Agricultural Biology and Biotechnology (IBBA), National Research Council (CNR), Via Bassini 15, Milan, Italy.

Methodologies, such as flow cytometry and computer assisted sperm analysis (CASA), provide objective, reproducible, rapid and multi-parametric evaluation of semen quality. In this study, semen samples collected from six stallions were analysed for viability (by propidium iodide), chromatin stability by sperm chromatin structure assay (SCSA) and mitochondrial membrane potential by JC-1 using flow cytometry. Total and progressive motility, average path velocity (VAP), curvilinear velocity (VCL) and straight-line velocity (VSL) were determined by CASA system. The cytofluorimetric analysis provided results with low intra-assay variability respect to motility analysis by CASA system. The data on viability and mitochondrial assessment were rather uniform between stallions. The SCSA was able to distinguish potential fertility levels between stallions. In fact statistical differences were found between stallions especially for %-DFI and SD-DFI parameter. The %-DFI parameter was negatively correlated with VCL parameter. The higher repeatability of %-DFI parameter respect to those of other SCSA parameters confirms the importance of this parameter notoriously related to fertility. In conclusion, the simultaneous assessment of different functional sperm parameters, by flow cytometry and CASA, may be allow to obtain detailed and repeatable evaluations of sperm quality in the stallion, usually not considered in breeding selection programs.

Key words: Flow cytometry, stallion, sperm quality.

INTRODUCTION

Pregnancy and foaling rates are considered true indexes of fertility in the horse. However, both of them are retrospective and dramatically influenced by factors extrinsic to the stallion, such as mare's reproductive capacity and breeding management. In many circum-stances, a prospective test is desired so that, likely, subfertility can be identified before a stallion embarks on his breeding career (Colenbrander et al., 2003). Constraints in horse breeding – small number of fertilized mares per ejaculate/ stallion and tremendous variations in mare management and insemination – did not allow carrying out trials similar to bovine species concerning artificial insemination techniques. For these reasons, the fertility trials of the horse are of little value and laboratory tests could help in accurately predicting the fertility of commercially produced stallion semen (Colenbrander et al., 2003). Thus, considerable effort has been invested in identifying

---

*Corresponding author. E-mail: fiorenza.minervini@ispa.cnr.it.

Abbreviations: PI; Propidium Iodide, **AO;** Acridine Orange, Δψm; Inner Mitochondrial Membrane Potential, **CASA;** Computer Assisted Sperm Analysis, **VAP;** Average Path Velocity, **VSL;** Straight-line Velocity, **VCL;** Curvilinear Velocity.

markers for functional sperm capacity that can more accurately predict the fertility of a stallion, such as ubiquitin and seminal plasma proteins ubiquitination (Sutovsky et al., 2003; Barrier-Battut et al., 2005). Given the limitations of the standard Breeding Soundness Examination in predicting fertility or even in identifying all seriously subfertile stallions, many different approaches have been investigated in the hope of finding a relatively straight-forward and inexpensive test closely correlated with fertility (Neild et al., 2005). The limitation of this approach is that most of tests only assess a limited number of attributes that a sperm must possess to fertilize oocyte. For this reason, the combined use of different tests could offer more promise for reliable assessment of functional attributes of sperm quality (Colenbrander et al., 2003; Rodriguez-Martinez, 2003; Gillan et al., 2005; Mocé and Graham, 2008).

The primary goal of all semen analyses is to determine the fertilizing potential of a semen sample, using rapid, automated and objective procedures although there is no laboratory assay that can be considered reliably correlated with fertility (Mocé and Graham, 2008). Most of laboratory assays are not objective and some of them, such as hypo-osmotic swelling test, morphology and motility assessed by microscopic observation, require also long procedures (Graham, 2001; Kuisma et al., 2006). Flow cytometry (FC) allows to simultaneously measure multiple sperm attributes which can be combined in a predictive equation (Graham, 2001; Kuisma et al., 2006).

Indeed, the flow cytometric analysis of cells is objective since it measures the amount of fluorescent stain in an unbiased manner and evaluates tens of thousands of cells in about a minute (Graham, 2001). The common cytofluorimetric parameters considered for sperm quality assessment are viability, chromatin structure stability and mitochondrial function (Graham, 2001; Gillan et al., 2005).

Membrane integrity is a fundamental requisite for sperm viability and for the success of fertilization. Viable spermatozoa are defined as cells that possess an intact plasma membrane. Different viability assays assess the integrity of different plasma membrane compartments (Mocé and Graham, 2008) that can be evaluated by microscopic or cytofluorimetric approach after cell staining (Garner et al., 1986; Magistrini et al., 1997; Merkies et al., 2000; Love et al., 2003). Many techniques, based on the use of viable (fluorescein diacetate and SYBR-14) and non-viable (propidium iodide – PI - and eosin-nigrosin) dyes allow to detect sperm viability (Martinez-Pastor et al., 2004). Sperm viability assessments by using nucleic acid stains (such as SYBR-14 and PI) are considered to be less variable than enzyme-based stains (such as carboxyfluorescein diacetate – CFDA- and fluorescein diacetate) and sperm DNA is believed to be a more appropriate cellular target due to its sustainability and staining uniformity (Gillan et al., 2005).

The sperm chromatin structure assay (SCSA), a procedure originally designed by Evenson et al. (1994), is based on the metachromatic properties of the fluorescent probe acridine orange (AO) in relation to the DNA structure. Acridine orange emits green or red fluorescence when intercalates to double-stranded DNA (native) or single-stranded DNA (damaged DNA structure), respectively. After acid treatment, the increased susceptibility of sperm cells to in situ DNA denaturation can be evaluated by FC. Sperm chromatin stability is related to the content of P1 and P2 protamines (Corzett et al., 2002). In the horse, the occurrence of both protamines (Andrabi, 2007) induces lower chromatin stability compared with species containing only P1 (Evenson et al., 2002). Additional factors, such as seminal plasma volume, storage time and extender type, could influence the intrinsic susceptibility of equine chromatin structure (Love, 2005). The integrity and structural stability of equine sperm DNA is associated with fertility (Love and Kenney, 1998).

Mitochondrial status plays an important role in determining sperm cell competence because of its relationship with the energetic status of the cell and motility and has been related to fertility (Martinez-Pastor et al., 2004). Rhodamine 123 and MitoTracker fluorochromes have been used to evaluate mitochondrial function of spermatozoa, but no distinction can be made between spermatozoa exhibiting different respiratory rates (Gillan et al., 2005). On the contrary, the potentiometric dyes, such as JC-1 probe, permit a distinction between spermatozoa with poorly and highly functional mitochondria by the determination of the inner mitochondrial membrane potential ($\Delta\psi m$) since this molecule can exist in two different states (aggregates or monomers), each with a different emission spectra. If the JC-1 molecule remains in the monomeric form, it will fluoresce in green (that is, $\Delta\psi m^{low}$) after passage through mitochondrial membrane. If it converts into the JC-1 aggregate form, it will fluoresce in orange (that is, $\Delta\psi m^{high}$) (Reers et al., 1991; Mocé and Graham, 2008). The reliability and the sensitivity of JC-1 staining on stallion spermatozoa were assessed by FC by Gravance et al. (2000) and Mari et al. (2008).

Sperm motility is a very important parameter for sperm quality evaluation and it can be assessed by using Computer Assisted Sperm Analysis (CASA). The CASA system is based on the capture of successive microscopic images which are then digitized. Motile spermatozoa are identified in successive images, thus allowing to follow their trajectories. Finally obtained trajectories are processed, allowing the establishment of several kinetic characteristics. Sperm motility assessment by CASA is more objective and rapid than the microscopic evaluation although the inter-assay and intra-assay variability were similar for these two assay methods (Mocé and Graham, 2008). The CASA system was used for the evaluation of equine spermatozoal motility by several authors (Jasko et al., 1991; Varner et al., 1991; Ball et al., 2001).

The aim of this study was to assess fresh sperm quality in the stallion by means of several, objective and rapid laboratory assays and to correlate in order to select the parameter useful for a future prediction of fertilizing potential of semen sample. In particular viability, SCSA and mitochondrial potential were assessed by FC and motility by CASA system.

MATERIALS AND METHODS

Materials

MitoProbe JC-1 assay kit (Molecular Probes). Propidium iodide (Sigma). Acridine orange (Sigma). Semen extender INRA 96 (IMV Technologies).

Semen collection

In this study, during a breeding season (from 14 February to 14 July), six mature stallions, routinely used in artificial insemination programs and located in Equine Reproduction Centre ("Pegasus" Bari, Italy) were used. During the experiments, horses were kept in boxes on straw and received two times a day two Kg of oat and five Kg of hay. Water was *ad libitum*. All experiments were carried out in June and July 2007. The management of stallions and the collection of semen samples were performed in accordance with health and welfare regulations in force.

Semen was collected by using an artificial vagina (Missouri model) with an in-line gel filter. Gel-free semen volume was measured with a graduate cylinder, spermatozoa concentration was assessed photometrically and motility parameters were estimated by using CASA system. Subsequently, semen was diluted with INRA 96 at 37°C, to a final concentration of about $20 - 25 \times 10^6$ spermatozoa/ ml (Gravance et al., 2000). The extended semen was chilled for 1 h in commercial portable containers, to maintain viability during transport to the flow cytometer laboratory. From each stallion two semen samples were collected and analysed in two independent days, once a month.

Apparatus

Flow cytometric analyses were carried out by using a FACS Calibur (BD, Milan, Italy) equipped with 15 mW air-cooled Argon laser. After acquisition, data were evaluated by using Cell Quest v. 3.3 (BD, Milan, Italy) and Win List v. 5.0 (Verity Software House Inc., Topsham, ME, USA) software.

Sperm viability

Sperm viability was assessed by using PI that can enter the cell and stain the nucleus when the plasma membrane is damaged. The protocol of Papaioannou et al. (1997) with some modifications was used. Aliquots of extended semen, at the final concentration of 2×10^6 spermatozoa/ml, were transferred into 5 ml culture tubes (352052 BD Falcon, Becton Dickinson, Milan, Italy) and supplemented with 1 ml of PI solution (2 μg/ml). The tubes were gently mixed and further incubated for 30 min in the dark at room temperature. Each sample was analyzed by FC after acquisition of 20,000 total events at a flow rate of ~200 cells/s. An analysis gate was applied in the FCS/SSC dot-plot to restrict the evaluation to sperm cells and to eliminate small debris and other particles from analyses. Propidium iodide, excited at 488 nm, was read with 650/13 nm

bandpass emission filter (FL-3) by using a logarithmic histogram. Viable spermatozoa with intact plasmalemma (PI-negative) were observed in the second decade while membrane-damaged cells (PI-positive) were visible in the fourth decade. Cell viability was expressed as percentage of PI-negative cells. For each stallion, sperm viability was analyzed on two different ejaculates.

Sperm chromatin structure assay (SCSA)

The SCSA protocol used in this study was described by Benzoni et al. (2008). Fresh semen was diluted into 1 ml of TNE buffer (0.01 M Tris-HCl, 0.15 M NaCl, 1 mM EDTA and pH 7.4) at a final concentration of 2×10^6 spermatozoa/ml. In crushed ice, aliquots of semen (0.2 ml) were mixed with 0.4 ml of a low pH detergent solution (0.15 M NaCl, 0.1% Triton X-100, 0.08 N HCl and pH 1.4) for a partial *in situ* DNA denaturation. After 30 s, cells were stained by adding 1.2 ml of AO work solution (6 μg/ml in 0.1 M citric acid, 0.2 M Na_2HPO_4, 1 mM EDTA, 0.15 M NaCl, pH 6.0). After 3 min, cytometric analysis was carried out at a flow rate of ~200 cells/s and total acquisition of 20,000 events for each sample. Acridine orange was excited at 488 nm and its red fluorescence was read at 650/13 nm bandpass emission filter (FL3), while its green one was read at 530/30 nm bandpass emission filter (FL1). In order to standardize all analyses, FC was calibrated with semen collected from a stallion control with regular reproductive function and known SCSA parameters. The FC was adjusted such that the mean of FL1 and FL3 fluorescences were set on histograms with 1024 channel resolution, at 439 and 110 channels, respectively. Flow cytometer calibration was checked at each start-up of the FC and after every 5 samples. The instability of sperm structure chromatin was assessed by using three different parameters of DNA Fragmentation Index (DFI), as described by Benzoni et al. (2008): the mean ($\overline{X}$) DFI, calculated selecting those cells to the right of the main population on a dot-plot of red (FL3) versus green (FL1) fluorescence; the percentage (%) and the standard deviation (SD) DFI, calculated by using αt parameter (which expresses the ratio between FL3/FL3+FL1) obtained by WinList 5.0 software. For each stallion, four determinations were carried out on each ejaculate.

Mitochondrial function assessment

A 200 μM JC-1 stock solution in DMSO was prepared prior to use. For each sample, 1×10^6 sperm cells were suspended in 1 ml of warm PBS and JC-1 solution (2 μM final concentration) was added. Cells were incubated for 30 min at room temperature. For each sample, a total of 20,000 events were analyzed at a flow rate of ~200 cells/s. As suggested by the protocol, in order to confirm the JC-1 sensitivity to changes in membrane potential, carbonyl cyanide 3-chlorophenylhydrazone (CCCP = 50 μM final concentration) was used as membrane potential disruptor. Emission filters of 535 and 595 nm were used to quantify the population of spermatozoa with green ($\Delta\psi m^{low}$ "inactive mitochondria") and orange ($\Delta\psi m^{high}$ "active mitochondria") fluorescence, respectively. Dot-plot for FL-1 (green) and FL-2 (orange) fluorescence was used. In order to determine the population of cells with $\Delta\psi m^{high}$, the percentage of cells which concentrate JC-I into aggregates was considered. Because of technical problems, a single ejaculate for each stallion was assessed.

Motility determination

The CASA analysis was performed by using HTM-IVOS Sperm Analyzer v. 12.3 (Hamilton Thorne, Beverly, MA, USA). Briefly, the analysis was performed in a 10 μm Leja chamber at 37°C. Setting for cell detection (capture of 45 frames; 60 Hz) were: minimum

Table 1. Viability, sperm chromatin structure assay (SCSA) parameters, mitochondrial function and motility parameters of fresh spermatozoa collected from six stallions.

Parameters	Stallion 1	Stallion 2	Stallion 3	Stallion 4	Stallion 5	Stallion 6
Sperm viability (%)	93.5± 1.85	96.5 ±1.85	99 ± 1.85	97 ± 1.85	99.75 ± 2.77	98.75 ± 2.77
SCSA						
$\overline{X}$ -DFI	128 ± 9.63	121 ±9.63	134 ± 9.63	135.5 ± 9.63	141 ± 14.45	143 ± 14.45
%-DFI	16.5 ± 0.86a	17 ± 0.86b	29.5 ± 0.86c	32 ± 0.86c	4.25 ± 1.29d	8.25 ± 1.29e
SD-DFI	64 ± 2.59a	49.5 ± 2.59b	76.5 ± 2.59c	54 ± 2.59ab	83.25 ± 3.89ce	79.25 ± 3.89cf
Mitochondrial potential Δψm high (%)	63	52	58	54	70	74
Motility						
Total motility (%)	87.5 ± 8.05a	86.5 ± 8.05a	63 ± 8.05a	65.5 ± 8.05a	88.12 ±12.08a	95.12 ± 12.08a
Progressive motility (%)	30 ± 8.12a	26.5 ± 8.12a	22.5 ± 8.12a	28.5 ± 8.12a	39.37 ±12.19a	18.37 ± 12.19a
VAP (μm/s)	125.5 ± 8.59a	118 ± 8.59a	114.5 ± 8.59a	94.5 ± 8.59a	104.87 ± 12.89a	128.87 ±12.89a
VSL (μm/s)	77 ± 9-6a	72.5 ± 9-6a	69 ± 9.6a	66.5 ± 9.6a	69.5 ± 14.4a	61.5 ± 14.4a
VCL (μm/s)	229 ± 13.67a	227 ± 13.67a	187.5 ±13.67a	189 ±13.67ab	209.87 ± 20.51a	263.87±20.51ac

LS means ± SD. a,b,c,d,e,f, LSmeans in the same row without common subscript differ (P < 0.05).

contrast 70, minimum cell size 4 pixel, path velocity (VAP) cut-off =50 μm/s and straightness (STR) threshold = 75%. The determination was carried out on 3000 spermatozoa (about 400 cells for each field with a total of 7 fields). Sperm kinetic assessment was based on the determination of the percentage of total motile cells, progressively motile cells, average path velocity (VAP-μm/s), straight-line velocity (VSL–μm/s) and curvilinear velocity (VCL–μm/s) of motile cells. For each stallion, two ejaculates were collected with an interval of one month and analyzed.

Statistical analysis

Data were analysed by using the Statistical Analysis System package (SAS, 1998). General linear model (GLM) procedure was undertaken to assess the effect of age, breed and collection date on the semen quality parameters: motility, viability, SCSA parameters and mitochondrial function assessment. Student's t test was used for comparison between least square means (LSmeans). Partial Correlation Coefficient among the variables were calculated on the residuals of the GLM analysis.

Repeatability of SCSA parameters was computed from the components of variance obtained by MIXED procedure on 6 stallions from which two or more measurements were performed. In the model collection date and breed are considered as fixed effect, the animal as random effect.

RESULTS

Parameters of sperm quality for each examined stallion are shown in Table 1. In Table 2, the correlations between parameters are reported, in Table 3 repeatabilities of SCSA parameters.

Viability assessment

The total mean viability was of 96.9 ± 2.64 (mean ± SD) and ranged from 92 to 99%. No statistical correlation between viability and other sperm parameters was found.

SCSA

The mean values of SCSA parameters are shown in Table 1. The total $\overline{X}$ –DFI parameter ranged between 115 to 146 with a mean of 131.8 ± 10.63. The %-DFI parameter ranged between 7 to 35 with a mean of 20.84 ± 9.54. The total SD-DFI parameter ranged between 44 to 86 with a mean of 65.6 ± 14.2. Statistical differences between stallions were observed for %-DFI and SD-DFI parameters. Concerning %-DFI parameter, stallion 5 and 6 showed values significant lower than those of the other stallions. Concerning SD-DFI parameter, stallion 2 had the lowest values. Significant negative correlations was seen between %-DFI and VCL (r = -0.98; p = 0.01).

Repeatability values of %-DFI was high (0.91) showing high time-consistency of this trait. Repeatabilities for the other SCSA parameters were moderate to low: 0.47 and 0.11 for $\overline{X}$ –DFI and SD-DFI respectively.

Mitochondrial function assessment

As observed in Figure 1a, the FL1-FL2 dot-plot, obtained after gating spermatozoa in FSC/SSC dot-plot, shows two cell populations: one with high green fluorescence and low orange fluorescence, (which is considered as the population with Δψm^{low} "inactive mitochondria"), and another one with high green fluorescence and high orange fluorescence, (which is considered as the population with Δψm^{high} "active mitochondria").

After the addition of a Δψm disrupter CCCP, a decrease on Δψm was recorded as only one population with high green fluorescence and low orange florescence

Table 2. Partial correlation coefficients from the residuals of General Linear Model (r-value) between viability, sperm chromatin structure assay (SCSA) parameters, mitochondrial membrane function and motility parameters in fresh equine spermatozoa.

Parameters		Mitochondial potential $\Delta\psi m^{high}$ (%)	SCSA $\overline{X}$-DFI	DFI %	SD-DFI	Motility (CASA) Tot (%)	Prog (%)	VAP (µm/s)	VSL (µm/s)	VCL (µm/s)
Sperm viability (%)		0.4 p = 0.7	0.10 p = 0.8	0.4 p = 0.3	-0.24 p = 0.8	0.45 p = 0.5	0.66 p = 0.3	0.38 p = 0.6	0.55 p = 0.4	-0.04 p=0.9
Mitochondrial potential $\Delta\psi m^{high}$ (%)		/	0.8 p = 0.3	0.9 p = 0.08	0.4 p = 0.75	1 p = 0.08	0.8 p = 0.3	-0.3 p = 0.7	-0.8 p = 0.3	-0.9 p=0.07
SCSA	$\overline{X}$-DFI	/	/	0.76 p = 0.2	0.93 p = 0.06	0.93 p = 0.06	0.63 p = 0.3	-0.13 p = 0.8	0.25 p = 0.7	-0.67 p=0.3
	%-DFI	/	/	/	0.81 p = 0.18	0.65 p = 0.3	0.13 p = 0.8	-0.71 p = 0.2	-0.36 p = 0.6	-0.98 p = 0.01
	SD-DFI	/	/	/	/	0.74 p = 0.25	0.33 p = 0.6	-0.36 p = 0.6	-0.03 p = 0.9	-0.71 p = 0.3
Motility (CASA)	Tot (%)	/	/	/	/	/	0.83 p = 0.1	0.07 p = 0.9	0.47 p = 0.5	-0.57 p = 0.4
	Prog (%)	/	/	/	/	/	/	0.59 p = 0.4	0.87 p = 0.1	-0.05 p = 0.9
	VAP (µm/s)	/	/	/	/	/	/	/	0.91 p = 0.08	0.77 p = 0.2
	VSL (µm/s)	/	/	/	/	/	/	/	/	0.44 p = 0.5

Table 3. Repeatabilities of sperm chromatin structure assay (SCSA) parameters.

		Repeatability
SCSA	$\overline{X}$-DFI	0.11
	%-DFI	0.91
	SD-DFI	0.47

was found (Figure 1b). The mean of $\Delta\psi m^{high}$, derived from the analysis of all stallions, was of 61.83 ± 8.81 (range 52 - 74). No correlation between $\Delta\psi m^{high}$ and other sperm parameters was found.

Motility determination

Mean values for total motility was 77.9% ± 13.84 (range 54 - 94), 26.8% ± 8.97 (range 15 - 42) for progressive motility, 113.7 µm/s ± 14.1 (range 83 - 128) for VAP, 70 µm/s ± 9.27 (range 52 - 81) for VSL and 213 µm/s ± 28 (range 168 - 264) for VCL (Table 1). As already discussed, a high (r = -0.98) negative correlation between VCL and %-DFI was found (Table 2).

DISCUSSION

Viability is the most common parameter evaluated for sperm quality by FC. The determination of sperm viability by using FC after PI staining produced similar results between stallions. Our sperm viability results, obtained by a single dye (PI), were higher than results reported in literature by dual DNA staining (SYBR-14/PI) (Merkies et al., 2000; Aziz et al., 2005). The PI staining can not omit from flow cytometric analyses (that is, gated out) particles without nuclei into viable cell (PI negative) percentage (Mocé and Graham, 2008), causing an overestimation of sperm viability. However, there was low correlation between viability and other spermatic functions, also considering low variation of sperm viability found among the six examined stallions. The lack of correlation between viability and mitochondrial function is not in agreement with the observations by Papaioannou et al. (1997) and could be probably related to the different probe (Rhodamine 123) used for mitochondria activity assessment.

Concerning the assessment of sperm chromatin quality, although significant differences on SCSA para-meters were found between stallions, the discrimination between fertile and subfertile stallions should be made by the following SCSA parameters described by Love et al. (2002): $\overline{X}$-DFI (199 ± 10 vs 228 ± 13), %-DFI (14 ± 4.4 vs 35 ± 8) and SD-DFI (45 ± 6.2 vs 60 ± 13). The %-DFI parameter is strictly associated with stallion fertility and it is able to discern fertility levels in the horse, as described by Love and Kenney (1998). In fact, ejaculates of stallions 1, 2, 5 and 6 could be considered excellent or

a

b

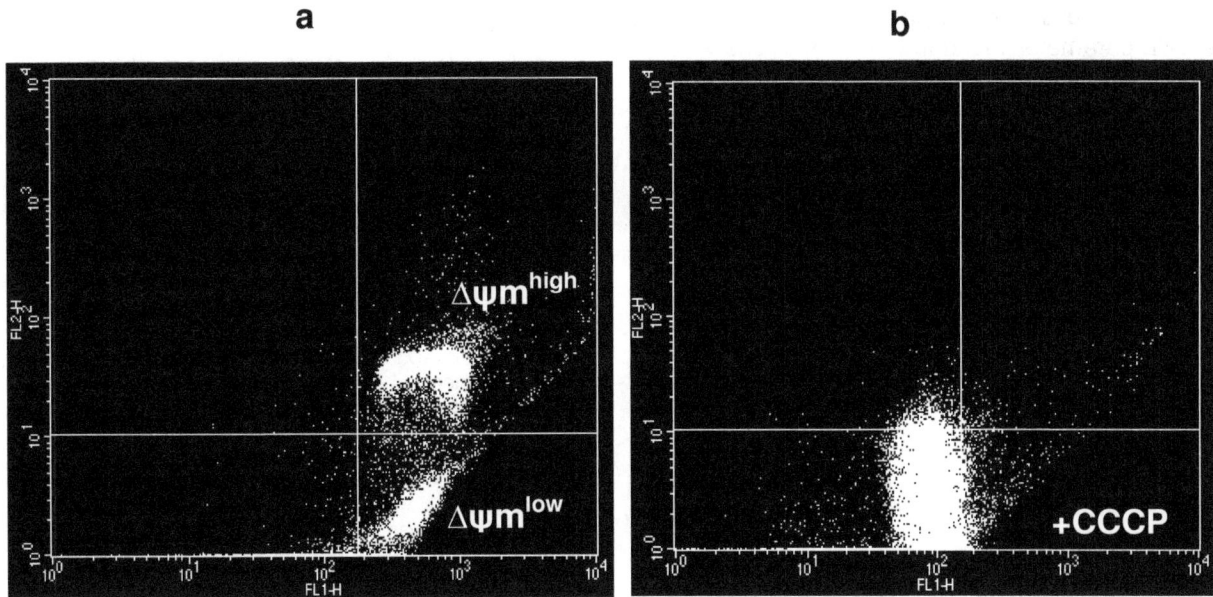

Figure 1. Flow cytometric dot plots of inner mitochondrial membrane potential ($\Delta\psi m$) by using JC-1 probe in equine spermatozoa. a) The sperm population with $\Delta\psi m^{high}$ and $\Delta\psi m^{low}$ is indicated in the upper and lower right quadrant, respectively. b) A single population with $\Delta\psi m^{low}$ after addition of $\Delta\psi m$ disrupter CCCP.

good. While $\bar{x}$–DFI is the least sensitive indicator of fertility because a higher degree of denaturability must occur in the whole spermatozoal population before this variable changes (Love and Kenney, 1998). In our study all ejaculates tested showed values of $\bar{x}$–DFI parameter included in fertile range defined by Love et al. (2002). As reported by Love and Kenney (1998), the SD-DFI parameter, which reflects the degree of denaturability, is not highly correlated with fertility parameters in the stallions. Our SD-DFI values, which were higher in all stallions, could not be considered as indicator of subfertility.

In our study, the overall fertility rate (defined as percentage of pregnant mares at 14 - 21 days after AI) for the whole season of stallions used was comprised between 78 and 89%. The integrity of sperm chromatin is important not only for fertilization but also normal embryonic development (D'Occhio et al., 2007). It has been proposed that alterations in chromatin structure may affect the rate of decondensation, a prerequisite for male pronucleus formation during fertilization and thereby, disrupt embryo development. Indeed, in humans, a reduced sperm chromatin stability has been related to recurrent abortion and it is possible that a similar mechanism explains why, while fertilization rates in horses appear to be very high, the early pregnancy loss rate is high (Colenbrander et al., 2003).

The negative correlation found between some SCSA and motility parameters in stallions is in agreement with related results reported by Giwercman et al. (2003) in human field. Moreover, the contemporaneous acquisition of motility and DNA reorganization around protamine

molecules in the same organ (epididymis), could explain the negative correlation, found in our study, between SCSA and some motility parameters, as supposed in human field by Giwercman et al. (2003).

Mitochondrial membrane potential is a sensitive indicator of the functional status of mitochondria and it plays an important role because of its relationship with the energetic status and motility of spermatozoa (Martinez-Pastor et al., 2004). The lipophilic fluorescent probe, JC-1, has been validated in the assessment of bull and stallion spermatozoa using FC (Garner et al., 1997; Gravance et al., 2000) and provides a more rigorous estimate of metabolic function than rhodamine 123 or MitoTracker (Gravance et al., 2000). The percentages of spermatozoa with active mitochondria ($\Delta\psi m^{high}$), found in our study, were homogeneous among the six examined stallions and in accordance with results reported by Mari et al. (2008). The lack of relation between mitochondrial function and kinetic parameters, found in our study, was in agreement with data reported also in other species and could be explained by the involvement of many factors in sperm motility, not only mitochondrial status (Martinez-Pastor et al., 2004; Mari et al., 2008; Volpe et al., 2008).

The CASA system is one of the simplest and most reliable methods for studying sperm motility. The results of this processing are reflected in a series of parameters which precisely define the exact movement of each individual spermatozoon (Quintero Moreno et al., 2003). Conflicting opinions concerning relationship between motility and fertility in stallion were reported by Jasko et al. (1991) and Kuisma et al. (2006) although the spermatozoa motility and the quality of motility are still the most reliable estimates in practice. The total motility and the VSL parameter found in our six stallions were in agreement with data reported by Jasko et al. (1991), Varner et al. (1991), Aurich et al. (2007) and Mari et al. (2008). Concerning VCL parameter, our samples are in agreement with Love et al. (2005). The VAP values found in our study were similar to data reported by Mari et al. (2008) although this velocity parameter is not considered a reliable and repeatable parameter to analyze sperm quality, because of its high inter-assay variability (Kirk et al., 2005). The progressive motility values found in our study were similar to results found by Mari et al. (2008).

Other authors report, concerning this parameter, discordant results (Jasko et al., 1991; Varner et al., 1991; Albrizio et al., 2005). These discordant data could be related to different factors, such as stallion, ejaculation frequency, season, extender used for semen dilution, CASA parameters setting (threshold setting, specimen concentration, video digitalisation rate), as reported by several authors (Jasko et al., 1991; Davis and Katz, 1992; Brinsko et al., 2000; Ball et al., 2001; Sieme et al., 2004).

In conclusion, our study reported simultaneous evaluation of different functional sperm parameters by using objective methodologies, useful tools for complete sperm quality assessment in order to improve stallion's fertility. In general, fertility in stallions used in breeding programs is lower and more variable than in other farm animal species, primarily because current selection procedures are based on pedigree, looks and/or athletic performances, with little consideration of fertility or fertility potential (Neild et al., 2005). The modest consideration of stallion fertility could be greatly improved by the introduction of objective methodologies such as FC and CASA system.

As reported by Kirk et al. (2005), the flow cytometric data exhibited less intra-assay variability than did motility assays and the high repeatability of %-DFI parameter found in this study by using FC is in agreement with this statement. The sperm motility and chromatin integrity are significantly associated. Spermatozoa with excellent chromatin structure but without motility could be infertile or motile and normal spermatozoa with damaged DNA and chromatin could have an impaired functions, as reported by Giwercman et al. (2003) on human field. It is suggested that both parameters, obtained by objective methods (such as FC and CASA) can complement each other.

Further studies need to be carried out in order to establish the degree of reliability and variability of sperm functional parameters in relation to some factors, such as breed, seasonal influence, geographic location etc.

REFERENCES

Albrizio M, Guaricci AC, Maritato F, Sciorci RL, Mari G, Calamita G, Lacalandra GM, Aiudi GG, Minoia R, Dell'Aquila ME, Minoia P(2005). Expression and subcellular localization of the μ-opioid receptor in equine spermatozoa: evidence for its functional role. Reproduction 129:39-49.

Andrabi SMH (2007). Mammalian sperm chromatin structure and assessment of DNA fragmentation. J. Assist. Reprod. Genet. 24: 561-569.

Aurich C, Seeber P, Müller-Schlösser F (2007). Comparison on different extenders with defined protein composition for storage of stallion spermatozoa at 5 °C. Reprod. Dom. Anim. 42:445-448.

Aziz DM, Ahlswede L, Enberges H (2005). Application of MTT reduction assay to evaluate equine sperm viability. . 64: 1350-1356.

Ball BA, Medina V, Gravance CG, Baumber J (2001). Effect of antioxidants on preservation of motility, viability and acrosomal integrity of equine spermatozoa during storage at 5 °C. Theriogenoly. 56: 577-589.

Barrier-Battut I, Dacheux JL, Gatti JL, Rouviere P, Stanciu C, Dacheux F, Vidament M (2005). Seminal plasma proteins and semen characteristics in relation with fertility in the stallion. Anim. Reprod. Sci. 89: 255-258.

Benzoni E, Minervini F, Giannoccaro A, Fornelli F, Vigo D, Visconti A (2008). Influence of in vitro exposure to mycotoxin zearalenone and its derivatives on swine sperm quality. Reprod. Toxicol. 25: 461-467.

Brinsko SP, Rowan KR, Varner DD, Blanchard TL (2000). Effects of transport container and ambient storage temperature on motion characteristics of equine spermatozoa. Theriogenoly. 53:1641-1655.

Colenbrander B, Gadella BM, Stout TAE (2003). The predictive value of semen analysis in the evaluation of stallion fertility. Reprod. Domest. Anim. 38: 305-311.

Corzett M, Mazrimas J, Balhorn R (2002). Protamine 1: protamine 2 stoichiometry in the sperm of eutherian mammals. Mol. Reprod. Dev. 61:519-527.

Davis RO, Katz DF (1992). Standardization and comparability of CASA instruments. J. Androl. 13: 81-86.

D'Occhio MJ, Hengstberger KJ, Johnston SD (2007). Biology of sperm chromatin structure and relationship to male fertility and embryonic survival. Anim. Reprod. Sci. 101: 1-17.

Evenson DP, Larson KL, Jost LK (2002). Sperm chromatin structure assay: its clinical use for detecting sperm DNA fragmentation in male infertility and comparisons with other techniques. J. Androl. 23:25-43.

Evenson DP, Thompson L, Jost L (1994). Flow cytometric evaluation of boar semen by the sperm chromatin structure assay as related to cryopreservation and fertility. Theriogenoly. 41: 637-651.

Garner DL, Pinkel D, Johnson LA, Pace MM (1986). Assessment of spermatozoal function using dual fluorescent staining and flow cytometric analyses. Biol. Reprod. 34:127-138.

Garner DL, Thomas CA, Joerg HW, DeJarnette JM, Marshall CE (1997). Fluorometric assessment of mitochondrial function and viability in cryopreserved bovine spermatozoa. Biol. Reprod. 57: 1401-1406.

Gillan L, Evans G, Maxwell WMC (2005). Flow cytometric evaluation of sperm parameters in relation to fertility potential period, Theriogenology 63: 445-457.

Giwercman A, Richthoff J, Hjøllund H, Bonde JP, Jepson K, Frohm B, Spanò M (2003). Correlation between sperm motility and sperm chromatin structure assay parameters. Fertil. Steril. 80: 1404-1412.

Graham JK (2001). Assessment of sperm quality: a flow cytometric approach. Anim. Reprod. Sci. 68:239-247.

Gravance CG, Garner DL, Baumber J, Ball BA (2000). Assessment of equine sperm mitochondrial function using JC-1. Theriogenoyly. 53: 1691-1703.

Januskauskas A, Johannisson A, Rodriguez-Martinez H (2003). Subtle membrane changes in cryopreserved bull semen in relation with sperm viability, chromatin structure, and field fertility. Theriogenoly 60: 743-758.

Jasko DJ, Lein DH, Foote RH (1991). The repeatability and effect of season on seminal characteristics and Computer-Aided Sperm Analysis in the stallion. Theriogenoly 35: 317-327.

Kirk ES, Squires EL, Graham JK (2005). Comparison of in vitro laboratory analyses with the fertility of cryopreserved stallion spermatozoa. Theriogenoly 64: 1422-1439.

Kuisma P, Andersson M, Koskinen E, Katila T (2006). Fertility of frozen-thawed stallion semen cannot be predicted by the currently used laboratory methods. Acta. Vet. Scand. 48: 14-22.

Love CC, Brinsko SP, Rigby SL, Thompson JA, Blanchard TL, Varner DD (2005). Relationship of seminal plasma level and extender type to sperm motility and DNA integrity. Theriogenoly 63: 1584-91.

Love CC, Kenney RM (1998). The relationship of increased susceptibility of sperm DNA to denaturation and fertility in the stallion. Theriogenoly 50: 955-972.

Love CC (2005). The sperm chromatin structure assays: a review of clinical applications. Anim. Reprod. Sci. 89: 39-45.

Love CC, Thompson JA, Brisko SP, Rigby SL, Blanchard TL, Lowry VK, Varner DD (2003). Relationship between stallion sperm motility and viability as detected by two fluorescence staining techniques using flow cytometry. Theriogenoly 60: 1127-1138.

Love CC, Thompson JA, Lowry VK, Varner DD (2002). Effect of storage time and temperature on stallion sperm DNA and fertility. Theriogenoly 57: 1135-1142.

Magistrini M, Guitton E, Levern Y, Nicolle JCI, Vidament M, Kerboeuf D, Palmer E (1997). New staining methods for sperm evaluation estimated by microscopy and flow cytometry. Theriogenoly 48: 1229-1235.

Mari G, Rizzato G, Merlo B, Iacono E, Seren E, Tamanini C, Galeati G, Spinaci M (2008). Quality and fertilizing ability in vivo of sex-sorted stallion spermatozoa. Reprod. Dom. Anim. DOI: 10.1111/j.1439-0531.2008.01314.x.

Martinez-Pastor F, Johannisson A, Gil J, Kaabi M, Anel L, Paz P, Rodriguez-Martinez H (2004). Use of chromatin stability assay, mitochondrial stain JC-1, and fluorometric assessment of plasma membrane to evaluate a frozen-thawed ram semen. Anim. Reprod. Sci. 84:121-133.

Merkies K, Chenier T, Plante C, Buhr MM (2000). Assessment of stallion spermatozoa viability by flow cytometry and light microscopic analysis. Theriogenoly. 54:1215-1224.

Mocé E, Graham JK (2008). In vitro evaluation of sperm quality. Anim. Reprod. Sci. 105:104-118.

Neild DN, Gadella BM, Aguero A, Stout TAE, Colenbrander B (2005). Capacitation, acrosome function and chromatin structure in stallion sperm. Anim. Reprod. Sci. 89:47-56.

Papaioannou KZ, Murphy RP, Monks RS, Hynes N, Ryan MP, Boland MP, Roche JF (1997). Assessment of viability and mitochondrial function of equine spermatozoa using double staining and flow cytometry. Theriogenoly 48: 299-312.

Quintero-Moreno A, Miró J, Teresa Rigau A, Rodríguez-Gil JE (2003). Identification of sperm subpopulations with specific motility characteristics in stallion ejaculates. Theriogenoly 59:1973-1990.

Reers M, Smith TW, Chen LB (1991). J-aggregate formation of a carbocyanine as a quantitative fluorescent indicator of membrane potential. Biochem. 7: 4480-4486.

Rodriguez-Martines H (2003). Laboratory semen assessment and prediction of fertility: still utopia? Reprod. Domest. Anim. 38: 312-318.

Sieme H, Katila T, Klug E (2004). Effect of semen collection practices on sperm characteristics before and after storage and on fertility of stallions. Theriogenoly 61: 769-784.

Sutovsky P, Turner RM, Hameed S, Sutovsky M (2003). Differential ubiquitination of stallion sperm proteins: possible implications for infertility and reproductive seasonality. Biol. Reprod. 68: 688-698.

Varner DD, Vaughan SD, Johnson L (1991). Use of a computerized system for evaluation of equine spermatozoal motility. Am. J. Vet. Res. 52: 224-230.

Volpe S, Leoci R, Aiudi G, Lacalandra MG (2008). Relationship between motility and mitochondrial functional status in canine spermatozoa. Proceedings of 6th ISCFR (Eds England G, Concannon P, Schäfer-Somi G,) Vienna (Austria), pp. 287-288.

17

Climate change and the abundance of edible insects in the Lake Victoria Region

Monica A. Ayieko[1]*, Millicent F. O Ndong'a[1] and Andrew Tamale[2]

[1]Maseno University, P. O. box 333-40105, Maseno, Kenya.
[2]Busoga University, P. O. box 154 Iganga, Uganda.

Global warming is adversely affecting the earth's climate and its profound effects are virtually on all ecosystem. Every living animal will be affected in one way or another by climatic changes and insects being an integral biotic component of nearly all ecosystems are not an exemption. However, the various ways by which change will occur is yet to be determined by scientists. Insects being an integral biotic component of nearly all ecosystems will be affected by the change in a variety of ways not yet determined by scientists. This partial review and empirical observation paper discusses how edible lake flies in Lake Victoria and termites in the lake region are responding to climate change and how they are likely to impact on entomophagy and gastrophagy as part of food chain among the riparian communities. The dynamics of the insect population have been observed by several households that collect the insects for domestic purposes. The focus is given to the lake flies (Ephemeroptera and Diptera), termites (Isoptera) and formicidae ants (Hymenoptera) which form part of livestock and human feed. Several factors of climate change are identified and discussed in relation to how they influence insect abundance. Ability to respond successfully to challenges requiring a lot of collaboration across different fields of study is solicited. It requires understanding of all stakeholders, how they will be affected by climate change and strategic adaptive measures open to all. Analysis of impact on humans' livelihoods with specific focus on developing countries is discussed. The interrelationship in the metamorphosis of entomology and entomophagy in food production in the region is proposed.

Key words: Lake flies, termites, climate change, edible insects, entomophagy.

INTRODUCTION

There is increasing evidence that the earth's climate is undergoing change largely due to human activities. It is estimated that the global climate change will have profound impacts on virtually all ecosystems. Insects being an integral biotic component of nearly all ecosystems will be affected by the change in a variety of ways not yet determined by scientists. Studies point out that insect population is likely to increase with the changing climate (Saunders, 2008). Given that Lake Victoria basin is inhabited by the greatest diversity of insects, the climate change is set to influence both plants and insects alike. Insects have a unique symbiotic relationship with plant life (Fleshman, 2007). Plants and insects that may not adapt to new changes will have to give way to the survivors and these survivors will have to re-group at different suitable habitats. Lake Victoria is the second largest fresh water lake and supports the largest fresh water fishery in the world. Over the past 75 years, the lake has become eutrophic due to deforestation, increased agriculture and urbanization, (Verschen et al., 2002). The climate change is already having an impact on the food security of the large population of persons within the Lake Victoria basin.

Entomophagy studies in several parts of the world indicate that consumption of insects is gaining ground not only in rural Africa but in many parts of the western societies (Huis, 2002; Huis, 2003; Saunder, 2008; Banjo et al., 2006; Ayieko et al., 2010). With the realization of the pending food shortage particularly in developing coun-tries, the consumption of insects is sure to increase (Meyer-Rochow, 2009). As such, global climate change and the increasing food insecurity in many parts of the developing world may put insects on the menu for many families in many communities. This will particularly be

---

*Corresponding author. E-mail: monica_ayieko@yahoo.com.

so with the reduction in supply of conventional food items such as fish, meat, poultry or plant produce of which production is fast spiralling down with the climate change. Data and information on climate change show that agriculture and fisheries sectors are also being affected to a significant degree thereby influencing the already constrained food situation in Africa (Fleshman, 2006; Saunder, 2008). Although effects of climate change will be far from uniform around the world, the lake region is likely to be more vulnerable because of its biodiversity and the size of the population of people living in the area. Furthermore, the climate change is also having an impact of fisheries in the lake. The shift in the ecosystem due to climatic change is changing the resource base of fishery with far reaching consequences for local livelihoods. This change in the ecosystem of Lake Victoria is creating new opportunities in the livelihoods of the people around the lake. Many are attracted to take up fishing and related activities as observed in most fish landing beaches in Kenya. Climate change that has been observed to enhance the abundance of the insects is thus of an added advantage.

This paper therefore discusses how the edible lakeflies in Lake Victoria and the elate termites in the region are responding to climate change and how it is likely to impact on entomophagy and gastrophagy as part of food security among the riparian communities. The focus is given to the lake flies (Ephemeroptera and Diptera), and termites (Isoptera) and formicid ants (order Hymenoptera) which are normally collected within the lake region for domestic purposes such as feeding family members and poultry because of their unique food value. Other species are in use for traditional medicine and witchcraft (Ayieko and Oriaro, 2008; Banjo et al., 2006; Huis, 2002; Huis, 2003).

THE EDIBLE INSECTS IN THE LAKE VICTORIA BASIN

Chaoborid and Chironomid larvae had been reported to be the most common insect larvae in Lake Victoria in the 1950s (Macdonald, 1956). Birds, human and many aquatic organisms feed on these lake flies. Of similar significance are the mayflies which traditional medicine practitioners have indicated to be of unique uses (Ayieko and Oriaro, 2008) and both bird and fish species feed on. In the 1980s, the lake's food web changed from a complex fauna with many species to a simplified food web dominated by three fish species (tilapia, *Oreochromis niloticus*, the nile perch, *Lates niloticus* and a cyprinid, *Rastrineobolus argetea*) and one shrimp, *Ciridina niloticu* (Barel et al., 1985; Gouds et al., 2005).

This increased biomass of shrimps and lakeflies in the Lake Victoria region could be attributed to increased eutrophication and this provides additional feed for the nile perch and other species in the lake region. The elate termites are important as human and poultry feed (Ayieko, 2007; Ayieko and Oriaro, 2008). Grasshoppers are equally a delicacy, hence are used for human feeding. These insects form part of the food chain in the lake ecosystem that provides valuable food nutrients such as protein, vitamins and essential minerals in animal nutritional requirement. These are nutrients which are normally found to be lacking in poor communities within the lake region (Ayieko, 2007; Banjo et al., 2005). Insects are not only important for human nutrition, but also for other aquatic organisms in the lake which help to maintain the riparian ecosystem. However, with the changes in the benthic macrofauna community of the lake, the dynamics of the population and distribution of these insects may change.

DYNAMICS OF THE EDIBLE INSECT POPULATION IN THE VILLAGES

The dynamics of the insect population have been observed by several households that collect the insects for home consumption. A total of 11 different groups of villagers surrounding the lake on the Kenyan side have reported an increase in insects. These households have collected lake flies and elates termite for a long period of time. (Ayieko, 2007; Ayieko and Oriaro, 2008). The villagers noted that lately there have been plenty of elate termites and edible lake flies. The edible lake flies are commonly referred to as "sam" by the locals living along the lake. Field observations indicated that there is a gradual increase in the observed abundance and frequency of swarming of the lake flies and elate termites relative to 5 to 10 years back. Based on our observations and that of the residents of Lake Victoria region, it was confirmed that the edible insects are more available than before. The lake flies and the elate termites have exhibited a prolonged period of emergence than residents had expected as a result of protracted period of rainfall. A case in point is the period between March and November, 2008 and 2009 when the villagers of the lake region collected edible insects throughout the season. Similarly, 12th and 13th May 2009 villagers in Kisumu East and West and Vihiga districts (in western Kenya) experienced a heavy emergence of elate termites which was fairly distinct from the normal swarming. In a normal situation, these insects would be collected from the onset of rains in March till early May only, but 2009 season yielded a surprisingly abundance of edible insects in the region.

CLIMATE CHANGE AND EDIBLE INSECT POPULATION

Several factors of climate change have been identified to influence reproduction of insects. Dunn and Crutchfield (2006), Heegaard et al. (2006), Both et al. (2006),

Bale et al. (2002), Hunter (2001), Landsberg and Stafford, 1992) briefly outlined some of the factors that influence insect population. Weather conditions precipitated by climate change favour the increased emergence of insects. Moisture and temperature play a significant role in insect ecology. Climate change influence insect population by influencing benthic fauna and its biodiversity that sup-ports the insects.

In East Africa, particularly in Uganda, the long-horned grasshopper, Orthoptera *Raspolia nitidula* Scopoli is a delicacy when in season. This insect is highly associated with forests (Owen, 1973) which are not yet affected by the current climate change. However, its emergence and swarming is highly dependent on the onset of the rain season which is currently unpredictable due to climate change. Nevertheless, this could affect it indirectly because it feeds on the seasonal lash growth of forest trees (Wellington and Trimble, 1984; Martinat, 1987). The unpredicted seasonal changes could affect the temporal distribution of the emergence and the quantity thereof Insects have been reported to be sensitive to temperature increases and are known to be more active in higher temperatures (Schindler, 1980; Both et al., 2006; Dunn and Cutchfield, 2006; Heergaard et al., 2006). Patrick Durst, the FAO's Senior Forestry Officer, comment in an interview with Media Global on February 2008, that there is an expectation that if the climate change is warming, there will in fact be more insects. Indeed the observed increase in temperatures with the current climate change has favoured increased reproduction of the many other insects as observed by Dunn and Crutchfield (2006).

Several pointers confirm increased emergence of lake flies and elate termites among other insects of economic importance in the country side. For example, several termite mounds that have been known to yield only one swarm of elate termites annually are realizing more than one or two emergence in a year. Other termite mounds that had been dormant or giving just a few elates for several years are back in high production. Villagers are also witnessing emergence of elate termites in new places, indicating increased under the ground activities of the termites. Most likely this could be as a result of high temperatures on the physiology of the developing juveniles causing them to develop faster than the usual rate of development (Taylor, 1981). Studies have indicated that high temperatures also stimulate high fecundity in female insects consequently giving rise to a large number of individuals at emergence (Rattle, 1985).

Moisture availability and variability have also been shown to be a major determinant of insect habitats. In the recent past, weather conditions have kept termite mounds moist much longer in certain areas than in other years. This has also encouraged high reproduction in the termites and this has resulted in large populations of elates warming to form or founding new colonies. Swarming and migration in insects can also be triggered by changes in ambient temperature (Rabb and Kennedy, 1979). Insects also respond to change in their thermal enviro-nment through migration, adaptation, or evolution (Dunn and Crutchfield, 2006). As such, the insects are able to adopt faster and widely spread to other areas to survive the climate changes, thereby increasing their availability to human consumers and other predators such as birds and fish in the lake region.

INSECT HARVESTING AND CLIMATE CHANGE

Patrick Durst has also commented during the interview with Media-global that when crops fail due to adverse weather changes such as draught and flooding, most insects will survive the harsh conditions relative to other food sources such as plants (Saunders, 2008) because they tend to be hardy to such conditions. Polyphagous insects often utilize alternative food plants in times of scarcity because they can access and consume other suitable plant among the remaining plants that can survive dry conditions. Thus, polyphagous edible insects can be harvested sustainably for domestic consumption and be incorporated into the family diet requirement as sources of protein. This observation is therefore an encouragement to collection of edible insects for domestic use during such a time as this of climate change. Saunders (2008) suggested that commercialization and marketing of edible insects could create money-making opportunities and add key nutrients to the diet of many vulnerable populations. Saunders notes that given the rapidity with which many edible insects are able to reproduce imply that most insects do not have any real problems or threat to extinction. Edible insects are thus an ideal mini livestock. For example, lake flies are not likely to compete with other food production due to their natural habitat and choice of feeding habitat at the lake bottom because the lake flies nymphs inhabit the bottom of the lake. Whereas high ambient temperatures negatively impact on agricultural production on land, the same increases reproduction of edible insects which is an advantage for entomophagy (Coviella and Trumble, 1999). This increases in the frequency of emergence of insects as the developmental time for each generation becomes gradually shorter and shorter. The spatial distribution of elate termites in arid and semi-arid areas is likely to be improved by moisture availability and warmer temperatures which accompany climate change.

The major constraint in consumption of edible insects has been sustainability in supply. However, with careful considerations, the observed increased frequency of emergence of edible insect populations should promote the supply side economy of entomophagy.

WATER QUALITY AND AQUATIC EDIBLE INSECTS

As alluded to, human activities have heavily contributed

towards the current climate change and global warming through their indiscriminate release of greenhouse gases. The emission of carbon gases which heavily contributes to the greenhouse gases which affects the PH of water. The high temperature also has an effect on dissolved oxygen in water. This in turn affects oxygen availability for both plants and insects that use water from the lake as their habitat (Elzinger, 1997). For example, the author explains that lake flies such as stoneflies and mayflies need plenty of dissolved oxygen in the water. Inadequate oxygen has negative impact on insect population. Population of mayflies has been reduced in certain areas of Lake Victoria due to pollution (Muli and Mavuti, 2001). These researchers noted changes in the lake ecosystem and that there is spatial distribution of the fauna attributed to the industrial effluence discharged from the lake catchments. Kisumu town (the Winam gulf) has witnessed reduced swarming of lakeflies within the past several years due to water pollution by the industrial effluence. The benthic macrofauna community of Winam gulf indicate changes from an Oligochaeta and Insecta classes of invertebrates dominated community to the present one now dominated by Phyla Mullusca and Oligochaeta due to pollution. Such changes have, even further reduced emergence of mayflies in certain areas of the lake.

Insects are well suited for use in environmental impact assessment because they respond to environmental perturbation in characteristic fashion, (Rosenberg et al., 1986) and the condition of the waters often can be monitored to a certain degree, through observing characteristics of some aquatic insect species. Certain species of insects can be used as bio-indicators of the status of the environmental conditions such as toxicity, bioaccumulation and bioavailability of environmental contaminants (Walton, 1989). As stated above, studies show that the presence of some edible aquatic insects in the lake can be used to monitor the environmental quality of the water (Schindler, 1980; Muli and Mavuti, 2001). Elzinger (1997) reported that stonefly nymphs require high concentration of dissolved oxygen and relatively pure waters. Their presence with a high diversity of other aquatic species indicates a measure of clean water quality. On the other hand, the presence of some chironomid larvae only indicated water pollution. Despite the above observations some villagers still, unknowingly, harvest the insects for domestic purposes.

Scientists have shown how chironomids may act as a proxy for water quality and temperature (Heegard et al., 2006; Both et al., 2006). Studies show that fossils of chaoborids and chironomids are widely used by palaeolimnologists as indictors of past environmental changes, including past climatic changes. The Environmental Change Research Centre at the University College London (UCL) also observes in their web sites that the midges are sensitive indicators of environmental changes. Chironomid larvae preserved in the lake sediments can be used to make quantitative reconstructions of palaeoenvironmental change. The role of these lake flies in studying the environmental changes cannot be overstated.

Water renewal rate also play a role in population of the insects in Lake Victoria since the productivity and trophic state of lakes is a balance between nutrient supplies and water renewal (Schindler, 2009), which cause washout of nutrients. For example, Schindler (1980) says that lakes in high rainfall areas have a faster water renewal than lakes in more arid climates. Schindler (2009) further comments that several paleoecological studies such as Lewis et al. (2001), Croley and Lewis (2006), McCormick (1990), Hari et al. (2006), Winder et al. (2009), Rood et al. (2005), Schindler and Donahue (2006), Rood et al. (2008) have all shown how slight changes in the balance between precipitation and evapotranspiration can dramatically change conditions in lakes. Of late, the lake region has experienced erratic high rainfalls, higher water run-off in rivers and uncontrollable water pathways causing flooding in several areas such as Budalangi and Nyando districts in Kenya. The sporadic high rainfall pattern within the lake region has brought about changes in the physico-chemical properties of the lake water. The change in physico-chemical properties that defines the chemistry and biology of the lake has significantly influenced the population of the lake flies at certain points.Climate change determines the temperatures in which various species and communities of fauna and flora survive and it controls the rates of chemical weathering and evaporation, which in turn determine chemical concentrations of minerals in the water (Schindler, 1980). Schindler records that the release of sulphur and nitrate oxides from the burning of fossil fuels and sulphur con-taining ores are major contributors to the acidification of soft waters. Wild fires during extended dry spells in the lake catchments areas such as the one of March 2009 in the Mau Forest of Kenya to some extent contributed to acidification of the lake waters. Emission of carbon gases from such activities on the Nandi hills and surrounding Mau forests only add to weather changes within the lake basin. Pickford (2006) has reminded us that the harvest of elate termites flourished in the Namibian savannah and coastal regions as evidenced by the presence of numerous termite mounds. But when a massive climate change occurred, millions of years ago all the mounds were buried by sand dunes. Thus, indicating that such extinctions of insects are also possible with climate change. It is indeed a wake-up call to experience the current abundance of insects with the present climate change.

POTENTIALS OF EDIBLE INSECTS FOR LIVESTOCK FEEDING

The lack of appropriate livestock feeds at affordable

prices has greatly contributed to the slow pace of development in aquaculture and agriculture particularly in East Africa. Research on aquaculture has shown that the demand for fish is increasing at a rate of 2.5% per annum. This rate is expected to increase despite the declining fish supply from the lakes, a factor exacerbated by the growing processing plants targeting the export markets which is likely to be destroyed by human activities (Barel et al., 1985; Gouds et al., 2005).

Feed used in commercial livestock production must contain all essential nutrients at adequate levels to meet total nutritional requirements of livestock for normal growth and development. Fish are generally fed on diets containing 28 to 32% protein, while for most livestock feeds, 15 to 25% protein are generally used (Donald et al., 2002; Edwin et al., 1998). *Rastrineobola argentea* has been the major animal protein source for most livestock species that allows high protein digestibility with amino acid profile, which closely matches the fish and poultry dietary requirements. On the other hand, Haplochromines that would provide the necessary feed quality has other competing need that is, it is used as food for human and also as a medicament. This has left the nutritionist with no choice other than seek for alternative feed sources for the livestock. The use of lake flies and grasshoppers in the livestock diets can greatly boost the macro and micro minerals to help curb on the deficiency diseases which result from the lack of sodium, potassium, cobalt, iodine, iron, copper, manganese, zinc, selenium and vitamin E, which are rampart nutritional challenges in livestock feeding. Grazing animals rely on pasture for their nutrition. The trace element content in pasture is measured in parts per million (ppm). There should be at least 10 ppm of copper, but only 0.25 ppm of iodine, 0.10 ppm of cobalt and a mere 0.03 ppm of selenium is enough for animals to grow well. If the soil in which the pasture grows lacks these trace elements, feed supplement for the livestock is necessary for their growth. Farmers choose what to feed their livestock, especially cattle, based on availability and the cost of feeding. Cattle feeds tend to be bulky and costly such that farmers often do not take into consideration nutritional value of the feed. This has led to nutrition deficiencies, hence diseases in various livestock species. Among the most important malnutrition is nutrient intake deficiency of macro and micro nutrients. Such deficiencies often predispose the livestock to various types of ill health conditions broadly referred to as metabolic diseases because it affects the animal ability to convert food into energy and growth. Nutrient availability in edible insects has been shown to be adequate to provide part of essential source of vitamins and minerals. The lake flies, termites and grasshoppers which are found in abundance with climate change provide additional option of the essential macro and micro minerals in feed supplements. Studies show that the high food nutrients in grasshopper and lake flies which range between 62 and 66% may help

in alleviating the negative balance of herd health and improve udder health common in cattle. For example, studies done by Brand et al. (2001) on nutrition and clinical mastitis found that vitamin E, A and Beta carotene and selenium had positive effects on the health of cattle especially on the improvement of udder health. Supplementation of livestock feed can be obtained by adding selected insects which are made abundant with climate change.

CONCLUSION AND RECOMMENDATIONS

The amounts and pattern of emergence of insects during the onset of rainy season has been influenced by the changes in the weather patterns. This paper described how the climate change is impacting on edible insects. Periodicity of emergence of the insects has been influenced by the unpredictability of the onset of the rain season and other weather activity on which the emergence of the insects depend on but these influences are yet to be explored. It is expected that the climate change will have both positive and negative impacts on collection and utilization of edible insects for human consumption and livestock feeds formulation. The status of food security of the marginal areas around the Lake Victoria will depend on how scientists interpret and manage the climate change outcomes.

Ability to respond successfully to challenges as a result of climate changes requires a lot of collaboration across different fields of study. It requires understanding on the part of stakeholders, how they will be affected by climate change, and strategic adaptive measures open to all. Analysis of the status and impact of insect population change on humans' livelihoods with specific focus on developing countries is critical. Specifically, understanding the interrelationship in the metamorphosis of entomology and entomophagy in food production in the region is paramount in interpreting and managing the climate changes outcome on the lakeflies and the elate termites in order to understand and adopt appropriate measures for utilization of these edible insects.

The reducing number of swarming aquatic insects in certain areas is attributed to environmental pollution as a result of the global warming. The balance of the insect population and other consequences are subject to studies. Therefore, this discussion extrapolates scientific reasoning for the increase based on recorded documents. The changes could be considered a wake-up call for researchers to pay closer attention to insect population (Fleming and Volney, 1995; Fleshman, 2007; Fraser, 2006). This could be an opportuned time for increased entomophagy as an available means for insect control and also improved food security when other sources of food items decrease.

The relationship between insects, plants and climate may elicit new kinds of behaviour yet to be determined. It is interesting to note that researchers have tended to record potential climate change effects of insects on human and animal health (Petzoldt and Seaman, 2008; Marigi et al., 2005; Marigi and Wairoto, 2005). This is particularly so, due to threats of increased transmission of pathogens. Resulting dynamics of effects of climate change on insects and plants may not necessarily be due to any single player but to their interactions. In any case, it is not certain that such interaction will maintain ecosystem stability, or lead to biodiversity instability (Bredenh, 2008; Connor, 2008; Coviella and Trumble, 1999). This is an issue yet to be researched more for future effective management. The interrelationship between moisture and temperature and the insect population may create new insect dynamics in the tropics with widespread influence on riparian communities. New things do happen with change! For example, Dunn and Crutchfield (2006) say that due to climate change, pine beetles have been noted to do things which have not been recorded before. The insects are attacking younger trees and attacking timber in altitudes they have never been before. What the future holds for edible insects is unknown. Formal research may help to unearth even additional phenomena. Different environmental factors may influence the rates of change in the composition and activities within and among groups of insects (Harringgton, 2005; Harrington et al., 2001; Harvel et al., 2002). Multi-disciplinary proxy studies comprising of several independent groups of the insects can be used to reconstruct past environmental conditions and the future sustainability of the edible insect fauna.

The role played by insects as source of feeds for livestock cannot be overstated. This will go a long way in providing livestock feed concentrate to some of the animals that now compete for the same sources of feed. The current abundance of insects may be such an opportunity presenting itself. In view of the above changes and the interdependence of many aquatic species, there is need to investigate the biology and ecology of the edible insects and their impacts in general on the benthic fauna. This could spell the beginning of insect farming for food security! The bottom line is how we interpret the increased population of the edible insects. Is it a boon or just a wake-up call not yet answered?

REFERENCES

Ayieko MA (2007). Nutritional value of selected species of reproductive isopteran and phemeroptera within the ASAL of Lake Victoria basin. Discovery. Innovation. 19(2): 126-130.

Ayieko MA, Oriaro V (2008). Consumption, indigenous knowledge and cultural values of the lakefly species within the Lake Victoria region, Afri. J. Environ. Sci. Technol. 2(10): 282-286.

Ayieko MA, Oriaro V, Nyanmbuga IA (2010). Processed products of termitesand lake flies: Improving entomophagy for food security within the Lake Victoria region, Afr. J. Food. Agric. Nut. Devel., 10(2): 2085-2098.

Bale JS, Masters GJ, Hodkinson ID, Awmack C, Bezemer TM, Brown VK, Butterfield JA, Buse JC, Coulson J, Farrar JEG, Good R, Harrington S, Hartley TH, Jones RL, Lindroth MC, Press I, Symrnioudis AD, Watt WJB (2002). Herbivory in global climate change research: direct effects of rising temperatures on insect herbivores. Global. Change. Biol., 8: 1-16.

Banjo AD, Lawal OA, Sngonuga EA (2006). The nutritional value of fourteen species of edible insects in Southern Nigeria. Afr. J. Biotechnol., 5(3): 298-301.

Barel CDN, Dorit R, Greenwood PH, Fryer G, Hughes N, Kawanabe PBN, McConel HRH (1985). Destruction of fisheries in African lakes. Nat., p. 315.

Bazzaz F (1990). The response of natural ecosystems to the rising CO_2 levels. Ann. Rev. Ecol. Syst., 21: 167-196.

Both C, Bouwhuis S, Lessells CM, Visser ME (2006). Climate change and population decline in a long distance migratory birds, Nat., 441: 81-83.

Brand JP, Noordhuizen TM, Schukken YH (2001). Herd health and production management in dairy practice. 3rd Ed. Wageningen. Netherlands.

Bredenh E (2008). Impact of a dam on benthic macroinvertebrates in a small river in a biodiversity hotspot: Cape Floristic Region, South Africa. J. Insect. Conserv. pp, 297-307

Connor S (2008). Insects 'will be climate change's first victims'. The Independent, (Tuesday, 6 May 2008). Retrieved on 16th June, 2010 from http://www.independent.co.uk/news/science/insects

Coviella C, Trumble J (1999). Effects of elevated atmospheric carbon dioxide on insect plant interactions. Conserv. Biol., 13: 700-712.

Donald PMC, Edwards RA, Greenhalgh MCA (2002). Animal nutrition. 6th Ed. Prentice hall. Ashford colour press limited. Goisport. United Kingdom.

Dunn D, Crutchfield JP (2006). Insects, Trees, and Climate: The bioacoustic ecology of deforestation and entomogenic climate change. Santa Fe Institute Working paper New Mexico, 06: 12-30.

Edwin HR, Meng HL, Martin WB (1998). Feeding Catfish in Commercial Ponds. Southern Regional aquaculture Centre. SRC Publication, p. 181.

Elzinger RJ (1997). Fundamentals of Entomology. 4th Ed. Chap 10-Insects Plants and Humans. Prentice-Hall, New Jersey.

Fleming RA, Volney WJA (1995). Effects of climate change on insect defoliator population processes in Canada's boreal forest: Some plausible scenarios. J. Water. Air, Soil. Poll., 82(1-2): 445-454.

Fleshman M (2007). Climate change: Africa gets ready. Planning how to deal with higher temperatures and shifting weather, Africa Renewal, United Nations, (website), 21(2): 1-8.

Fraser H (2006). Climate Change and Insects. OMAFRA. Retrieved on 16th June, 2010 from http://www.omafra.gov.on.ca.

Fraser MC, Bergeron JA (1991). The Merck Veterinary Manual: A hand book of diagnosis, therapy and disease prevention and control for the veterinarian.7th Ed. Merck and co Inc. New Jersey USA.

Gouds WPC, Witte F, Wanink JH (2005). The shrimp caridina nilotica in Lake Victoria (East Africa) before and after the Nile perch increase. Hydrobiologia, 00.

Harrington H (2005). Impacts of climate change on insect populations: Understanding and predicting climate change impacts. Rothamsted Research, Harpenden, UK. Retrieved on 16th June, 2010 from http://www.edinburgh.ceh.ac.uk.

Harrington R, Fleming R, Woiwood IP (2001). Climate change impacts on insect management and conservation in temperate regions: can they be predicted, Agric. For. Entomol., 3: 233-240.

Harvell CD Mitchell CE, Ward J, Altizer S, Dobson AP, Ostfeld RS, Samuel MD (2002). Climate Warming and Disease Risks for Terrestrial and Marine Biota. Science, 296: 2158-2162.

Heegaard E, Lotter AF, Birks HJ (2006). Aquatic biota and detection of climate change: Are these consistent aquatic ecotones, J. Paleolimnol., 35(3): 507-518.

Huis van A (2003). Insects as food in Sub-Saharan Africa. Insect Sci. Appl. 23(3): 163-185.

Huis vA (2002). Medicinal and stimulating properties ascribed to arthropods and their products in Sub-Saharan Africa. Insects in oral literature and traditions. Elizabeth Motte-Florac and Jacqueline M.C. Thomas, (Eds). Paris Louvai, Peeters-SELAF Ethnosciences, pp. 367-382.

Hunter MD (2001). Effects of elevated atmospheric carbon dioxide on insect-plant interactions. *Ag. For. Entomol.* 3: 153-159.

Landsberg J, Stafford SM (1992). A functional scheme for predicting the outbreak potential of herbivorous insects under global atmospheric change. Aust. J. Bot., 40: 565-577.

Martinat PJ (1987). The role of climatic variation and weather in forest insect outbreaks. Insect outbreaks. (Barbosa, P., and Schults, J.C., Ed.). Academic Press, London, pp. 241-268.

Macdonald WW (1956). Observation on the biology of chaoborids and chironomids in Lake Victoria and on the feeding habits of the Elephant-Snout fish kannume Forsk. J. Anim. Ecol., 25(1): 36-53.

Marigi SN, Wairoto JG Ambenje PG, Gikungu D (2005). Climate change and implications on Public Health Care. Climate Network Africa Impact Bulletin, (August 2005), pp. 12-15.

Marigi SN, Wairoto JG (2005). Climate Change and Associated Recent Impacts in Africa. Climate Network Africa Impact Bull, (February 2005), pp. 4-9.

Meyer-Rochow VB (2009). Entomophagy and its impact on culture of the world: The need for a multidciciplinary approach. Cultural impact of entomophagy and ethnoentomophagy. FAO- press Books, New York.

Muli JR, Mavuti KM (2001). The benthic macrofauna community of Kenyan waters of Lake Victoria. *Hydrobiologia*, 458: 83-90.

Owen DF (1973). Man's environmental predicament: Introduction to human ecology in tropical Africa. London: Oxford University Press, pp. 132-136.

Petzoldt C, Seaman A (2008). Climate Change Effects on Insects and Pathogens. Retrieved on 16th June, 2010 from http://www.climateandfarming.org.

Pickford M (2006). A termite tale of climate change. *Quest.*, 2(3): 28.

Rabb RL, Kennedy GG (1979). Movement of highly mobile Insects: Concepts and Methodology in Research. North Carolina State University, Raleigh.

Rattle HT (1985). Temperature and insect development. Environmental Physiology and Biochemistry of insects. (Hoffman KH Ed.), Springer-Varlag, Berlin, pp. 33-65.

Rosenberg DM, Danks HV, Lehmkuhl DM (1986). Importance of insects in environmental impact assessment. Retrieved on 8th June, 2010 from http://www.springerlink.com J. Environ. Manage. 10(6).

Saunders A (2008). FAO serves up edible insects as part of food security solution. *Mediaglobal*, (February, 2008), United Nations Secretariat, New York, FAO Rome.

Schindler DW (1980). Aquatic ecosystems and global ecology. Fundermentals of aquatic ecology. 2nd Ed. (Barnes RSK, Mann KH, Eds.). Blackwell Science.

Shetty P (2009). Climate change and insect-borne disease: Facts and figures. Science and Development Network. Retrieved on 16th June, 2010 from http://www.scidev.net.

Taylor F (1981). Ecology and Evolution of physiological time in insects. Am. Nat., 117: 1-23.

Verschen D, Thomas J, Hedy JK, Edington DN, Leavitt PR (2002). The chronology of human impact on Lake Victoria, East Afr. Proc. *R. Soc.* London, B 269.

Walton BT (1989). Insects as indicators of toxicity, bioaccumulation and bioavailability of environmental contaminants. J. Environ. Toxicol. Chem., 8(8). John Wiley and Sons, Inc. Retrieved on 8th June, 2010 from http://www.setacpeachnewmedia.com.

Wellington WG, Trimble RM (1984). Weather. Ecological Entomology in Huffacker CB, Rabb RL Eds.), John Wiley, New York pp. 399-425.

18

When DNA sequences and microsatellites loci tell the story of field groundnut infestation by *Caryedon serratus* Ol. (Coleoptera, Chrysomelidae, Bruchinae)

Mbacké Sembène[1*], Awa Ndiaye[1], Khadim Kébé[1], Ali Doumma[2], Antoine Sanon[3], Guillaume K. Kétoh[4], Laurent Granjon and Jean-Yves Rasplus [5]

[1]Faculty of Science and Technology, University Cheikh Anta Diop, P. O. Box 5005 Dakar, Senegal.
[2]Faculty of Science, University Abdou Moumouni Niamey, P. O. Box 10662, Niamey, Niger.
[3]Laboratory of Entomology University of Ouagadougou, Burkina Faso.
[4]Laboratory of Applied Entomology, Faculty of Science, University of Lomé, P. B. 1515, Lomé, Togo.
[5]NRA – UMR 1062 CBGP (INRA / IRD / Cirad / Montpellier SupAgro), Campus international de Baillarguet, CS 30016, 34988 Montferrier-sur-Lez, France.

The first infestations of stored groundnuts by the seed-beetle *Caryedon serratus* were reported in this country at the turn of the 20th century. This bruchid has a wide distribution in Africa, from Senegal to South Africa and in southern Asia. Native hosts of *C. serratus* in Senegal include *Bauhinia rufescens*, *Cassia sieberiana*, *Piliostigma reticulatum* and *Tamarindus indica*, all of which belong to the legume subfamily Caesalpinioideae. Molecular marker, DNA sequences and microsatellites loci polymorphism were used to investigate the mechanisms of first groundnut infestation by *C. serratus*. Sequence analysis of ribosomal DNA nuclear (ITS1) and mitochondrial coding DNA (Cytochrome b) reveal several biotypes in Senegal, with restricted past and/or present gene flow between each other. Samples typically clustered according to host plant, except for groundnut and *P. reticulatum*, which clustered together. Polymorphic microsatellites loci confirm the allelic proximity between *P. reticulatum* and groundnut *C. serratus*. These strains, genetically very close, begin however to diverge but the number of migrants between them keeps relatively important. Historical hypothesis of the first groundnut infestation in West Africa is also debated in this study.

Key words: groundnut, *Caryedon serratus*, *Piliostigma reticulatum*, groundnut, infestation, ITS, Cyt. B, DNA sequences, microsatellite loci.

INTRODUCTION

The disruption of biological communities creates several problems for people and can greatly affect productivity and profitability of agricultural systems. Also, conservation of food stored by the appearance of species or biotypes from other regions or agro-ecological zones creates several problems, but by increase populations of native species favored by global change. Nowadays, production of groundnut is provided to more than 85% of small farms in Africa and Asia. The conservation of crops ensures the availability of food resources which is a key factor in the food security of a country. Unfortunately, agricultural production is generally seasonal, while the consumer needs are extended throughout the year. Thus, the establishment of an adequate policy to keep the plant populations from the risks of food shortages during the off-season farming is the most important thing that developing countries should achieve. In this context, particular emphasis should be placed on controlling insect pests of crops in stocks because damage caused by insects can lead to financial loss, hunger and risks associated with intoxication consumption or damaged goods. As such, crops should be treated with pesticides

*Corresponding author. E-mail. mbacke.sembene@ird.fr.

(Zuoxin et al., 2005). In countries like those in the Sahel, where the dry season lasts most of the year, storage of crops is a matter of survival.

Groundnut (*Arachis hypogaea*) is an herbaceous, annual legume, originating in Peru. Currently, it is cultivated in sub-tropical savanna regions in sub-saharan Africa, USA, Middle East and Asia. It was introduced into Africa from Brazil by the Portuguese at the end of the 15th century (Adrian and Jacquot 1968; Perhaut, 1976). The grain contains 38 - 50% oil and is also rich in protein. It is consumed by humans as grain, flour, paste and oil. Groundnut cake and foliage are used as animal fodder and the oil is also used in soap production.

In Africa, the first infestations of stored groundnuts by the bruchid *Caryedon serratus* (Olivier) were reported in this country at the turn of the 20th century (Davey, 1958; Delobel, 1995). *C. serratus* has a wide distribution in Africa, from Senegal to South Africa and in southern Asia (Johnson, 1986). Its larvae feeds on the seeds of wild Caesalpiniaceae belonging to four genera: *Bauhinia, Cassia, Piliostigma* and *Tamarindus* (Borowiec, 1987). The groundnut seed-beetle is responsible for important weight losses in stored groundnuts, thereby reaching about 83% in four months storage in Senegal (Ndiaye, 1991). About 60 years after its first record as a pest of groundnut in West Africa (Roubaud, 1916), it has become a major primary groundnut pest in Central Africa (Matokot et al., 1987) and Asia (Dick, 1987). It was also recorded in Central and South America in the seeds of ornamental *Bauhinia*. Its larvae bore through groundnut hulls and favour attacks by secondary pests. In the same time, they favour the spread of *Aspergillus flavus*, a mould which produces a toxic substance, aflatoxin (Gillier and Bockelée-Morvan, 1979). Groundnut infestation by the seed-beetle puts the question of the mechanisms by which *A. hypogaea*, a plant of the family Fabaceae, became part of the insect's range of hosts.

Food-plant selection and larval development researches (Robert, 1985) as well as morphometric (Sembène and Delobel, 1996), allozyme variation (Sembène et al., 1998; Sembène and Delobel, 1998) and DNA sequences (Sembène et al., 2008 and 2010) provided support to the hypothesis that a certain amount of genetic isolation exists between populations with different feeding habits, and in particular, between groundnut-feeding and Caesalpiniaceae-feeding forms. In this study, we are particularly interested in the historic and present modalities of the infestation of the groundnut dried in fields. Our primary question was whether groundnut-infesting populations of *C. serratus* represent a newly formed host race of the species. In evolutionary biology, it is evident that the differentiation in host races is a phenomenon widely discussed to phytophagus insects. Indeed in a lot of case, it was observed after the introduction of a new plant. We are also interested in knowing the intensity of gene flow and migrants between the beetles on *Piliostigma reticulatum* and those on groundnut dried in the field.

MATERIALS AND METHODS

C. serratus sampling

Beetles were sampled from eggs, larvae or pupae on/inside pods collected from the different host trees between 2005 and 2010. Samples were collected in Bignona (16°13'W, 12°48'N), Fimela (16°41'W, 14°08'N), Keur Baka (15°57'W, 13°56'N), Linguere (15°25'W, 15°88'N), Ouarak (16°04'W, 15°33'N) and Thies (16°56'W, 14°48'N). Samples were named after their host plant species and geographic origin (Table I). Pods were collected from the trees between January and May. Groundnut samples were collected from the field during drying. Adult beetles were genetically analysed almost immediately after they emerged from the pods. In all, a total of 1200 individuals of both sexes corresponding to 30 different populations with 40 individuals per population were genetically scored in the study.

DNA extraction and PCR

DNA was extracted, amplified and sequenced with standard protocols described elsewhere (Sembène et al., 2010). Each sequence was obtained from the DNA of a single seed-beetle. The abdomen, elytra and antennae were kept apart to avoid contamination by fungi and nematodes and to permit subsequent morphological observations.

A partial Cytochrome B (Cyt. B) end region was PCR-amplified using the primers CB1 (5' – TATGTACTACCATGAGGACAAATATC - 3') and CB2 (5'-ATTACACCTCCTAATTTATTAGGAAT - 3'). The ITS1 ribosomal DNA was targeted for PCR, amplified and sequencing with primer CIL (5' GCGTTCGAARTGCGATGATCAA 3') and CIU (5'GTAGGTGAACCTGCAGAAGG3').

For both markers, PCR amplification were performed in 25 µl reaction volume 2.5 µl enzyme buffer supplied by the manufacturer, 2.5 mM $MgCl_2$, 0.6 unit of Taq polymerase (Promega), 17.5 pM of each primer, 25 nM of each DNTP and 1µl of DNA extract. After an initial denaturation step at 92°C for 3 min, reaction were subjected to 35 cycles for 1 min at 92°C, 1 min 30 s at 48°C and 1 min at 72°C. PCR products were loaded on a 1.3% agar gel. The PCR band was cut and then purified with Quiaquick PCR gel purification kit (Qiagen) and directly sequenced on an Abi 373 automated sequencer using TaqFS and Dye-labeled terminators (Perkin-Elmer).

Radioactive polymerase chain reaction (PCR) amplifications of microsatellite loci were carried out on 10 µl as previously described in Estoup et al. (1995). Ten microsatellite loci were scored: M193, M2149, M66, M625, M836, M97, M984, M12o mega, M13113 and M1425. As such, primer sequences and annealing temperatures were used for each locus (Sembène et al., 2003).

Sequence alignment and phylogenetic analyses

Sequence alignment was performed using ClustalW (Thompson et al., 1994) as implemented in BioEdit. Alignment of coding sequences (Cyt. B) was unambiguous as no gap event was detected. For ITS1, we proceeded to several multiple alignments using ClustalW under different gap opening and gap extension coast. Aligned sequences were finally entered in McClade 3.06 (Maddison and Maddison, 1992) for subsequent treatments.

The resulting data set was used to calculate Kimura 2-parameter genetic distances between the haplotypes. G-tests (log-likelihood ratio test) were performed to test for genetic distance homogeneity among hosts at the same site and among sites for the same host(Sokal and Rohlf, 1981).

Phylogenetic relationships were reconstructed with PAUP $4*_B8$ (Swofford, 2001) using the maximum parsimony method (MP). The MP analysis was carried out with the heuristic search option with 50

Table 1. Population sampling: origin, host plant, abbreviations and numbers of scored individuals (between parentheses).

Sites and coordinates	Sampling plants, abbreviations of populations and number of scored individuals				
	A. hypogaea	*B. rufescens*	*C. sieberiana*	*P. reticulatum*	*T. indica*
Bignona (16°13'W, 12°48'N)	Abi (40)	Bbi (40)	Cbi (40)	Pbi (40)	Tbi (40)
Fimela (16°41'W, 14°08'N)	Afi (40)	Bfi (40)	Cfi (40)	Pfi (40)	Tfi (40)
Keur Baka (15°57'W, 13°56'N)	Akb (40)	Bkb (40)	Ckb (40)	Pkb (40)	Tkb (40)
Linguere (15°25'W, 15°88'N)	Ali (40)	Bli (40)	Cli (40)	Pli (40)	Tli (40)
Ourack (16°04'W, 15°33'N)	Aou (40)	Bou (40)	Cou (40)	Pou (40)	Tou (40)
Thies (16°56'W, 14°48'N)	Ath (40)	Bth (40)	Cth (40)	Pth (40)	Tth (40)

In Bignona, Abi was obtained from *A. hypogaea*, Cbi from *C. sieberiana*, Pbi from *P. reticulatum* and Tbi from *T. indica*. In Fimela, sample Afi was obtained from *A. hypogaea*, Bfi from *Bauhinia rufescens*, Cfi from *Cassia sieberiana*, Pfi from *P. reticulatum* and Tfi from *T. indica*. In Keur Baka, Akb was obtained from *A. hypogaea*, Ckb from *C. sieberiana*, Pkb from *P. reticulatum* and Tkb from *T. indica*. In Ouarak, Aou from *A. hypogaea*, Bou from *B. rufescens*, Pou from *P. reticulatum* and Tou from *T. indica*. In Linguere sample Ali was obtained from *A. hypogaea*, Bli from *Bauhinia rufescens*, Cli from *Cassia sieberiana*, Pli from *P. reticulatum* and Tli from *T. indica*. In Thies: Ath from *A. hypogaea*, Pth from *P. reticulatum* and Tth from *T. indica*.

random stepwise taxon addition replicates, using the branch swapping tree bisection-reconnection (TBR) option. A bootstrap procedure (1000 iterations with the same option of heuristic search) was used to establish the score of each node (Felsenstein, 1985) by retaining group compatible with the 50% majority rule consensus. A strict consensus tree was computed whenever multiple equal parsimonious trees were obtained.

The molecular clock hypothesis was tested following Possada and Crandall (2001) by comparing the log likehood scores of the bests trees obtained with molecular clock enforced and no enforced. The Shimodaira-Hasegawa test (Shimodaira and Hasegawa, 1999) using reel Bootstrap with 1000 replications was used to test for a significant difference between the score of the different trees obtained.

Analyses were also conducted using the distance-matrix method with the Neighbour-Joining (NJ) algorithm (Saitou and Nei, 1987) on Kimura 2- parameter distances with PAUP 4*B8 (Swofford, 2001). The robustness of inferences was assessed through bootstrap resampling (1000 replicates). The consistency index CI and the retention index RI (Farris, 1994) were also calculated. The two analyses were made separately with Cyt.B and the ITS1. The partition homogeneity test (Farris, 1994), as implemented in PAUP was used to determine the appropriateness of combining both partial Cyt.B and ITS1 genes into a single analysis with the same options. The Wilcoxon's signed-rank test (Templeton, 1983) was applied to compare the statistical significance of the best tree produced by each tree reconstruction method to one another. Two out groups were used in our study: *Bruchidius atrolineatus* which belongs to the family Bruchidae and *Caryedon gonagra* obtained from *Bauhinia variegata* from Egypt (BvC). *C. gonagra* (F.) is a species which may be synonymous with *C. serratus* (Delobel et al., 2003).

Polymorphic microsatellite data

Quantitative analysis of microsatellite data, tests for Hardy-Weinberg equilibrium and genotypic linkage disequilibrium were computed using Genepop V1.2 (Raymond and Rousset, 1995). In both cases, Fisher's exact test available in Genepop V1.2 was used after pooling the data by sample for a given locus and by locus for a given sample. G-tests (log-likelihood ratio test) were performed to test allele frequency homogeneity among hosts at the same site and among sites for the same host (Sokal and Rohlf, 1981). Fstat V1.2 (Goudet, 1995) was used to compute Weir and Cockerham's (1984) estimators f and θ of F-statistics (Wright, 1931) in order to

evaluate sub-structuration among hosts and sites. Parameter f (consanguinity coefficient) corresponds to Wright's F_{IS}. Parameter θ (degree of genetic differentiation between populations) corresponds to Wright's F_{ST}. Multiple test significance was assessed using Fisher exact test. The numbers of migrants (F1 and F2) between *C. serratus* feeding on the native hosts and the groundnut form is evaluated using Strcture.exe. Relationships among populations based on gene frequencies were displayed using the neighbour-joining algorithm (Saitou and Nei, 1987) and the mean chord distance (Cavalli-Sforza and Edwards, 1967). This distance is one of the most efficient for obtaining the correct tree topology under both the stepwise mutation model (SMM) and the infinite allele model (IAM) (Takezaki and Nei, 1996). Node stability was evaluated using 1000 bootstrap replicates (Hedges, 1992) re-sampling loci and/or individuals and majority-rule consensus trees were obtained using procedures "SEQBOOT" and "CONSENSE" in Felsenstein's (1993) PHYLIP V3.57c package. Swofford and Berlocher (1987) frequency parsimony program (FREQPARS) was used to build trees based on a modified Wagner algorithm (Farris, 1970) and also to compare the different topologies obtained. A strict consensus tree was computed after NJ and FREQPARS analysis. Tree of individuals (only with samples reared from *P. reticulatum* and groundnut) were constructed with Population.exe using the DAS (shared allele) distance (Chakraborty and Jin, 1993) using the distance-matrix method with the neighbour-joining algorithm. The robustness of inferences was assessed through bootstrap re-sampling (1000 replicates). For this tree, a classification index was calculated to establish the validity of the clusters of individuals on tree branches according to their membership to a known population. An individual is 'well classified' when it is attached to his group of origin (Estoup et al., 1994).

RESULTS

Alignment and genetic distance

We obtained 518 bp of the partial Cyt. B gene in 30 *C. serratus* populations. The alignment was straightforward and involved no insertions/deletions. The sequences could be unambiguously aligned and showed 22 different haplotypes due to 51 polymorphic sites. Of these sites, 94% were parsimony informative. The number of nucleotide differences in pairwise comparisons of *C. serratus* populations ranged from 0 to 16.1% due to a large part of *C. serratus* sampled on *C. sieberiana* and the others. Within the same host species, the number ranged from 0 to 0. 02%.

Sequences obtained from ITS1 domain were 954 bp in *C. serratus* feeding on *C. sieberiana* and 879 bp feeding on the others including, in both cases 6 bp in 18 S and 49 bp in 5.8 S. Only 68.2% of the total could be aligned between both groups. Of these sites, 35.9% were variable and 89% of these were parsimony informative. The number of nucleotide differences in pairwise comparisons *C. serratus* populations ranged from 0 to 34, 6%. Within the same host species, the number ranged from 0 to 0. 03%.

Among the *C. serratus*, the genetic distances measured on the total alignment of Cyt.B + ITS1 (1418 bp) ranged between 0 and 0.204, but clearly fall in two classes: One group comprises haplotypes sampled in *A. hypogaea, B. rufescens, P. reticulatum* and *T. indica* from 0 to 0.032 and a second group gathers the haplotypes raised from *C. sieberiana* from 0.16 to 0.204; separating *C. serratus* into two major groups. The original data set was reanalysed without *C. serratus* sampled in *C. sieberiana*. Within host genetic distance decreased and was low: 0 - 0.004 for "groundnut", 0 - 0.002 for "Bauhinia", 0 − 0.002 for "Piliostigma" 0 − 0.002 for "Tamarindus". Genetic distance between hosts (0.036) decreased but remained significant (G-test p < 0.01). Genetic distance between localities was non-significant.

Phylogenetic relationships

The maximum parsimony (MP) analysis on Cyt.B nucleotide data yielded 32 equally parsimonious trees that were 274 steps long (CI = 0.882; RI = 0.785). The same methods on the ITS1 data set yielded 7 equally parsimonious trees, 263 steps long (CI = 0.967; RI = 0.992). Finally, analysis of the combined data set yielded 178 steps long (CI = 0.932; RI 0.988). Similar patterns of relationships were obtained with Neighbour-joining (NJ) analysis. Samples typically clustered according to host plant, except for groundnut and *P. reticulatum*, which clustered together. *C. sieberiana* samples were clearly separated from all other samples and showed high bootstrap values. Bootstrap values were all over 50%. Separation according to host plant is clear. For each data set, the topology obtained with MP and NJ methods were compared using the Shimodaira-Hasewaga test and the Wilcoxon's signed-rank test. These tests did not support the significant difference between the trees. The strict consensus tree of MP and NJ trees is presented in Figure 1 for the pooled data.

Polymorphic microsatellites

The microsatellite loci M12omega failed to sustain

Figure 1. Phylogenetic relationships among nucleotide sequences of the pooled data (partial cytochrome b and ITS1 genes) of 30 specimens of *C. serratus* populations. This tree is the consensus between maximum parsimony (MP) consensus tree and neighbour-jonning (NJ) consensus tree. The numbers above/under branches are the means of MP and NJ % bootstrap values (1,000 replicates). *Bruchidius atrolineatus* (BrA) an *Caryedon gonagra* (BVC) are the out group.

reliability. M66 and M13113 proved to be monomorphic. Only M625, M836, M97, M984 M193, M2149 and M1425 showed scoreable polymorphic loci.

The number of alleles varied from 4.8 to 6.9 (mean 5.85). The difference between allele frequencies at a given locus was lower between geographically distant samples from the same host plant than between sympatric samples from different host plants. All individual populations were in Hardy-Weinberg equilibrium at all loci with the exceptions of Ali and Afi. Deviation was due to a deficiency of heterozygotes (0.23 < f < 0.67). Over all loci, deviations from Hardy-Weinberg expectations occurred in populations from the same locality (p < 0.01).

Table 2. *C. serratus*, θ at various hierarchical levels of differentiation estimated for the host plants (a) the localities (b). Significance of deviation from zero of θ values was tested using $\chi^2 = 2N(\theta)(k-1)$ for k alleles (see Table 1) and s populations with (k-1)(s-1) df. * P < 0.001.

| Locus | Bignona | Fimela | Localities | | | | | |
			Keur Baka	Linguere	Ouarak	Thies	Between localities	Between samples
M1425	0.332*	0.308*	0.384*	0.403*	0.336*	0.364*	0.008	0.312*
M625	0.318*	0.316*	0.314*	0.381*	0.317*	0.312*	0.007	0.291*
M836	0.321*	0.321*	0.326*	0.310*	0.282*	0.303*	0.005	0.262*
M97	0.398*	0.432*	0.415*	0.475*	0.430*	0.418*	0.002	0.389*
M984	0.390*	0.419*	0.392*	0.452*	0.380*	0.382*	0.006	0.358*
M2149	0.312*	0.324*	0.342*	0.352*	0.299*	0.315*	0.006	0.349*
M193	0.242*	0.298	0.247*	0.288*	0.251*	0.249*	0.007	0.301*
Overall	0.355*	0.368*	0.362*	o.405*	0.344*	0.358*	0.006	0.321*

| Locus | Host plants | | | | | |
	A. hypogaea	B. rufescens	P. reticulatum	T. indica	Between host	Between samples
M1425	0.007	0.004	0.024	0.004	0.359*	0.312*
M625	0.017	0.016	0.003	0.009	0.336*	0.291*
M836	0.005	0.022	0.007	0.007	0.314*	0.262*
M97	0.009	0.007	0.059	0.007	0.447*	0.389*
M984	0.063	0.012	0.010	0.012	0.412*	0.358*
M2149	0.007	0.006	0.008	0.009	0.447*	0.349*
M193	0.006	0.004	0.006	0.002	0.412*	0.301*
Overall	0.011	0.006	0.009	0.006	0.368*	0.321*

Finally, over all loci and samples, the probability of deviations from Hardy-Weinberg expectations was highly significant (χ^2 = 246.7; ddl = 154). No linkage disequilibrium was found among any pair of microsatellite loci for any sample (P > 0.05). The large positive f value indicated a lower number of heterozygous individuals relative to that expected when data was pooled for all populations.

In "*Bauhinia*" and "*Tamarindus*" samples, allele frequencies were on the whole homogeneous. "*Groundnut*" and "*Piliostigma*" samples showed signify-cant or nearly significant heterogeneity and were globally homogeneous. Comparable results were obtained from the analysis of θ values within and between host plants and within and between localities (Table 2). Genetic differen-tiation between hosts was highly significant (average θ = 0.368) and was due to a large part of "*Bauhinia*" samples, which differed from all other samples. All loci contributed to the genetic differentiation between hosts. Differentiation between sites was not significant.

Similar patterns of relationships between populations were obtained from neighbour-joining analysis and from Wagner parsimony analysis. The consensus tree generated by the two analyses is presented in Figure 2. Samples typically clustered according to host plant as results obtained and largely discussed by Sembène et al. (2008*)*. The individuals tree (Figure 3) show a slight divergence between *C. serratus* feeding on *P. reticulatum*

and those feeding on groundnut.

Results obtained with the "Structure.exe" reveal 12 migrants of *P. reticulatum C. serratus* to groundnut in the first generation. At the same time, 3 migrants left the groundnut lying on the residual seeds of *P. reticulatum* in its nature, and 23 infested the recently harvested groundnut. In the second generation, all these values increase approximately with 15%.

DISCUSSION AND CONCLUSION

It is evidence that a strong genetic differentiation clearly exists between *C. serratus* feeding on different host plant. Phylogenetic hypotheses and the relative genetic isolation between these populations are best explained by the fact that they feed on different host plants. These host race differentiation and host race formation mechanisms are largely debated in Sembène et al. (2008 and 2010).

In this study, we were particularly interested in under-standing the origin of dried groundnut infestation in the field. Our results show that Groundnut and *P. reticulatum C. serratus* populations are indistinguishable on the basis of sequence sets. These samples show however similarities and was genetically very close. High gene flow probably exists between the two populations, as period. The introduction of groundnut into the environ-

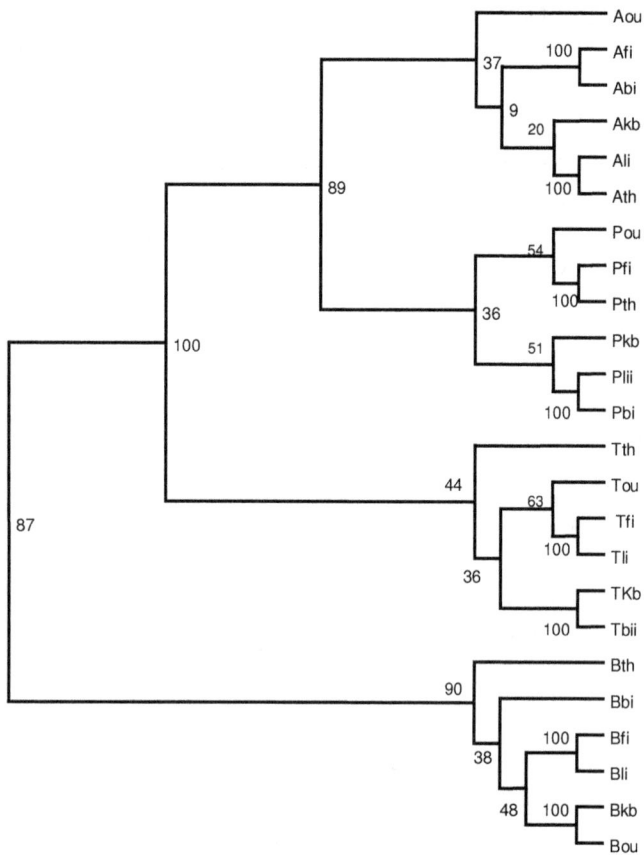

Figure 2. Relationships among *C. serratus* samples from different sites and host plants (see Table 1). *C. sieberiana* samples are excluded. This tree is a consensus between neighbour-joining consensus tree and FREQPARS consensus tree using 7 microsatellites loci. Node stability was evaluated using 1,000 bootstrapping replications re-sampling loci or populations. The numbers above branches are the neighbour-joining % bootstrap values (1,000 replicates).

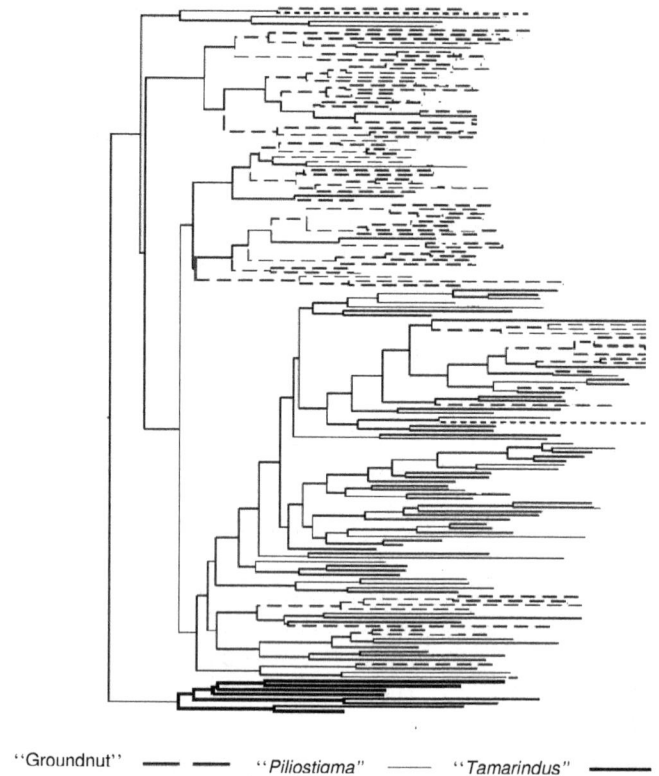

Figure 3. Relationships among "*Piliostigma reticulatum*" and "Groundnut" *C. serratus* individuals. This tree was constructed with the DAS (shared allele) distance using 7 microsatellite loci. It derives from 50% majority-rule consensus of 1,000 bootstrap replicates. "Tamarindus" individuals are the out group.

ment of wild plants supported the transfer of "Piliostigma" indicated by morphometric (Sembène and Delobel, 1996) and allozymic (Sembène and Delobel, 1998) analysis, and it may be hypothesized that beetles feeding on *P. reticulatum* were responsible for the initial infestation of groundnuts at the turn of the 20th century. Polymorphic microsatellites loci confirm and explain this hypothesis. Beetles feeding on *P. reticulatum* and those infesting the groundnut, although genetically very close begin to diverge. The number of individuals not being able to be brought back to the one or the other group is however 10 %. The rate of "well classified" is 90 % for each of its two stumps.

The depth of knots in the individuals tree between these two stumps and the big resemblance of the allele frequencies of the individuals belonging to these two origins show that both strain began to diverge only recently on the scale of time. This situation may be explained by the absence of wild pods during the rainy

on groundnut and the "cycling" of the groundnut stock in the residual pods of stores. Currently, adults of *C. serratus* which came from stores and those which emerged from residual seeds of *P. reticulatum* at the end of the dry season, when there were still some pods on trees, or from the few pods of *P. reticulatum* in decomposltion on the ground, were responsible for the re-infestation of new harvests. These beetles feed on pollen or nectar during the rainy season. In the presence of adequate food (pollen, water), beetles can survive 80 to 90 days in the laboratory (Delobel, 1989), and so in nature may also live 3 months or even longer. In periods of intense heat or strong rains (factors enhancing mortality of *C. serratus*) the beetles shelter, by negative phototropism, under the trees in litter. This finding points to the importance of residual populations in stores in the re-infestation of newly collected groundnuts.

Even though, we do not have references for estimate divergence dates on Cyt. B transversions, but molecular data make it possible today to propose how food preferences evolved in certain groups of insects and to identify the possible host plant ancestor of a studied group. In the case of groundnut infestation, we can suggest that *C. serratus* feeding on *P. reticulatum* were the origin of groundnut infestation at the beginning of the century, by allotrophy in Senegal. This population, undoubtedly, extended to infest groundnut harvests in most of Western Africa. Historical data reveal that it is in Western Africa (Senegal) that the first groundnut attacks by *C. serratus* were reported (Roubaud, 1916). Until the beginning of the 1970's, when damage in Congo was first reported (Delobel, 1989), this was the only area of the world where *C. serratus* infested groundnut stocks. Today, *C. serratus* is a pest of stored groundnuts from Senegal to Chad and southwards to the Central African Republic and Congo (Matokot et al., 1987). It has become a pest of groundnuts in Asia (Dick, 1987) and has colonized parts of the New World tropics in the seeds of tamarind and ornamental *Bauhinia* species (Johnson, 1966; Nilsson and Johnson, 1992). The present-day distribution of *C. serratus* as a pest of groundnuts may be explained by successive introductions of infested material, as was the likely cause in the Congo (Delobel and Matokot, 1991).

The groundnut crop was first stepped in Senegal without major difficulty, and was even free from plant health problems. In 1910, however, a sudden deterioration of commodities in the metropolis was reported. In 1912, Azemard, sub-inspector of agriculture Diambour (Senegal), draws the study's attention to the Senegalese head of agriculture report on the plant health of peanuts stocks and he echoed the concerns of traders with the subsequent receipt of stocks heavily damaged and diagnosed, following attacks by termites, millipedes of elaterides and secondary infestations by *Plodia interpunctella*, *Tribolium confusum* and *Oryzaephilus surinamensis*. To remedy the deterioration of stocks, Azemard advocates the elimination of grain "stuck or broken". Finally in 1913, Roubaud, a prominent expert on sleeping sickness, then head of laboratory at the Pasteur institute, was sent to Senegal, where he reported a significant amount of data. His list of insect pests in groundnut shell is much more comprehensive and more accurate than those previously published. It includes, in particular, groundnut bruchid (under the name *Pachymaerus acaciae* Gyll. However, this is the first time *Caryedon serratus* is associated with groundnut. The damage as indicated by Roubaud, may significantly prolong storage, but the author is very far from it according to the weevil importance he gives to termites, in that *C. serratus* grows slowly only in recent literature. In 1914, an increase in insect damage was attributed to *Ephestia cautella*. Eight years later, the weevil is still relatively uncommon in stocks, but passes unnoticed in stores mills of Bordeaux, which annually treat 80,000 tons of peanuts from Senegal (Feytaud, 1924). Yet more than 15 years after the first observations of echoes, it has become the main enemy of peanut stocks in Senegal in 1935 and Sagot Bouffil emphasize three crucial points concerning the mode of contamination in groundnut: (i) *C. serratus* from peanut thrives in tamarind seeds (ii) This

host is infested wilderness in nature by *C. serratus* (iii) Infestation of peanut field occurs during drying in windrows. Guiraud, especially in a study on the economics of groundnut cultivation, cites a pest of stocks, and this is the weevil, which, he says, "is more and more damage. These are particularly serious when they are made in stocks *of* foresight, because the seeds attacked are quickly incapable of germination "(Guiraud, 1938).

The preceding history highlights the difficulty of interpreting the evidence left by the early century entomologists. Our only certainty lies in the fact that *C. serratus* had already attacked the Senegalese groundnut in 1913 at least in some stocks. If we can assume that low levels of infestation have escaped in 1912 and 1913 to a non-specialist like Azemard, it is surprising that in 1910 Perez could not detect *C. serratus* had he been present in the stocks of Metropolitan mills. Indeed, the time limits imposed on stocks before their arrival in Metropole probably reached several months or one year; such delays could only encourage the growth of the weevil and the appearance of witnesses and compelling the obvious presence of *C. serratus*: pupa cocoons, outlet of larvae and adults. And it should be noted that ten years later, Bordeaux still does not know the groundnut bruchid, although its warehouses are continuously fed with peanuts from West Africa (Feytaud, 1924).

One can imagine two explanations for this apparent contradiction: The first is that there has been rejection by buyers of seed lots infested or sort particularly effective. But on one hand the chronic left no trace of such practices on the other hand, it is difficult to imagine that they could be effective as to obscure *C. serratus* metropolitan entomologists for so long. The second explanation is more plausible: It is likely that in 1913 the Senegalese groundnut infestation by the weevil was far from widespread and affected only a very limited number of production areas. That is why it escaped the successors of Roubaud. The geographically discontinuous nature of infection corresponds to a reality that had since been observed elsewhere, particularly in Congo.

It is difficult to specify the date of onset of weevil in Senegalese stocks because if we can date the onset of weevil in stocks for export around 1910, this does not mean that the passage populations of *C. serratus* on groundnut date precisely from this period. In fact, it perhaps became effective after multiple failed attempts that have occurred over several centuries. However, it is clear that the widespread infestation throughout West Africa took place in 15 or 20 years.

ACKNOWLEDGEMENTS

We thank A. Delobel and JF Silvain for helpful collaboration. This publication was made possible through the support provided by the IRD-DSF and by the international foundation of science (IFS). Facilities were provide by CBGP (Montpellier).

REFERENCES

Adrian J, Jacquot R (1968). Valeur Alimentaire de l'Arachide et de ses Dérivés. Collection Techniques Agricoles et Productions Tropicales. Maisonneuve and Larose, Paris, France, pp. 54.

Borowiec L (1987). The genera of seed-beetles (Coleoptera, Bruchidae). Polskie Pismo Entomol., 57: 3-207.

Cavalli-Sforza LL, Edwards SV (1967). "Phylogenetic Analysis: Models and Estimation Procedures" Evolution, 21(3): 550-570.

Chakraborty R, Jin L (1993). Determination of relatedness between individuals by DNA fingerprinting. Human Biol., 65: 875-895

Davey PM (1958). The groundnut bruchid, *Caryedon gonagra* (F.). Bulletin. Entomol. Res., 49: 385-404.

Delobel A (1989). Influence of pod peanut (*Arachis hypogaea*) and imaginal feeding on oogenesis, mating and spawning in the weevil Caryedon serratus. Entomologia Experimentalis Applicata., 52: 281-289.

Delobel A (1995). The shift of *Caryedon serratus* (Ol.) from wild Caesalpiniaceae to groundnuts took place in West Africa (Coleopter : Bruchidae*). J. Stored. Prod. Res., 31: 101-102.

Delobel A, Matokot L (1991). Control of groundnut insect pests in African subsistence farming. In: Fleurat-Lessard, F, Ducom, P. (Eds). Conference of Stored Product Protection, Bordeaux, pp. 1599-1607.

Delobel A, Sembène M, Fédière G, Roguet D (2003). Identity of groundnut and tamarind seed-beetles (Coleoptera: Bruchidae : Pachymerinae), with the restoration of *Caryedon gonagra* (F.). Annales de la Société Entomologiques de France, 39: 197-206.

Dick KM (1987). Losses caused by insects to groundnuts stored in a warehouse in India. Trop. Sci., 27: 65-75.

Estoup A (1995). Contribution of microsatellite markers to study genetic variability in two social insects, the honeybee (*Apis mellifera* L) and bumblebees (Bombus terrestris Latreille) of the colony to the species. Thesis University Paris Sud, Centre d'Orsay. p. 69

Estoup A, Solignac M, Cornuet JM (1994). Precise assessment of the number of patrilines and of genetic relatedness in honeybee colonies. Proc. R. Soc. Lond. B., 258: 1–7.

Farris JS (1994). Methods for computing Wagner trees. Systematic Zool., 34: 21-34.

Felsenstein J (1985). Confidence of limits on phylogenies: an approach using bootstrap. Evolution, 39: 783-791.

Felsenstein J (1993). PHYLIP (Phylogenetic Inference Package), version 3.5c. Department of Genetics, University of Washington, Seattle. p. 145.

Feytaud J (1924). Les insectes de l'arachide. Rev. Zool. Agric. Appl., 4: 85-92.

Gillier P, Bockelée-Morvan A (1979). Stock protection against peanut insect. Oilseeds, 3: 131-137.

Guiraud X (1938). The Senegalese peanut. Libr. Tech. Econ, Paris, p. 269

Hedges LV (1992) Modelling publication selection effects in meta-analysis. Stat. Sci., 7: 246–255.

Johnson CD (1966). *Caryedon gonagra* (Fab.) established in Mexico. Pan-Pacific Entomol., 42: 162-166.

Johnson CD (1986). *Caryedon serratus* (Olivier) (Bruchidae) established in northern and southern America with additional host and locality recorded from Mexico. Coleopt. Bull., 40: 264.

Maddison WP, Maddison DR (1992). MacClade: Analysis of Phylogeny and Character Evolution. Version 3.0. Sinauer. Associates, Sunderland, Massachusetts, pp. 398.

Matokot L Mapangou Divassa-S, Delobel A (1987). Evolution of populations *Caryedon serratus* (Ol.) in stocks of peanut in the Congo. Tropical Agric., 42: 69-74.

Ndiaye S (1991). The groundnut bruchid agroecosystem in central-western Senegal: Contribution to the study of contamination in the field and in the stocks of peanut (*Arachis hypogaea* L.) Caryedon serratus (Ol.) (Coleoptera -Bruchidae); role of legumes wild hosts in the cycle of this weevil. Thesis University of Pau and Pays de l'Adour. p. 96

Nilsson JA, Johnson CD (1992). New host, *Bauhinia variegata* L., and new locality records for *Caryedon serratus* (Olivier) in the New World. Pan-Pacific Entomol., 68: 62-63.

Olivier AG (1790). Methodical Encyclopedia, dictionary of insects. Vol 5. Pankouke, Paris.

Perhaut Y (1976) Oilseeds in West African countries into the Common Market Associates. Production, Trade and Transformation Products. Champion, Paris, France, p. 890.

Possada D, Crandall KA (2001). Selecting the Best-Fit Model of Nucleotide Subtitution, Evolution. 50: 580-601.

Raymond M, Rousset F (1995). Genepop (version 1.2), population genetics software for exact tests and ecumenicism. J. Heredity, 86: 248-249.

Robert P (1985). A comparative study of some aspects of the reproduction of three *Caryedon serratus* strains in presence of its potential host plants. Oecologia (Berlin), 65: 425-430.

Roubaud E (1916). Insects and degeneration of peanuts in Senegal. Memory Study Committee historical and scientific French West Africa, 1: 363-438.

Saitou N, Nei M (1987). The neighbor-joining method: A new method for reconstructing phylogenetic trees. Mole. Biol. Evol., 4: 406-425.

Sembene M, Delobel A (1996). Morphometric identification of populations of Sudan-Sahelian groundnut bruchid, *Caryedon serratus* (Olivier) (Coleoptera Bruchidae). J. Afr. Zool., 110: 357-366.

Sembène M, Brizard JP, Delobel A (1998). Allozyme variation among populations of groundnut seed-beetle C*aryedon serratus* (Ol.) (Coleoptera: Bruchidae) in Senegal. Insect Sci. Appl., 18: 77-86.

Sembène M, Delobel A (1998). Genetic differentiation of groundnut seed-beetle populations in Senegal. Entomologia Experimentalis et Applicata, 87: 171-180.

Sembène M, Vautrin D, Silvain JF, Rasplus JY, Delobel A (2003). Isolation and characterization of polymorphic microsatellites in the groundnut seed beetle, *Caryedon serratus* (Coleoptera, Bruchidae). Mole. Ecol. Notes, 3: 299-301.

Sembène M, Rasplus JY, Silvain JF, Delobel A (2008). Genetic differentiation in sympatric populations of the groundnut seed beetle, *C. serratus* (Coleoptera: Chrysomelidae): new insights from molecular and ecological data, Int. J. Trop. Insect Sci., 28(3), 168-177.

Sembène M, Kébé K, Delobel A, Rasplus JY (2010). Phylogenetic information reveals the peculiarity of *Caryedon serratus* (Coleoptera, Chrysomelidae, Bruchinae) feeding on *Cassia sieberiana* DC (Caesalpinioideae). Afr. J. Biotechnol., 9 (10): 1470-1480.

Shimodaira H, Hasegawa M (1999). Multiple comparisons of log-likelihoods with applications to phylogenetic inference. Mole. Biol. Evolution, 16: 1114-1116.

Sokal RR, FJ Rohlf (1981). Biometry. 2nd Ed. W.H. Freeman & Co, New York, San Francisco. 859 pp.

Swofford DL, Berlocher SH (1987). Inferring evolutionary trees from gene frequency data under the principle of maximum parsimony. Syst. Zool., 36: 293-325.

Swofford DL (2001). PAUP*. Phylogenetic Analysis Using Parsimony (*and other methods), Version 4. Sunderland, Massachusetts : Sinauer Associates.

Takezaki N, Nei M (1996). Genetic distances and reconstruction of phylogenetic trees from microsatellite DNA. Genetics 144: 389-399.

Templeton AR (1983). Phylogenetic inference from restriction endonuclease cleavage site maps with particular reference to the evolution of humans and the apes. Evolution, 37: 221-244.

Thompson LG, Mosley-Thompson, E, Davis M, Lin PN, Yao T, Dyurgerov M, Dai J (1994). ``Recent warming": ice core evidence from tropical ice cores with emphasis on central Asia, Global Planetary Change, 7: 145-156.

Weir BS, Cockerham CC (1984). Estimating *F*-Statistics for the analysis of population structure. Evolution, 38: 1358-1370.

Wright S (1931). Evolution in Mendelian populations. Genetics, 16: 97-159.

19

Study on prevalence of internal parasites in semi-intensive dairy production system of Sudan

Awad G. Mohammed[1], Atif E. Abdelgadir[2]* and Khitma H. Elmalik[2]

[1]Private sector, Khartoum, Sudan.
[2]Department of Preventive Medicine and Veterinary Public Health, Faculty of Veterinary Medicine, University of Khartoum, Sudan.

A cross sectional study was conducted in the dairy cattle of Al-Rodwan dairy project in Omdurman town during the three different seasons of the year. The results of the faecal examinations (n-290) showed that the prevalence of the internal parasites was 16, 8.42, and 7.36% for dry cool, dry hot, and wet hot season, respectively. The prevalence of coccidiosis was found to be 13, 4.21, and 2.10% for dry cool, dry hot, and wet hot season, respectively, while the prevalence of Fasciolosis was 1, 4.21, and 4.21% for dry cool, dry hot, and wet hot season, respectively. Statistically, no association between season and the prevalence of internal parasites ($P > 0.05$). A positive association ($P < 0.01$) between the milk yield and the occurrence of internal parasites was observed (infection with internal parasite reduce milk yield of the animal). Similarly, an association was recorded for the breed and age of the animal with infection of internal parasites ($P < 0.05$). Application of odds ratio (OR) indicated that breed was considered to be a protective factor (OR = 0.294), while age of the animal was considered to be a risk factor (OR = 3.638) for presence of internal parasites.

Key words: Al-Rodwan dairy project, internal parasites, season.

INTRODUCTION

The high needs for animal proteins in tropical countries, especially milk and milk products in recent years, oriented the producers to import high milk foreign breeds to meet human consumption. However, the high susceptibility of crossbred cattle to internal parasites is regarded as an important role in dissemination of health problems in dairy farms. The effects of parasitic diseases on livestock include mortality losses, condemnation of meat, weight loss, and depreciation of animal products and reduced resistance to other diseases as well as high expenditure on drugs. Moreover, helminthes of the gastrointestinal tract are a major cause of reduced productivity in ruminants throughout the world (El Bihari et al., 1974).

In Sudan, in spite of the large animal population still the dairy products do not satisfy the national demand. This is why the dairy industry faces many problems. For instance, parasitic diseases were known to exist in the country, way back as 1902. However, most cases reported were based on the tentative diagnosis, rarely supported by laboratory examination. In reviewing the available literature of Sudan veterinary services, no record was encountered while dealing with parasitic infestations or infections in dairy cattle in the different seasons of the year in Khartoum State. Therefore, the aim of this study was to provide basic information regarding the presence of internal parasites of dairy cattle in Omdurman district among different seasons (Karib, 1961).

MATERIALS AND METHODS

Study area

A cross sectional study was conducted in El-Rodwan Dairy Project which was located at the north western periphery of Omdurman town and regarded as one of the most important sites of the semi-

---

*Corresponding author. E-mail: atifvet@yahoo.com.

Table 1. Description of the study population in El-Rodwan dairy project during the three different seasons.

Unit	Season frequency (%)		
	Dry cool	Dry hot	Wet hot
Total No. of animals examined breed	100	95	95
local breed	11 (11%)	11 (11.58%)	10 (10.53%)
cross breed	89 (89%)	84 (88.42%)	85 (89.47%)
Age (years)			
<1	33 (33%)	28 (29.47%)	28 (29.47%)
1-3	4 (4%)	4 (4.21%)	40 (42. 10%)
> 3	63 (63%)	63 (66.32%)	27 (28.42%)
Milk yield (kg)			
< 4	39 (39%)	34 (35.79%)	34 (35.79%)
4-8	39 (39%)	39 (41.05%)	39 (41.05%)
>8	22 (22%)	22 (23.16%)	22 (23.16%)

intensive systems for milk production in Khartoum State. The climate of the study area is an arid type which is characterized by a wide range in daily and seasonal temperature. A temperature of 45°C may occur during the summer with hot dry weather and low humidity. During winter the weather is cool and dry with a mean daily temperature of 24°C. The maximum rainfall is from mid July to mid September, in this season there is an increase in relative humidity with a maximum of 68% in August.

Sampling

The study animals were cattle kept in El-Rodwan Dairy Project. For the purpose of sampling selection of pens was done according to cluster sampling method (two stage cluster) as described by Thrusfield (1995). Selection of clusters (10 pens) was done randomly and within each pen only 10% of the herds were sampled randomly. Description of the study population is shown in Table 1.

Faecal samples collection

Two hundred and ninety faecal samples were collected during the three seasons of the year. Accordingly, 100 faecal samples were collected during the dry cool season, 95 during the dry hot season, and 95 during the wet hot season). Samples were collected straight from the rectum of the animal or from the ground only if the animals were seen passing out their faeces. The faeces were then collected in plastic bags, labeled and immediately transferred to the laboratory for fecal examination (Angus and Todd, 1978).

Faecal examinations

Flotation method

This test was used to detect the presence of the eggs of nematodes and cestodes, as well as oocysts protozoa. Two to three grams of faeces were taken in a mortar and emulsified with 42 ml salt solution. The suspension was then poured through a tea sieve into a beaker to remove the large particles. The sieved suspension was then poured in a test tube. More of salt solution was added into the test tube until it was completely filled and then covered with a cover

slip. The cover slip was removed after 20 min and it was placed into a clean slide and examined under the microscope (Angus and Todd, 1978).

Sedimentation method

This test was used for detecting those eggs which do not float well in available flotation solutions. Those are the operculate eggs such as fluke infestation, Fasciola, Paramphistomes and Schistosoma. Two to three grams of faeces were taken in a mortar and emulsified with 42 ml tap water. They were grounded with pestle and mixed well. The suspension was then poured through a tea sieve into a beaker to remove the large particles. The sieved suspension was then poured in a centrifuge tubes and centrifuged at 1500 rpm for two min (this was the first wash). The dirty supernatant was poured off and re-suspended in water and centrifuged at 1500 rpm for two min. This was repeated four times till the supernatant fluid was clear. A bit of the deposit was taken and smeared on slide covered and examined under the microscope (Angus and Todd, 1978).

Data analysis

Stata 6.0 for Windows 98/95/NT was used for data analysis. Chi-square (χ^2) was used for assessing the statistical associations of various factors with presence of internal parasites. Logistic regression model was employed only for those factors which gave statistical significant by using chi-square (χ^2). Student t-test was employed to find out the relationship between milk yield and infection with internal parasites.

RESULTS

The results of microproscopic examinations from dairy cattle at El-Rodwan dairy project using flotation and sedimentation tests (n = 290) showed that the presence of internal parasites was 16% (n = 100) during the dry cool season, 8.42% (n = 95) during the dry hot season and 7.36% (n = 95) during the wet hot season.

Table 2. Prevalence of internal parasites infection detected during the study period.

Season	No. examined	Results			Over all prevalence n (%)
		Coccidia species (%)	*Fasciola* species n (%)	*Parmphistomum* species n (%)	
Dry cool	100	13 (13%)	1 (1%)	2 (2%)	16 (16%)
Dry hot	95	4 (4.21%)	4 (4.21%)	0 (0.0%)	8 (8.42%)
Wet hot	95	2 (2.10%)	4 (4.21%)	1 (1.05%)	7 (7.36%)

Prevalence of coccidiosis was observed during the three different seasons of the year given high prevalence of 13% (n = 100) during the dry cool season. While, prevalence of fasciolosis was low 1% (n = 100) during the dry cool season (Table 2). There was no effect of the season on the presence of internal parasites (P >0.05). A positive association (P <0.01) between the milk yield and the occurrence of internal parasites was observed. An association was obtained for the breed and age of the animal with infection of internal parasites (P <0.05). Application of odds ratio (OR) indicated that breed was considered to be a protective factor (OR = 0.294), while age of the animal was considered to be a risk factor (OR = 3.638) for presence of internal parasites.

DISCUSSION

The presence of internal parasites in El-Rodwan dairy project was mostly due to the poor hygiene in the pens resulting from crowd of animals in the center of the pen where there was a partial shade and this made it difficult to achieve thorough removal of animal dung. Moreover, animals were fed on fodder purchased from the market which increased the risk of infection with internal parasites as contamination with the infective stages can happen at any point as well as introduction of new animals particularly from areas known to be endemic for internal parasitic infections such as Gezira and White Nile areas.

Our study revealed that there was no relationship between the season and the occurrence of internal parasites. This result is not in line with a study conducted in Central Kenya by Waruiru et al. (2000) who stated that the season had a significant influence on the prevalence and intensity of helminth and coccidial infections in dairy cattle and they indicated that the higher intensity of infection with helminth and coccidia was found in the wet season. Similarly, the same authors stated that the total worm burden in the animals were highest during the rainy season and lowest during the dry season. This disagreement was attributed to the type of husbandry and management of the dairy cattle as most of these studies were conducted in pastoral production systems while our

study was done in closed or semi-intensive production system. A positive association was obtained between the age of the animal and the presence of internal parasites. This result was confirmed by Duval (1997) who stated that the age as well as the weight of animal determines susceptibility to infection with parasites. Young animals do not have strong immunity to parasitic infection during the first year in pasture. He also revealed that adult animals are much less susceptible to most parasites, unless they are in poor living conditions. A significant association was obtained for the breed of the animal and the presence of internal parasites. This result was confirmed in Uganda by Magona and Mayende (2002), who explained that infections with fasciola and gastrointestinal nematodes were higher in the exotic breed compared to the local breeds. Furthermore, another study conducted by Duval (1997) revealed that an animal which had never been exposed to infections with worms cannot develop resistance and immunity. Local breeds have strong ability to prevent the establishment or limit the subsequent development of parasitic infection due to the previous continuous exposure to worm infection.

However, the limited number of local breed included in this study could not be taken a definite reflection of breed susceptibility. Also the cross–bred animals have varying ratios of foreign blood. There was a highly significant association between the milk yield and the presence of internal parasites. This result was confirmed by Bliss and Todd (1976) who demonstrated increased milk production after treatment of dairy cows, thereby passing fewer than 10 epg of faeces. Also they demonstrated that milk production was suppressed in cows given 200.000 trichostrongylid larvae when the larvae were administered in the first 90 days of lactation.

In conclusion the results of this study showed that infections with internal parasites were common during the different seasons of the year in the selected dairy cattle in the study area. Many investigations on internal parasites in dairy cattle had been documented from different production systems in Sudan. Saad (2004) reported infection with fascioliasis in White Nile and Gezira states. He also stated that paramphistomiasis was common in White Nile and Gezira States. Based on results of this study it could be concluded that infections with internal parasites were prevalent in the dairy cattle of Al Rodwan dairy project. The seasonal variations had no influence on the prevalence of internal parasites in a closed dairy production system if other factors such as good management and adequate nutrition were controlled. Reduced milk production level is the most important feature for infection with internal parasites.

ACKNOWLEDGEMENT

The authors thank Mr. Ahmed Abdel Wahid for his keen help in the laboratory.

REFERENCES

Angus D, Todd D (1978). Veterinary Helminthology, 2nd ed. Butler and Tanner Ltd, Frome and London, Great Britain.

Bliss DH, Todd AC (1976). Milk production by Vermont dairy cattle after deworming, Vet. Med. Small Anim. Clin., 71: 1251-1254.

Duval J (1997). The control of internal parasites in cattle and sheep, MSc thesis, Mt Gill University, Canada.

El Bihari S, Gadir FA , Suleiman H (1974). Incidence and behavior of microfilariae in cattle. Sud. J. Vet. Anim. Husb., 15(2): 82-85.

Karib AA (1961). Animal trypanosomiasis in the Sudan. J. Vet. Sci. Anim. Hus., 2: 39-46.

Magona JW, Mayende JS (2002). Occurrence of current trypanosomiasis, theileriosis, anaplasmosis and helminthosis in Friesian, Zebu and Sahiwal cattle in Uganda, Onderstepoort J. Vet. Res., 69: 133-140.

Saad AA (2004). Investigation of Diseases in Diary Cows in the White Nile and Gezira States, Sudan, MSc thesis, University of Khartoum.

Thrusfield M (1995).Veterinary Epidemiology, 2nd ed. Blackwell Science Ltd. UK.

Waruiru RM, Kyvsgaard NC, Thamsborg SM, Nansen P, Bogh HO, Munyua WK, Gathuma JM (2000). The prevalence and intensity of helminth and coccidial infections in dairy cattle in central Kenya, Vet, Res, Commun., 24: 39-53.

20

Phosphatase profile in *Manihot escculenta* induced neurotoxicity; role in neuronal degeneration in the brain of adult Wistar rats

O. M. Ogundele[1]*, E. A. Caxton-Martins[2], O. K. Ghazal[3] and O. R. Jimoh[2]

[1]Trinitron Biotech LTD, Science and Technology Complex, Abuja, Nigeria.
[2]Department of Anatomy, University of Ilorin, Kwara State, Nigeria.
[3]Unilorin Stem Cell Research Laboratory, Ilorin, Kwara State, Nigeria.

As a general trend, a change in cell activity and morphology is usually depicted as biochemical differentiation occurring structural differentiation, cell migration and even cell death. In cassava induced neurotoxicity, several substance has been identified to be naturally occurring in cassava and Cyanogenic glycosides or other phytotoxins which has been found to have neurotoxic effects; Scopoletin, afflatoxin and CYANIDE as described by Osuntokun (1981) and Ernesto et al. (2002). These substances elicit toxicity by accumulation over a period of time, or exposure to high concentrations from environmental contamination of water, food substance and sometimes occupational exposure. Cassava has been found to be neurotoxic as its cyanide component is capable of inducing oxidative stress by blocking cytochrome c oxidase (CcOX) and inhibition of other metalloenzymes. In this study we investigated the profile of acid phosphates (ACP) and alkaline phosphatase (ALP) in the brain tissue of adult wistar rats treated with varying dose of cassava diet for a period of 60 days. ACP serves as a biochemical marker for lysosomal activity while ALP indicates membrane transport and integrity in the neuronal architecture. The brain tissue were excised and homogenized in 0.25 M sucrose (Sigma: β-D-Fructofuranosyl-α-D-Glycopyranoside) and centrifuged in Multifuge 3SR+ by ThermoScientific. The supernatant was obtained and assayed for ACP and ALP change in optical density per minute.

Key words: Alkaline phosphatase, acid phosphatase, neurodegeneration, membrane, lysosomes, cassava, cyanide, cytochrome c oxidase.

INTRODUCTION

Cassava (*Manihot escculenta*) is a major food crop in the tropics and sub-tropics as it serves as a source of cheap calorie food (Osuntokun, 1981; Oke, 1979; De la cruz et al., 2009; Ernesto et al., 2002). In cassava endemic regions of Uganda, Tanzania and Niger, various neurological disorders have been reported to include Tropical Ataxic Neuropathy (TAN), spastic endemic paraparesis (Konzo), gradual loss of vision and other symptoms resembling those observed in parkinsonism (Osuntokun, 1981; El-Ghawabi et al., 2005; Soler-Martin et al., 2010). Cassava however contains several cyanogenic glycosides (Mathangi et al., 2000; Lee et al., 2009); In

the animal system, the major defense of the body against cyanide is the enzyme rhodanese (Tor-Agbidye et al., 1999) which is capable of converting cyanide (CN⁻) to thiocyanate (SCN), the major form in which cyanide is being excreted. This reaction will occur in the presence of the S-group present in thiosulphate and S-containing amino acid (SAA) like cysteine and tyrosine, cysteine has been found to react with free cyanide to generate 2-Iminothiazoldine-4-Carboxylic acid which could be found in saliva of cyanide intoxicated animals (Mathangi and Namasivayam, 2000).

Thus, animals fed on low protein diets will elicit more toxic effects compared to those fed on protein diets especially those containing SAA (Mathangi et al., 2000).

Cyanide in cassava is released either as free cyanide or HCN, the most reactive form of cyanide is CN but the state depends on the pH, salinity and temperature of the

*Corresponding author. E-mail: mikealslaw@hotmail.com.

medium (Chen et al., 2003; Lee et al., 2009). At pH 7.0, 99% of cyanide will exist as HCN, at pH 11 about 99% of cyanide will exist as CN while an equilibrium has been observed for pH range of 9.0 - 9.3 (Li et al., 2000). The neurotoxic effects of cassava can not be attributed to a single substance as it contains several substances capable of generating neurotoxicity (Ernesto et al., 2002; Denison et al., 2009; Dorea, 2003). The cyanide released for cassava generates oxidative stress by inhibiting Cytochrome c oxidase (CcOX) a terminal enzyme in the electron transport chain, Cyanide also inhibit energy production by binding to the three states of the Binuclear centre heme a3-CuB formed by combination of CcOX to molecular oxygen released form water, this will thus prevent production of ATP at complex IV (Bonfoco et al., 1995; Bathachanya and Tulsawani, 2008).

The cellular mechanism of this inhibition is associated with generation of heat, leakage of proton into then mitochondria matrix and conversion of 20% of molecular oxygen to ROS (reactive oxygen species) this includes superoxide ions and are measured in the as cytoplasm as superoxide dismutase (SOD) (Bove et al., 2005; Nelson, 2006). ROS reacts with accumulated NO at complex I and III to generate RNS, NO are naturally occurring endogenous modulators of cellular activity but if present in high levels could trigger toxic pathways or cell death, the mode of cell death observed in neurons for different toxicology experiments have been found to correspond with the level of accumulated NO and ROS (Gruetter et al., 2001; Lee et al., 2009). Several models have been used to describe the mode of cell death in neurons, Isom and Way (1984) reported that elevated levels of cerebral calcium initiate a caspase system in cell death while secondary autophagic bodies of lysosomes have been found in other experiments by Osuntokun (1981), Li et al. (2000) and Dorea et al. (2003). Although they measured lysosomal activity using β-glucoronidase as indicator, this results has been found not to be weight (Gunasekar et al., 1996) dependent and also variations has been found in different brain regions this however explains the models involving region specific cytotoxic pathways in the brain Isom et al., 1999; Bathachanya and Tulsawani, 2008).

The study describes the role of acid and alkaline phosphatase in the various cellular changes observed in the cells for various dose of treatment and the adopted mode of cell death observed in the cells of the brain.

MATERIALS AND METHODS

ALP assay kits (Sigma, Germany) and ACP Assay kit (Sigma, Germany), Sucrose (sigma Aldrich, Germany).

Tissue preparation

The occipital region, superior colliculus and lateral geniculate body of adult Wistar rats fed with 2.5, 10, 20 and 30 g of cassava per animal/ day orally for 60 days alongside a control group treated with

0.25 M sucrose. The tissues were homogenized in 0.25 M sucrose at 4°C and then centrifuged at 10,000 rpm for 20 min using multifuge 3SR+; the supernatant was collected and assayed using the substrate technique using the spectrophotometer (Jenway, 5550) (Enulat et al., 2010). The data was analyzed in SPSS 15.0 software to determine the analysis of variance.

Alkaline phosphatase

The working reagent is composed of magnesium chloride 0.625 mMol/L, alkaline phosphate 2 ml, p-nitrophenyl phosphate 50 mMol/L pH7.8. The reagent was linear up to 700 μl. A blank working reagent was used in a couvette and was discarded since the absorbance has exceeded 405 nm. 1000 μl of the working solution was mixed with 20 μl of the sample, the solution was incubated at 37°C for 30 min and the absorbance was measured at 60, 120 and 180 s respectively, after each 60 s the solution was mixed to check for change in optical density (ΔOD/min) using a dilution factor of 1:50, the ALP activity was then expressed as ΔOD/MinX2750 (Fishman and Baker, 1998; Wintola et al., 2010).

Acid phosphatase

ACP activity was determined using the substrate method described for ALP, an additional dye. The working reagent was freshly prepared and its composed of Acid phosphate10 mMol/L, fast red 6 mMol/L, the reagent was linear up to 150 μl. one tablet of acid phosphate was dissolved in freshly prepared citrate buffer pH 5.2, this reagent was found to be stable for 2 days at 2 - 8°C; 100 μl of the working solution was added to 10μl of the titrate solution (fast red) (Nachilas et al., 1989) then 100 μl of the sample was added, the mixture was then vortexed and incubated at 37°C for 6 min. Change in OD was measured at intervals of 60 s for 180 s. Activity was measured as a factor of ΔOD/MinX750 (Baker, 1998; Enulat et al., 2010; Volbracht et al., 2009).

RESULTS

In Group 1 which is a high dose treatment group, ALP activity followed a sinusoidal pattern with the activity at the 60 s being 0.862, it increases at 120 s 0.891 and then falls at 180 s, Group 4 showed a rise in ALP at 60 s an decrease from this initial value was observed at 120 and at 180 s ALP activity increased but did not get to value observed at 60 s, group 2 showed a decline in ALP (0.636, 0.576, 0.519). Group 3 showed ALP activity below the levels observed in the control (60 s: group 3 (0.016) and group 5/control (0.238). The high dose treatment groups (Group 1, 2 and 3) showed an irregular pattern in ALP activity. Comparing the factors for each of the groups, only Group 3 which received a moderate dose showed ALP activity below that of the control, also the activity observed in Group 4, low dose group is higher than the factor in Group 2, which suggests that the activity of ALP in a model system is dose dependent in such a way that extreme doses triggers increase in membrane activity and synthesis rather than moderate does which generate similar effects but below those elicited by the extreme doses (Table 1: Group 1 and 5). ACP activity in Group 1 is similar to ALP activity characterized by a rise then a fall at

Table 1. ALP activity.

Seconds Group	60	120	180	Average	Factor
1	0.862	0.891	0.808	0.854	2,348.25
2	0.636	0.576	0.519	0.577	1,586.75
3	0.016	0.013	0.086	0.115	316.25
4	0.796	0.667	0.676	0.713	1,960.75
5	0.238	0.236	0.273	0.249	684.75

Table 2. ACP activity.

Seconds Group	60	120	180	Average	Factor
1	0.279	0.500	0.309	0.309	272.75
2	0.555	0.364	0.291	0.403	302.25
3	0.350	0.301	0.295	0.315	236.25
4	0.373	0.388	0.327	0.363	272.15
5	0.460	0.475	0.533	0.491	368.25

180 s, Group 2 and 3 (Table 2) shows a decrease from 60 - 180 s while Group 4 has similar pattern to Group 1 again this further re-affirms the hypothesis that phosphatase activity are elicited by extremes doses rather than moderate doses as described in this study for ALP.

All groups gave a factor below the control value (368.25) and the level of activity in Group 1 and 4 are almost equal (272.75 and 272.15), respectively. We can, however, deduce from these result that; 1. Extreme exposure to cassava causes neurotoxicity by altering the activity across and synthesis of membranes (rise in ALP) and suppressing lysosomal activity as seen in decrease in ACP factor; 2. Moderate exposure initiate degeneration by increasing the level of ALP to a point 25% less that Extreme dose treatment and having less suppressing effects on the ACP activity compared to the extreme dose group which was found to have a greater ACP suppressing activity, thus, characterized by an apoptosis-necrosis continuum. Rise in ALP indicates necrotic activity while partial suppression of lysosomal activity; increased activity compared to the extreme dose groups (302.25) indicates tendency of physiological cell death rather than programmed cell death. these two models suggests that extreme doses (high or low) have a greater tendency of inducing cell death by necrosis while moderate doses induces cell death by partial apoptosis and Necrosis (Figures 1 - 3).

DISCUSSION

The role of phosphatase is completely opposite to that of kinase and phosphorylase, which add phosphate groups to proteins by the help of energy-supplying molecules ATP (adenosine triphosphate) (Baker, 1998). The addition of a phosphate group can set off a protein-protein interaction. This also can activate or deactivate the function of an enzyme. Phosphatase is an important constituent of many biological processes involving genetic transduction because it can regulate the proteins to which they are attached (Fishman and Baker, 1998; Bathacharya and Tulsawani, 2008; Wintola et al., 2010). The alkaline phosphatases are determined by at least three gene loci, which can be sharply distinguished one from another by their sensitivity to inhibition with various amino acids and peptides and by thermostability (Becker et al., 2000). Alkaline phosphatase is present in the brains of guinea pig, rat, mouse, hamster, squirrel, rabbit, cat, sheep, cow, tamarin, baboon, and man. The gene locus coding for alkaline phosphatase in all these brains is the liver/ bone/kidney locus, as indicated by thermostability studies and by inhibition studies with L-phenylalanine, L-homoarginine, and L-phenylalanylgly-cylglycine. The average brain alkaline Phosphatase activity is about 35% of the average for the livers and only 7.2 and 4.4% of the average kidney and placental activities, respectively (Ven-Watson and Ridgway, 2007).

During growth and development, brain alkaline Phosphatase activity decreases in the mammals studied. The amount of change is tissue- and species-dependent. Phosphatases are enzymes that act as catalysts in the hydrolysis of organic phosphoric acid. A few classes of phosphatase enzymes are found to be involved with many common physiological disorder (Van-Watson and Ridgway, 2007; Soler-Martin et al., 2009; El-Ghawabi and De Filipo, 2005). This fact indicates that phosphatases control many fundamental processes in cellular

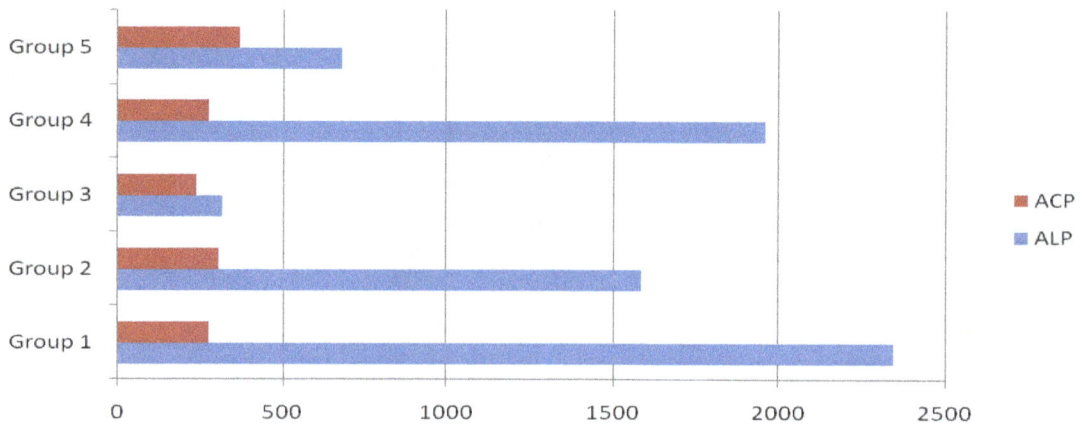

Figure 1. Curve demonstrating varying activity levels for ALP and ACP for each of the groups and withdrawal effects from group 6 - 8.

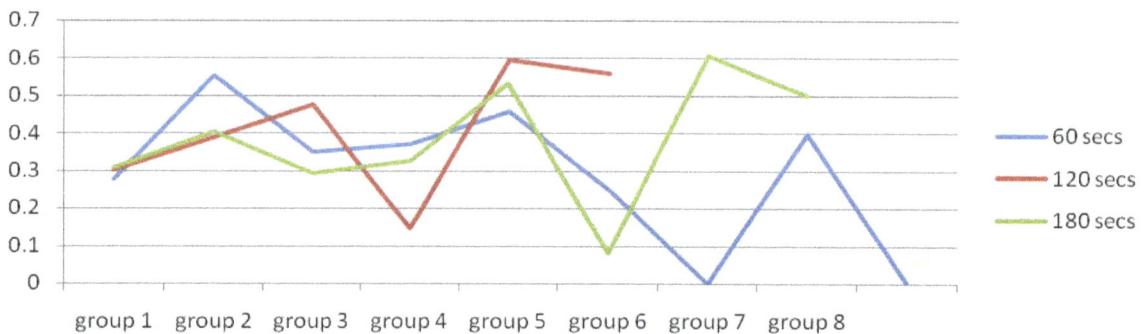

Figure 2. Curve demonstrating varying activity levels for ACP for each of the groups and withdrawal effects from group 6 - 8.

physiology. Phosphatases also can avert genetic changes. They are degenerated in response to DNA damage, thus preventing chromosomal abnormalities. Phosphatases also enhance the progression of cell cycle, in the brain, Phosphatase are found in multiple compartments of neuralgia and neuronal and play an important role in various neuronal functions (that is, pre-and-post synapses and gene expression). These functions are also important for maintaining the coordinated action of signaling cascades (Beckner et al., 2000).

As a general observation increase in size of the vacuolar spaces in the tissue conformed with an increase in activity of ACP (Figure 3) such that different regions of the brain reacts differently at the same treatment dose (Figure 3), in 3A (primary cortex) shows the presence of spaces around the cells while fibrous layer has become predominant in 3B (lateral geniculate body), while in 3C, presence of reduced metachromasia in the cells coupled with an increased cell diameter. After staining with Cresyl fast violet .This however, explains the variation in the cytotoxic pathway adopted by different brain regions. This was also similar to the findings of Solomonson et al. (1981) although they went further to examine these parameters at higher doses of cassava diet over a period

of 6 months but found out that the effects are too deleterious and irreversible on a long term basis. However, Osuntokun (1981) reported a rise in the level of β-Glucoronidase at low dose treatment as an indicator of lysosomal activity which was also found to be proportional to sulphur excretion. The pattern of change in the activity of the phosphatase enzymes is in a zigzag manner (Figure 1:ALP) and (Figure 2: ACP) as it explains the feedback mechanisms involved in the cytotoxic pathways (Isom et al., 1999), higher dose causes cell death and a decrease in the enzyme activity while lower doses will inhibit the ALP causing membrane malfunction and influx of the calcium ions (Di Filipo et al., 2008), internal build up of ROS and NO will activate caspase cascades which affects apoptotic pathways at moderate doses or stimulate necrosis at higher doses (Phrabakharan et al., 2007).

Available reports of toxicological studies lack information on the level of intake of cyanogenic glycosides or on the amount of hydrogen cyanide potentially released. No long-term toxicity or carcinogenicity studies were available. However, in vitro and in vivo genotoxicity were negative. Teratogenic and adverse reproductive effects attributable to linamarin (cassava) and hydrogen cyanide

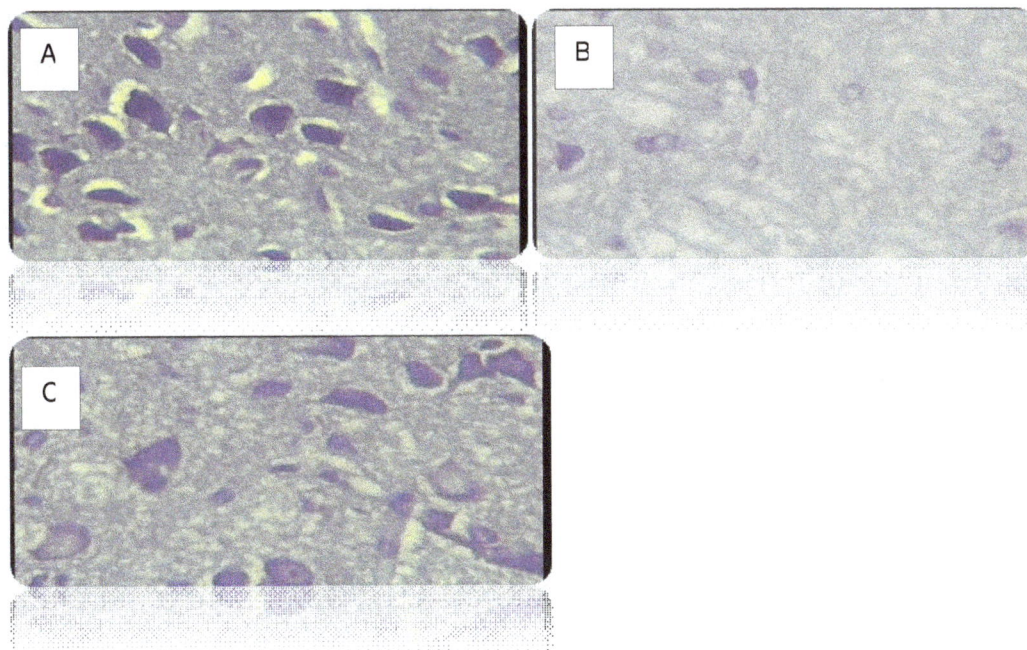

Figure 3. Structure of different parts of the brain A- Cortex, B- Lateral geniculate body and C- Superior colliculus showing different cellular changes to the same dose of treatment, this explains the variation in the cytotoxic pathway adopted by different brain regions.

were seen only at doses that also caused maternal toxicity (Ernesto et al., 2002). The toxic effects of cyanide on the thyroid (via its metabolite thiocyanate) depend on the iodine status of the test animals, as indicated earlier. On the basis of epidemiological observations, associations have been made between chronic exposure to cyanogenic glycosides and diseases such as spastic paraparesis, tropical ataxic neuropathy, and goiter. However, these observations were confounded by nutritional deficiencies, and causal relationships have not been definitely established (Osuntokun, 1981; Oke, 1979; Mathangi and Namasivayam, 2000).

Traditional users of foods containing cyanogenic glycosides usually have a basic understanding of the treatment required to render them safe for consumption. However, some products are sold commercially and are consumed by people who may not be familiar with such procedures. The EPA (Environmental protection Agency, USA) recommended that guidelines be developed to provide reliable and sensitive methods for the analysis of these foodstuffs for hydrogen cyanide releasable from cyanogenic glycosides, in order to ensure that amounts in foods as consumed do not present a hazard. Because of a lack of quantitative toxicological and epidemiological information, a safe level of intake of cyanogenic glycosides could not be estimated. However, it was concluded that a level of up to 10 mg/kg hydrogen cyanide in the Codex Standard for Cassava Flour (Varone et al., 2008) is not associated with acute toxicity (Table 3).

Conclusion

Toxicity of cassava has been found to be initiated as function of membrane malfunction and lysosomal activity thus causing degeneration of the neurons of the brain, although the effect is dose dependent as matter of general effects such that extreme doses causing cell death by necrosis and moderate doses causing cell death by apoptosis-necrosis continuum rather than solely apoptosis or necrosis, the effect is how ever non-specific and of irregular pattern. The pattern adopted in each of the cytotoxic pathways will determine the mode of cell death adopted by the neurons.

ACKNOWLEDGEMENTS

Trinitron Biotech LTD, Mr. Gerry Nash, Rosita Menezes, Helen Odedairo, Dr. Arise Olusanya.

FOOTNOTES

Ogundele, Olalekan Michael (B.Sc, M.Sc); Reasearch and Development Department, Trinitron Biotech JV, LTD. Science and Technology Complex, Abuja.

Prof E.A Caxton-Martins (FASN) Department of Cell Biology, University of Ilorin, Nigeria.

Dr.O.R Jimoh (FMSN): Department of Cell Biology, University of Ilorin, Nigeria.

Dr O.K. Ghazal (B.Sc, MB.ChB, M.Sc): Department of Cell Biology, University of Ilorin, Nigeria

REFERENCES

Baker JR, HEW H, Fishman WH (1998). The use of a chioral hydrate formaldehyde fixative solution in enzyme histocheunistry. J. Histochem. Cytochern., 6: 244-250.

Becker NH, Goldfischer S, Shin W, Novikoff AB (2000). The localization of enzyme activities in the rat brain. J. Riophys. Biochem. Cytol., 8: (149-663, 19(10).

Bhattacharya R, Tulsawani R (2008). *In vitro* and *in vivo* evaluation of various carbonyl compounds against cyanide toxicity with particular reference to alpha-ketoglutaric acid. Drug Chem Toxicol., 31(1): 149-61.

Blomgren K, Leist M, Groc L (2007). Pathological apoptosis in the developing brain. Apoptosis Review, 12(5): 993-1010.

Bonfoco E, Krainc D, Ankarcrona M, Nicotera P, Lipton SA (1995). Apoptosis and necrosis: two distinct events induced respectively by mild and intense insults with NMDA or nitric oxide/superoxide in control cell cultures. Proc. Natl. Acad. Sci. USA, 92: 7162-7166.

Bove J, Prou D, Perier C, Przedborski S (2005). Toxininducedmodels of Parkinson's disease. NeuroRx., 2: 484-494.

Chen Kk, Rose Cl, Clowes G (2003). Comparative values of several antidotes in cyanide poisoning. Am. J. Med Sci., 188: 767.

de Haro L (2009). Disulfiram-like syndrome after hydrogen cyanamide professional skin exposure: two case reports in France. J Agromed. 14(3): 382-384.

de la Cruz Cosme C, Medialdea Natera P, Romero Acebal M (2009). Amyotrophic lateral sclerosis syndrome-plus and consumption of cassava (Manihot) . Is this a new presentation of the neurotoxic motor-neuron syndrome? Neurologia., 24(5): 342-343.

Denison TA, Koch CF, Shapiro IM, Schwartz Z, Boyan BD (2009). Inorganic phosphate modulates responsiveness to 24,25(OH)2D3 in chondrogenic ATDC5 cells. J. Cell. Biochem., 1; 107(1):155-62.

Di Filippo M, Tambasco N, Muzi G, Balucani C, Saggese E, Parnetti L, Calabresi P, Rossi A (2008). Parkinsonism and cognitive impairment following chronic exposure to potassium cyanide. Mov. Disord. 15;23(3): 468-470.

Dorea JG (2003). Fish are central in the diet of Amazonian riparians: should we worry about their mercury concentrations? Review, Environ. Res. 92(3): 232-244.

El-Ghawabi G, De Flipe J (2005). A correlative electron microscopic studies of basket cells. Neuro. Sci., 17: 991-1009.

Ennulat D, Magid-Slav M, Rehm S, Tatsuoka KS (2010). Diagnostic performance of traditional hepatobiliary biomarkers of drug-induced liver injury in the rat. Toxicol. Sci. 13. [Epub ahead of print].

EPA (1990). Summary Review of Health Effects Associated with Hydrogen Cyanide, Health Issue Assessment Environmental Criteria and Assessment Office, Office of Health and Environmental Assessment Office of Research and Development, US Environmental Protection Agency Research Triangle Park, North Carolina, USA 1023-1029, 1955.

Ernesto M, Cardosso AP, Nicala D, Mirone E, Massasa F, Cliff J, Haque MR (2002). Persistent Konzo and Cyanogenic toxicity from cassava in Northern Mozambique. Acta. Trop., 82(3): 357-362.

Fishman WH, Baker JR (1956). Cellular localization of ι-glucuromsidase in rat tissues. J. Histochem. Cytochein., 4: 570-587.

Hour SJ (1998). In General Cytochernical Methods, Ed. J. F. Daniehhi, Academic Press, Inc. New York.

Gamper N, Li Y, Shapiro MS. Structural requirements for differential sensitivity of KCNQ K+ channels to modulation by Ca2+/calmodulin. Mol Biol Cell. 2005 Aug; Epub 2005 May 18, 16(8): 3538-3551.

Gruetter R, Seaquist ER, Ugurbil K (2001). Am. J. Physiol., 281: E100–E112.

Hashem MA, Mohamed MH (2009). Haemato-biochemical and pathological studies on aflatoxicosis and treatment of broiler chicks in Egypt. Vet Ital. Apr-Jun; 45(2): 323-337

Isom GE, Gunasekar PG, Borowitz JL (1999). Cyanide and neurodegenerative disease. In *Chemicals and Neurodegenerative Disease* (S. C. Bondy, Ed.), pp. 101–129. Prominent Press, Scottsdale, AZ.

Isom GE, Way JL (1984). Effects of oxygen on the antagonism of cyanide intoxication: Cytochrome oxidase *in vitro*. Toxicol. Appl. Pharmacol., 74: 57-62.

JB, John S, Flora MV, Nenad Š, Pasko R. Notch regulates cell fate and dendrite morphology of newborn neurons in the postnatal dentate gyrus PNAS 2007 104 (51) 20558-20563.

Li R, Sonik A, Stindl R, Rasnick, D, Duesberg P (2000). Aneuploidy vs. gene mutation hypothesis of cancer: Recent study claims mutation but is found to support aneuploidy. Proc. Natl. Acad. Sci. U.S.A., 97: 3236–3241.

Mathangi DC, Namasivayam A (2000). Neurochemical and behavioural correlates in cassava-induced neurotoxicity in rats. Neurotox Res., 2(1): 29-35.

Mathangi DC, Namasivayam A (2000). Effects of Chronic cyanide intoxification on memory in albino rats. Food Chem. Toxicol., 38: 51-55.

Nachlas MM, Seligman AM (1989). The histochemical demonstration of esterase. J., Vat. Cancer Inst., 9: 415-425.

Nelson L (2006). Acute cyanide toxicity: mechanisms and manifestations Review. J. Emerg. Nurs. Aug; 32(4 Suppl): S8-11.

Oke OL (1979). Some aspects of the role of cyanogenic glycosides in nutrition. Wld. Rev. Nutr. Diet, 33: 70-103.

Osuntokun BO (1981).Cassava in diet, chronic cyanide intoxification and neuropathy in Nigerians. World Rev. Nutr. Dietetics, 36: 259-339.

Soler-Martín C, Riera J, Seoane A, Cutillas B, Ambrosio S, Boadas-Vaello P, Llorens J (2009). The targets of acetone cyanohydrin neurotoxicity in the rat are not the ones expected in an animal model of konzo. Neurotoxicol Teratol. 2010 Mar-Apr; 32(2): 289-294.

Solomonson LP (1981). Cyanide as a metabolic inhibitor. In Cyanide in Biology, Vennes L and, B., Conn, E.E., Knowles, C.J., Wesley, J and Wissing, F. Academic Press, London, New York, Toronto, pp. 11-18.

Tor-Agbidye J, Palmer VS, Spencer PS, Craig AM, Blythe LL, Sabri MI (1999). Sodium cyanate alters glutathione homeostasis in rodent brain: relationship to neurodegenerative diseases in protein-deficient malnourished populations in Africa. Brain Res., 820(1-2): 12-19.

Varone JC, Warren TN, Jutras K, Molis J, Dorsey J (2008). Report of the investigation committee into the cyanide poisonings of Providence firefighters. New Solut., 18(1): 87-101.

Venn-Watson SK, Ridgway SH (2007). Big brains and blood glucose: common ground for diabetes mellitus in humans and healthy dolphins. Comp Med., 57(4): 390-395.

Volbracht C, Penzkofer S, Mansson D, Christensen KV, Fog K, Schildknecht S, Leist M, Nielsen J (2009). Measurement of cellular beta-site of APP cleaving enzyme 1 activity and its modulation in neuronal assay systems. Anal. Biochem. Apr 15; 387(2): 208-220.

Wintola OA, Sunmonu TO, Afolayan A (2010). Toxicological evaluation of aqueous extract of Aloe ferox Mill. in loperamide-induced constipated rats. Hum. Exp. Toxicol. May 24. [Epub ahead of print].

21

Study of the wildlife acarology (Acari: Oribatida) in the palm groves of Biskra

Ghezali Djelloul* and Zaydi Djamel-Eddine

Departement Zoologie Agricole et Forestier, Ecole Nationale Supérieure Agronomique
El –Harrach, Alger –Algérie.

This study, conducted in the region of Biskra, allows to define eight species of oribatid living in two palm groves of 7 and 10 years, seven species in palm groves of one and three years and finally, three species in a palm grove of five years. No species has been found in the witness palm grove. The prevailing species in all these palm groves is *Scheloribates* sp. with 450 individuals. This study shows that man's influence can be beneficial to the natural environment. Indeed, installing an oasis helps both modify the Saharan landscape and create a specific biotope that allows ground-acari to thrive. This biotope is characterized by a multitude of ecological factors that are of a great importance for these acari, they are mainly the microclimate and the feeding support.

Key words: Oribatida, palm groves, Biskra, ecological factors, oasis ecosystem, nutritional substrate.

INTRODUCTION

Semi-arid, arid, and Saharan regions are characterized by soils with variable temperature degrees that can reach very high levels. These soils will provide unfavorable conditions for the development of mites. Date palms are very important components in these settings. Their spread in almost all oases made them an important element of the landscape of these regions. Date palms are formidable ecological barriers against desertification, and the establishment of an agricultural oasis ecosystem allows a recovery of economic and ecological environments, and a gradual recovery of land and dry areas. The introduction of this tree that is associated with other trees, vegetables, and forages, forms the agricultural oasis system. Such a system will typically create an environment where ecological and environmental conditions are much better. The diversity of plant resources in palm groves are a very important ecological factor (Dajoz, 1970).

Microclimate, irrigation, and soil amendments are used to continuously provide an environment that is conducive to the development of mites. Mites require a number of factors for a favorable environment for development to be

there. The abundance, species distribution, and community structure of arthropods depend on biotic and abiotic conditions of the environment (Tousignat and Coderre, 1992). Also, environmental factors determine the distribution and multiplication of soil mites (Vikram, 1986). Many studies have been made in these areas, particularly those related to diseases and parasites (Toutin, 1977), and date palm arthropod fauna (Achoura, 2010; Hamdi, 1992), but studies on soil fauna acarology, however, remain a very poorly studied area. In this study, the impact of oasis agroecosystem on the evolution of oribatid will be investigated.

MATERIALS AND METHODS

Region of study

The study was conducted in the region (Daira) of Tolga in the wilaya of Biskra on presenting palm plantation with different ages thus providing different microclimatic and edaphic characteristics very different. The Daira of Tolga covers 1,334.10 km^2 which is 6.20% of the total area of the prefecture (21,510 km^2), Utilized Agricultural Area (UAA) represents 8.89% of the total agricultural area (SAT) (9,250 ha, of which 74.87% is date palm). The rest consists of uncultivated land and grassland meadow that cover 91% of the SAT. Tolga is located 390 km south-east of the capital

*Corresponding author. E-mail: dj_ghezali@hotmail.fr.

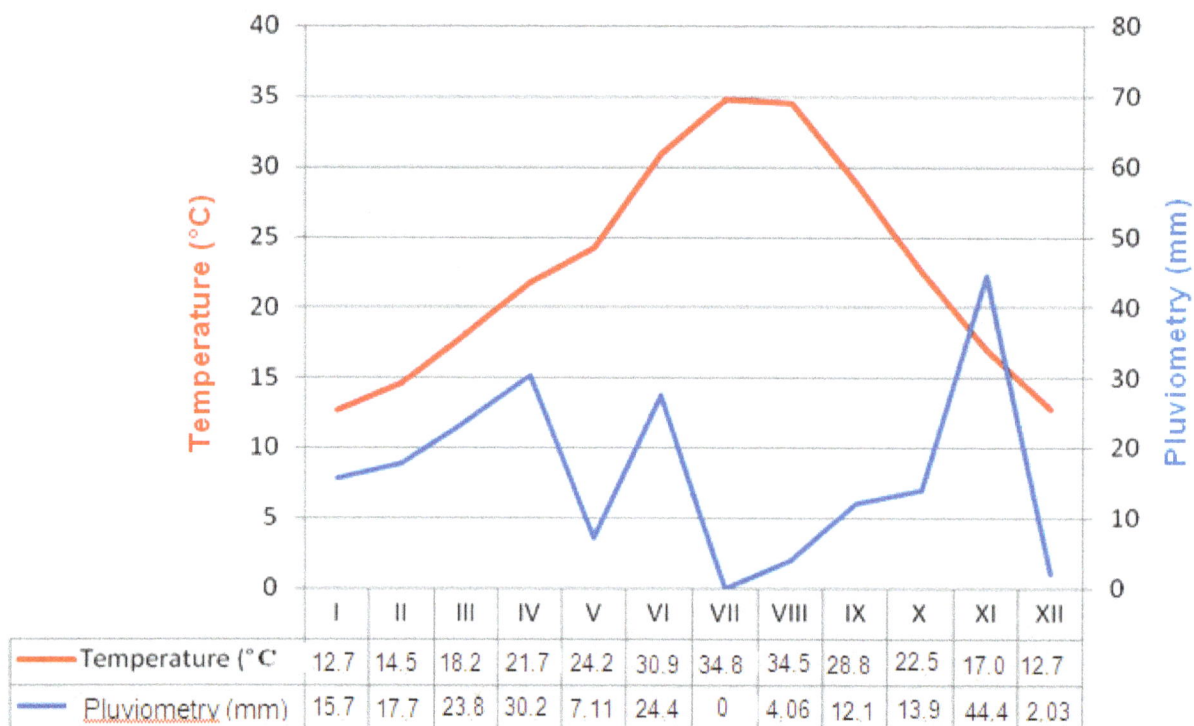

Figure 1. Climate data and ombrothérmic diagram of Biskra region in 2010.

Table 1. Relative humidity in the region of Biskra.

Month	I	II	III	IV	V	VI	VII	VIII	IX	X	XI	XII
H (%)	55.7	52.1	44.4	46.3	33.9	32.5	26.6	32	39.5	43.9	57.9	48.6

(Algiers) and 36 km northwest of the capital of Biskra wilaya. Its altitude is 128 m above sea level. It is characterized by cold winters, hot and dry in summer. Its geographical location makes it a region-oriented agro-Saharan Africa based on the vast areas of the oasis. The activity of date palm cultivation is an important component of this region, and it appears as one of the most important regions suitable for palms in Algeria. Daira of Tolga is well known worldwide for the high quality of its dates, including the noble variety "Deglet Nour."

Date palm crops

Manure fertilization is practiced by all farmers to maintain yields. However, the amendment of the manure is given in small amounts, below the needs of the generally poor soil. In the palm explored, these inputs are 33 kg for manure, sometimes within three years. The adverse effect of climatic and soil factors on date palm prosperity in Sahara are mainly a function of water availability. However, irrigation of palm is designed to ensure the amount of water required for normal development of the trees throughout the year and especially during the summer seasons in which demands are at their greatest. No prophylactic measure was provided: lack of maintenance, neglect of cleanliness in the majority of palm and chemical control dates which is rarely performed.

Climatic data

Results obtained show that the total rainfall during the year 2010 reached 198.88 mm (Figures 1 to 4). A maximum of 44.45 mm was recorded during the month of November and a zero value for the month of July. January is the coolest month with an average temperature of 12.7°C. July, in the contrary, is the hottest month with an average temperature of 34.8°C. As shown in Table 1, the average relative humidity of the air was greater than 57.9% during the months of November and was less than 26.6% for the month of July. According to the value of Q2 (13.96) calculated over ten years (2000 to 2010), the region of Biskra is localized in the Saharan bioclimatic to temperate winter (Tables 2 to 4).

Methods

Methods used are to count the quantity and quality of mite species in the various stations whereas the following samples were taken randomly. Ten samples were established for each station with three replicates and a total of thirty samples per station thus 180 per season x 2 seasons totaling 360 samples for the duration of experimentation. Each sample was taken using a square (15×15 cm) and a depth of 10 to 15 cm. Sampling was conducted during the months of April (spring season) and November (fall season),

Table 2. Bioclimatic zones of the study area.

Region	P (mm)	M (°C)	m (°C)	Q$_2$	Bioclimatic zones
Biskra	139.92	40.91	6.54	13.96	Saharan bioclimatic to temperate winter

P (mm): Pluviometry; M (°C): Maximum température; m (°C): Minimum température; Q$_2$: Rainfall quotient.

Table 3. Acarology fauna collected in the various the palms of Biskra.

Palmeraies	Species	Number (Autumn)	Number (Spring)	Total
Witness	**0**	**0**	**0**	**0**
	Oppia bicarinata	15	22	37
	Oppia neerlandica	1	12	13
P1 (1 year)	*Scheloribates* sp.	18	35	53
	Phthiracarus nitens	3	5	8
	Paleacarus sp.	2	3	5
	Galumna sp.	0	4	4
	Oppia bicarinata	15	16	31
P2 (3 years)	*Scheloribates* sp.	15	23	38
	Phthiracarus nitens	1	3	4
	Oppia bicarinata	49	35	84
	Scheloribates sp.	56	69	125
	Phthiracarus nitens	5	5	10
P3 (5 years)	*Paleacarus* sp.	48	52	100
	Galumna sp.	45	63	108
	Epilohmannia aegyptica	1	2	3
	Haplochthonius variabilis	15	5	20
	Oppia bicarinata	34	46	80
	Oppia neerlandica	22	41	63
	Scheloribates sp.	53	68	121
P4 (7 years)	*Phthiracarus nitens*	4	2	6
	Paleacarus sp.	34	53	87
	Galumna sp.	25	38	63
	Epilohmannia aegyptica	18	14	32
	Haplochthonius variabilis	11	23	34
	Oppia bicarinata	48	63	111
	O. neerlandica	38	54	92
	Scheloribates sp.	45	68	113
P5 (10 years)	*Phthiracarus nitens*	7	2	9
	Paleacarus sp.	12	9	21
	Galumna sp.	42	47	69
	Epilohmannia aegyptica	8	7	15
	Haplochthonius variabilis	17	39	56

and during the years 2010 and 2011. In the laboratory, the extraction of acarofaune is performed using Berlese funnels method and flotation method. Microarthropods were sorted and identified to the families through the identification key proposed by Balogh (1972). This determination was then refined at the specific level with a collection of M. Niedbala and M. Wauthy (Natural History

Table 4. Eigen values to the formation of axes.

Axes	Eigen values	% of inertia	Cumulative (%)
Axe1	0.112	49.966	49.966
Axe2	0.086	38.375	88.341

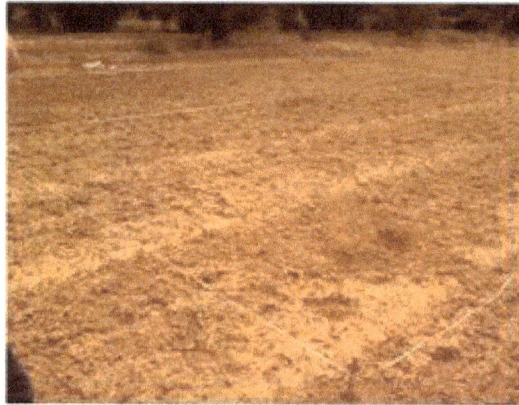

a) Control site   b) Palm groves (1 year)

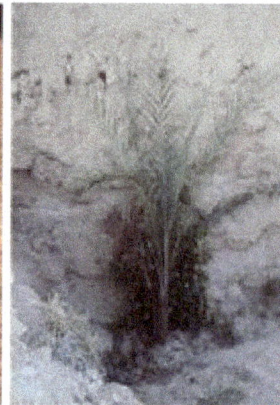

c) Palm groves (3 years)   d) Palm groves (5years)

e) Palm groves (7 years)   f) Palm groves (10 years)

Figure 2. Different stations (a, b, c, d, e, and f) selected for this study.

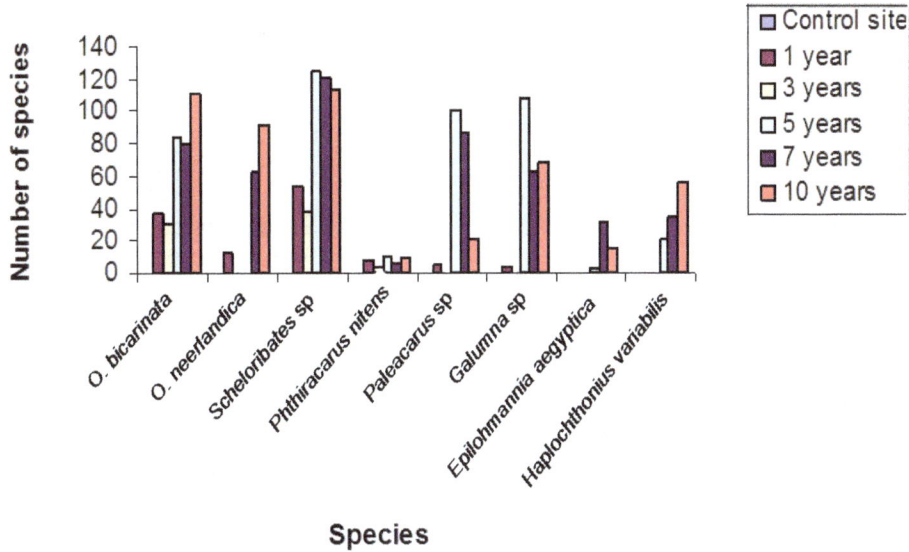

Figure 3. Evolution of species number according to palm age.

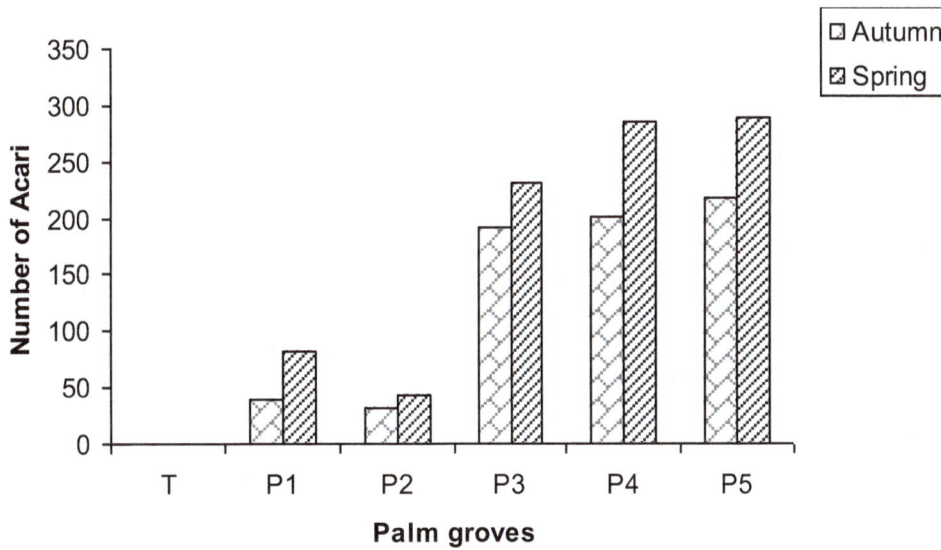

Figure 4. Number of mites collected in full and spring according to the different palm groves, the value of Khi² (G² Wilks) is equal to 0.024 which means that the distribution of mites on the basis of stations is highly significant.

Museum of Brussels, Belgium).

RESULTS

Factorial analysis of correspondence

The correspondence analysis (AFC) is used to describe the relationship between mite species and the different stations on the one hand and also between species. It is based on the absence (0) and presence (1) of mite species in the different stations. In the interpretation of the AFC, species and statements with relatively high concentrations were used in the calculations. The interpretation of an axis was done on the elements that provide to the axis the largest contributions to the axis that explained the maximum of inertia of this axis. However, the elements that contribute most to the construction of the axis are those that deviate from their origin, therefore they will present the highest coordinates. Digital processing was focused on the analysis of statements giving frequencies of eight species in the stations representing the five stations of the region of Tolga.

The contribution of the axis 1 is in the range of 49.96%

Table 5. Contribution of the different stations in the formation of axes.

Palm groves	Axe 1	Axe 2
P1	3.23	33.71
P2	0.82	46.56
P5	58.88	0.22
P7	0.00	15.10
P10	37.07	4.41

and that of axis 2 is 38.37% (4). Thus, the 1×2 factorial design alone accounts for 88.34% of the total inertia. In addition to the best plane of projection of all the elements it is, it will be very interesting to interpret the factorial design. For the formation of axis 1, the station P1 contributes 3.23%, that of the station P2 with 0.82% (Table 5). The third station P3 with 58.88%, the fourth with 0.00% and the fifth with 37.07%. For axis 2, stations 1 and 2 contribute 33.71 and 46.56%, respectively. The third with 0.22%, fourth with 15.10% and fifth with 4.41%.

Interpretation of factorial designs for stations 1 and 2 (palm planting age dependence)

For graphs interpreting proximity between points and major plants and the role of each point in determining an axis by contributions examination, must be considered (Saporta and Bouroche, 1980). Also, the eigenvalues can be described as part of the information explained by the different axes. Elements that have the highest contributions are the most explanatory to the main axis considered. A representation of the projection of the stations in the factorial designs 1 and 2 is shown in Figure 5. The position of stations reflects the affinities and the correlations between various plantations ages. According to axes 1 and 2, there are five subdivisions of point clouds. The first is located on the positive part of axis 1 which is formed by the station P5. The second and third, meanwhile, are on the positive part of axis 2 which includes the stations P1 and P3. The fourth station is represented by station P10, which is localized on the negative part of axis 2 and the fifth is represented by station P7, which is in the negative part of axis 1. Thus, we see that there is an affinity between the stations P1 and P3 on the one hand, and stations P7 and P10 on the other. These stations (P7 and P10) show a very rich species and individual numbers. Stations P1 and P3 represent the first years of plantation where they show a low wealth of species and individual numbers.

Interpretation of the axes 1 and 2 species composition

According to axes 1 and 2, there are 8 groups:

Group 1: *Epilohmania aegyptica pallida*
Group 2: *Haplochthonius variabilis*
Group 3: *Oppia neerlandica*

These three species are confined much more in the stations P7 and P10 station

Group 4: *Scheloribates* sp.
Group 5: *Phthiracarus nitens*

These two species are much more confined to P1 and P5 stations

Group 6: *Oppia bicarinata*

This species is common to all stations

Group 7: *Galumna* sp.
Group 8: *Paleacarus* sp.

These last two species mark their presence, especially at stations P5 and P7 station.

DISCUSSION

The present study was conducted in a palm groves ecosystem that was used to identify eight species of oribatid inhabiting palm plantations, whose ages were 7 and 10 years. There were 6 and 7 species in palms of 1 and 3 years, and 3 species in palms of 5 years. The number of species range from 0 in control, up to 125 individuals in P3 station (5 year palms). *Scheloribates* sp., with a total of 450 individuals, seems to be the most dominant species in all palm stations. This wealth, even if it is real low compared to other environments, including the Northern regions where ecological conditions are significantly better (Davet, 1996; Fekkoum, 2010) constitutes a very interesting index. Indeed, this wealth can show that oasis ecosystems in relation to the surrounding areas, offer a different biotope, where ecological conditions are much better. Lincoln et al. (1982) noted that the species, communities, or organizations can provide clues; and the presence or absence of certain species, provide information on the environmental quality. Indeed, the presences of oribatid species in these environments show a significant change in these groves, in particular point of view, the microclimate and nutritional support.

Thiele (1977) stressed that the distribution and activity of arthropods, are largely determined by climatic factors; such as temperature and humidity, as well as nutritional substrate (Travé, 1963). The control site, where no species were found, shows that the Saharan environment is an unfavorable environment for the development of mites. This is partly due to prevailing climatic factors and land degradation, which is characterized by the complete

Graphique symétrique   (axes F1 et F2 : 88.34%)

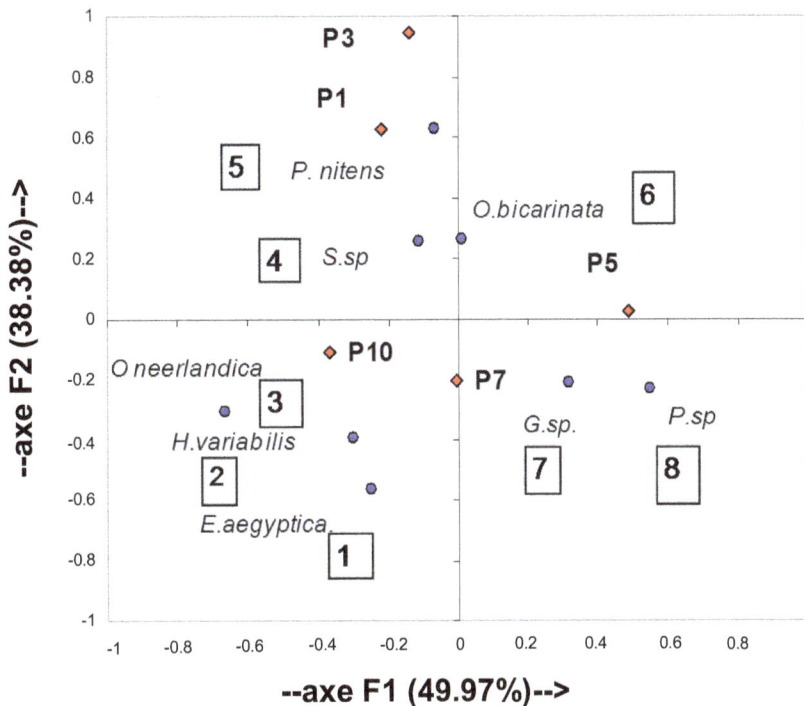

Figure 5. Factorial analysis of correspondence.

absence of plant cover that determine, to a large extent, the presence of oribatids.

The amendments and irrigation, and the microclimate created by palm trees create the requirements of mites that makes the environment more favorable. Tousignat and Coderre (1992) showed that species abundance and community structure of arthropods depend usually on biotic and abiotic factors of the environment. Also, Athias and Cancela (1976) showed that rainfall remains an important factor for soil fauna. Nef (1971) showed that moisture is also prominent as the temperature; it profoundly influences the soil fauna by adjusting the intensity, location, and activity of individuals involved in the numerical changes of microfauna. Furthermore, Temperature can induce a change in the community structure of Oribatid towards this factor (Webb et al., 1998; Gergocs and Hufnagel, 2009). The richness and complexity of communities trace the historical and biogeographical events of the medium and bio-ecological factors available (Lincoln et al.,1982).

These communities show intra-and interspecific relationships, on one hand, and their relationship with the environment, on the other. However, they can learn about the integrity or degree of impairment of the environment, thus constituting a basis for studies of ecosystems and their evolution. According to the presence of soil organisms, particularly acarofaune, depends directly on the nutritional substrate (Travé, 1963). It was noted that

Oribatid group had fundamental characteristics that can indicate the different environmental changes (Gergocs and Hufnagel, 2009). These characteristics are widely mentioned (Lebrun and Van Straalen, 1995; Behan, 1999; Gulvik, 2007). According to these authors, Oribatid behavior can be used to indicate human impact effects in an environment. This phenomenon of human impact has always a harmful effect on the environment.

We can say that this implementation of the oasis ecosystem, even as the work of man, has to change the landscape typical of the Saharan, by creating a particular habitat. While the objective of this change was a matter of survival for indigenous people, the combined effort of all generations throughout the ages has to create a microenvironment that, not only addressed the need of the population, creates an ecologically favorable environment for a variety of species that find it a suitable refuge. Indeed, the particular nature of date palm and climatic requirements, essential for growth should be noted; and the environment of the palm should be made a very special habitat (Benziouche et al., 2010).

Conclusion

This study on oribatid mites in the oasis ecosystem in Algeria, being the first, shows the effect of man's activities, where the negative effect on the environment can sometimes be useful. This can be generalized in the areas that have undergone degradation. But this can only be done through careful study and multidisciplinary research in order to create favorable habitats for all species of animals, and especially species that are actively involved in soil environments, to improve soil conditions by their action of biodegradation.

REFERENCES

Achoura A, Belhamra M (2010). Aperçu sur la faune arthropodologique des palmeraies d'El-Kantara. Courier du Savoir- N° 10, pp. 93-101.

Athias HC, Cancela JP (1976). Micro-arthropodes édaphiques de la aillais (Fontaine –Bleu). Composition et distribution spatio-temporelle d'un peuplement en placette à litière de Hêtre pure (Acarien et collembole) Rev. Ecol. Biol. Soil, 13(2): 315-329.

Balogh J (1972). The oribatid genera of the world, Akademic Kiado edition, Budapest, p. 188 + 71 planches.

Behan-Pelletier VM (1999). Oribatid mite biodiversity in agrosystems: Role For bioindication Agriculture, Ecosyst. Environ., 74: 411- 423.

Benziouche SE, Chehat F (2010). La Conduite du Palmier Dattier Dans les Palmeraies des Zibans (Algérie) Quelques éléments d'analyse. Eur. J. Sci. Res., 42(4): 644-660.

Bouroche JM, Saporta G (1980). L'analyse des données. Que sais-je ? Paris: Presses Universitaires de France. Dajor R. 1970 : Précis d'écologie. Ed. Dunod, Paris, pp. 357.

Gergocs V, Hufnagel L (2009). Application of Oribatid mites as indicators. Appl. Ecol. Environ. Res., 7(1): 79-98.

Gulvik ME (2007). Mites (Acari) as indicators of soil biodiversity and land use Monitoring: A review. Pol. J. Ecol., 55(3): 415-440.

Hamdi H (1992). Etude bio écologique des peuplements orthoptérologiques des dunes fixées du littoral algérois. Thèse, magister, I.N.A. El-Harrach. pp. 166.

Lebrun Ph, Van Straalen NM (1995). Oribatid mites: Prospects for their use in ecotoxicology. Exp. Appl. Acarol., 19: 361-379.

Lincoln R, Rosshall G, Clark PF (1982). Soil inhabiting arthropods as indicators of Environmental quality. Acta biologica Hungarica, 37(1): 79-84.

Nef L (1971). Comparaison de l'efficacité de différentes variantes de l'appareil de Berlese-tulgren. Centre Rech. Biol. Bookryl. genk, pp. 179-199.

Tousignat S, Coderre D (1992). Niche partitioning by soil mites in a recent Hardwood plantation in southern Quebec, Canada. Pedobiologia 36: 287-294.

Toutin G (1977). Eléments d'agronomie saharienne, de la recherche au développement. Ed. INRA. Paris. pp. 276.

Travé J (1963). Ecologie et biologie des Oribates (Acariens) saxicoles et arboricoles. Vie et milieu, Suppl. 14: 1-267.

Vikram M (1986). Soil inhabiting: Arthropods as indicator of environmental quality. Acta Biol. Hungarica., 37(1): 79-84.

Webb NR, Coulson SJ, Hodkinson ID, Block W, Bale JS, Strathdee AT (1998). The effect of experimental temperature elevation on population of cryptostigmatic mites in high arctic soils. Pedobiologia, 42(4): 298-308.

22

Biological characteristic of an embryonic fibroblast line from Xiaoshan chicken for genetic conservation

Weijun Guan [#1], Dianjun Wang[#3], Chunyu Bai[1 2], Minghai Zhang[2], Changli Li[1], Yuehui Ma[*1]

[1]Institute of Animal Science, Chinese Academy of Agricultural Sciences, Beijing, 100193, PR China.
[2]College of Wildlife Resources, Northeast Forestry University, Harbin, 150040, PR China.
[3]Department of Pathology, Chinese PLA General Hospital, Beijing, 100853, PR China.

A fibroblast line from embryos of Xiaoshan chicken was established successfully by direct culture of explants and cryopreservation techniques. The cells were morphologically consistent with fibroblasts, and the population doubling time (PDT) was about 46.72 h. According to karyotyping and G-banding, the diplontic cells with 78 chromosomes accounted for 98.58±1.27% of the total cells. The cells were tested for microbial contamination and they were free of infections from bacteria, fungi, viruses and mycoplasmas. There were no cross-contamination from other cell lines as revealed by isoenzyme polymorphism analysis of lactate dehydrogenase (LDH) and malate dehydrogenase (MDH). Three fluorescent protein genes were transfected into the Xiaoshan chicken embryonic fibroblasts and the transfection efficiencies of these genes were between 12.72 and 35.89%. All the tests showed that the quality of the cell line conformed to the quality criteria of the American type culture collection (ATCC). This work had not only preserved the precious genetic resources of Xiaoshan chicken, but also explored a new protocol to preserve the endangered animal breeds.

Key words: Xiaoshan chicken, fibroblasts, genetic conservation, biological characteristics.

INTRODUCTION

Animal genetic resources are an important element of biodiversity. Genetic diversity of China livestock and poultry has being seriously shrinking, and some high-quality species are on the edge of imminent extinction as the exacerbation of environmental pollution and the development of modern animal husbandry. While some government departments, organizations and experts appeal for conservation and management of the livestock and poultry genetic resources, but there is still a massive loss committed. If these genetic resources have not been preserved in any forms before their extinction, not only the genetic resources will lost evermore, but also it becomes impossible to investigate the unknown cell and molecular mechanisms in regard to the extinct livestock

and poultry to regenerate them through somatic cell cloning. Therefore, it is of urgent need to employ practical measures to conserve endangered species (Guan et al., 2002). At present, preservation in terms of individual animals, semen, embryos, genomic libraries and cDNA libraries are all available methods. In addition, modern cloning techniques have made somatic cells an attractive resource for conserving animal genetic materials (Wu, 1999). Thus, establishment of animal cell line can not only save the genetic resources on cell level, but also provide a precious experiment material for cell biology, genomics, post-genomics and embryonic engineering.

Xiaoshan chicken, also known as Yue chicken, originated in Jiangsu Province of China. It is characterized with big fatty body, high fecundity and tender flesh. As one of the local chicken breeds in China, Xiaoshan chicken has enjoyed a high reputation ever since the Yue State during the Sping and Autumn Period. Xiaoshan chicken was listed in 138 national protected species by Chinese government in 2006. This research successfully constructed a qualified fibroblast line through primary explantation and programmed cryopreservation.

---

*Corresponding author. E-mail: weijunguan301@gmail.com, yuehui_ma@hotmail.com.

# These two authors contributed equally to the work.

The genetic resources of this valuable local breed have thus been permanently preserved in the form of somatic cells, and this technique system will provide technical and theoretical support for conservation of other animal genetic resources at cell level.

MATERIALS AND METHODS

The 8-day old embryos of Xiaoshan chicken in this research were provided by chicken breeding farm of Chinese Academy of Agricultural Sciences, Beijing, China.

MEM (Gibco, USA), special grade fetal bovine serum (Biochrom, German), DMSO (Sigma, USA), Hoechst 33258 (Invitrogen, USA), Lipofectamine 2000 (Invitrogen, USA), polyacrylamide gel (Sigma, USA).

Isolation and culture of Xiaoshan chicken embryonic fibroblasts

The Xiaoshan chicken eggs incubated for 8 days were sterilized using alcohol swabs, and then the embryos were isolated and washed three times with phosphate buffered saline (PBS). The embryos were cropped into pieces of 1 mm^3 in size and seeded onto the surface of a tissue culture flask, and cultured inverted at 37°C in a humidified atmosphere, 5% CO_2 for 3 to 4 h. Modified Eagle's medium (MEM) containing 10% fetal bovine serum (FBS) was added into the flask. The medium was refreshed after 2 to 3 days. The cells were harvested at 80 to 90% confluence using 0.25% trypsin (m/v) solution and were separated into culture flasks at the ratio of 1:2 or 1:3 (Guan et al., 2005; Zhou et al., 2004).

Cryopreservation and recovery

The cells in logarithmic growth phase were harvested and counted with a hemocytometer. The viability before freezing was checked by the CellTiter-Blue® Cell Viability Assay (Promega, USA). The harvested cells were resuspended in freezing medium containing 40% DMEM, 10% dimethyl sulphoxide (DMSO) and 50% FBS to a final concentration of (3 to 4) ×10^6 viable cells/ml. The cell suspension was dispensed into sterile plastic cryovials labeled with species, gender, freezing serial number and the date. First, the sealed cryovials were kept at 4°C for 20 to 30 min to allow the DMSO to equilibrate. Then, cryovials were transferred to a commercially available freezing kit (Nalgene, USA), and refrigerated at -80°C overnight (a process in which temperature decreased at a rate of 1°C min). Whereafter, the cryovials were transferred to liquid nitrogen (LN2) for long term storage (Werners et al., 2004).

When recovering the cells, the cryovials were taken out from liquid nitrogen, and rapidly thawed in 42°C water bath, which were subsequently transferred to culture flask containing 90% MEM and 10% FBS with a straw. The suspension was pipeted gently to make it well-distributed, and then cultured at 37°C with 5% CO_2. The medium was refreshed after 24 h.

Estimation of cell viability by trypan blue exclusion test

Viabilities before freezing and after recovery were determined using trypan blue exclusion test. The number of non-viable cells was determined by counting 1000 cells (Qi et al., 2007).

Growth dynamics

Cells were plated onto 24-well plates at the density of approximately 1.5×10^4 cells per well. The cells were cultured for 7 days and counted every day (3 wells each time). The mean cell counts at each time point were then used to plot a growth curve, based on which the PDT was calculated (Qi et al., 2007; Hirofumi et al., 2006).

Microbial detection

Detection of bacteria and fungi

The cells were cultured in complete MEM media free of antibiotics and observed for the presence of bacteria and fungi at 3 days after subculture according to the method described by Doyle et al., 1990).

Detection of mycoplasmas

According to protocol of the ATCC, the cells were cultured in antibiotic-free medium for at least 1 week, and then fixed and stained with Hoechst 33258 (Simpson, 2003).

Detection of viruses

Hay's hemadsorption protocol was used to examine the samples for cytopathogenesis using phase contrast microscopy (Hay et al., 1992; Wu et al., 2008).

Karyotyping and chromosome analysis

Chromosome spreads were prepared, fixed and stained following standard methods (Costa et al., 2005). Cells were harvested when 80 to 90% confluent, and subjected to hypotonic treatment and fixed, then the chromosome numbers were counted from 100 spreads under an oil immersion objective upon Giemsa staining. Relative length, centromeric index and kinetochore type were calculated according to the protocol described by Sun et al. (2006).

Isoenzyme polymorphisms

Isoenzyme patterns of lactate dehydrogenase (LDH) and malate dehydrogenase (MDH) were identified by polyacrylamide gel electrophoresis (PAGE). The protein extraction solution (0.9% Triton X-100, 0.06 mmol NaCl: EDTA in mass ratio 1:15) was added to the harvested cells at the density of 5×10^7/ml. Then the suspension was centrifuged and stored in aliquots at -70°C. Equal volumes of 40% sucrose and 2.5 ml loading buffer were added to the sample. The PAGE apparatus was used with the electrophoretic buffer and was changed into Tris-glycin (pH 8.7), the gel buffer was prepared into discontinuous system using two kinds of Tris-citric acid buffer at different concentrations: 0.078 mol/L (pH 8.9) and 0.017 mol/L (pH 6.8). Electrophoretic mobility was defined by numbers and intensity of enzyme bands, as well as the distance of the band migrating from the point of origin. Different mobility patterns were reflected by the relative mobility front (RF), which was calculated as the ratio of the distance of the isozyme migration to that of bromophenol blue.

Expression of fluorescent protein genes in Xiaoshan chicken fibroblasts

According to the methods described by Wu et al. (2008) three plasmids carried fluorescent protein genes (pDsRed1-N1, pEGFP-N3 and pEYFP-N1) (BD Bioscieces Clontech product, USA) were transfected into the Xiaoshan chicken fibroblasts using Lipofectamine 2000. The tranfected cells were observed at 24, 48, 72, 96, 1 week, 2 weeks and 1 month after transfection. Expression of the three fluorescent protein genes was observed under a confocal microscope (Nikon TE-2000-E, Japan) with excitation wavelengths of 543, 488, 488 and 543 nm respectively. In each group 10 visual fields were captured to take pictures and to calculate the transfection efficiencies which was formulated as the ratio of positive cell numbers to the total cell numbers (Wu et al., 2008). At 48 h after transfection, DAPI staining was adopted to locate the nuclei. By trypan blue and DAPI staining, the viability and apoptosis rate of the transfected cells were detected to analyze the influence of exogenous gene transfection on the cells.

RESULTS

Morphological observation of Xiaoshan chicken fibroblasts

Cells sprouted out from small tissue pieces at 1 day after being plated on the bottom of tissue culture flasks, and then it continued to proliferate and were subcultured when 80 to 90% confluent within about 3 days. The cells displayed typical fibrous and fusiform morphology with centrally located oval-shaped nuclei. However, there existed some ones that were morphologically epithelial cells. The fibroblasts grew rapidly and replaced the epithelial cells gradually after 2 to 3 passages, and then a relatively purified fibroblast line was obtained. The viabilities of Xiaoshan fibroblasts before freezing and after recovery evaluated through trypan blue exclusion tests were 98.36±1.41 and 91.17±0.92% respectively ($p<0.5$) (Figure 1A, B and C).

Growth dynamics

The growth curve of Xiaoshan chicken fibroblast before cryopreservation and after recovery displayed typical "S" shape (Figure 2). Lag phase of approximately 24 h was observed after the cells were plated, which was corresponding to the recovery period against trypsin damage. Afterwards, they proliferated rapidly and entered the exponential growth phase until the stationary phase after about 5 days. From day 5, growth stagnated and the population began to degenerate gradually. The PDT calculated from the curve was about 46.72 h.

Microorganism detection

The medium was clear all the time and no abnormalities could be observed under the microscope. The results indicated that the Xiaoshan chicken fibroblasts were free of bacterial contamination. Mycoplasma detection suggested that the fibroblasts were free of mycoplasmas. Would there be abundant punctiform and filiform blue fluorescence in the nucleoli, it could be concluded that the cells were contaminated by mycoplasmas (Figure

1D) just as the positive result of Li et al. (2009). As is shown by the cytopathogenic evidence and the hemadsorption test, tests for virus contamination were all negative as well.

Karyogram and chromosome number

The chromosome number of diploid Xiaoshan chicken is 78, including 9 pairs of macrochromosomes and 30 pairs of minichromosomes. The sex chromosome type is ZZ(♂)/ ZW(♀) (Figure 3). The parameters including relative length, centromere index and kinetochore type were shown in Table 1. In this experiment 100 representative spreads at metaphase of passage 3 to 5 were observed under the microscope to count the chromosome numbers, and the mean proportion of diploid cells was 98.58±1.27%.

Isoenzyme polymorophisms

The profiles of isoenzyme polymorphisms might be characteristic in a specific species or tissue (MacLeod et al., 1999). Polymorphism analysis of isoenzymes is currently a standard method for the quality control to identify cell lines and to prevent interspecies contamination. Isoenzyme patterns of LDH and MDH were obtained using vertical slab non-continuous PAGE, stained by Coomassie brilliant blue and compared with those from other species (Figure 4). The LDH RFs were in the order of LDH5, LDH4, LDH3, LDH2, LDH1 (Figure 4A); while the two bands of MDH were Mitochondrial MDH (m-MDH) and cytosolic MDH(s-MDH) (Figure 4B). The three chicken breeds had their own characteristic bands, each band with a different migration rate (Tables 2 and 3). The results indicated that there was no cross-contamination from other cell lines.

Expression of exogenous genes in Xiaoshan chicken fibroblasts

The three fluorescent genes, EGFP, EYFP, DsRed1 were all highly expressed at 24 h according to the optimized conditions. The transfected cells increased gradually within 48 h (Figures 5 and 6), and the transfection efficiencies were listed in Table 4. The number of transfected cells decreased gradually as time passed by. However, fluorescence still could be observed in a few cells at 2 weeks after transfection.

DISCUSSION

Xiaoshan chicken fibroblast line was successfully established from 62 embryo samples by adherent culture, and was cryopreserved in 184 cryovials within 5 passages

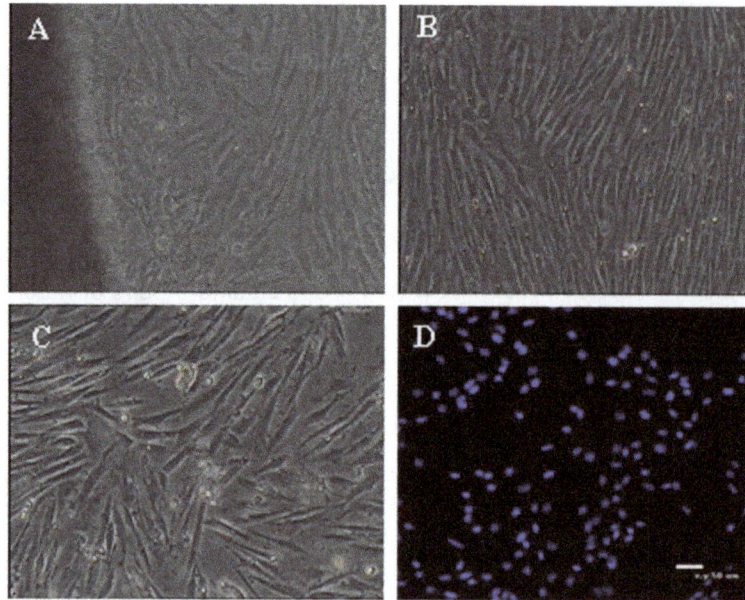

Figure 1. Morphology of Xiaoshan chicken fibroblasts A. Primary cells grew out from the embryo explants; B. Cells before cryopreservation; C. Cells at 24 h after recovery from cryo-storage. D. Xiaoshan chicken fibroblasts stained with Hoechst 33258. Bar = 100 um.

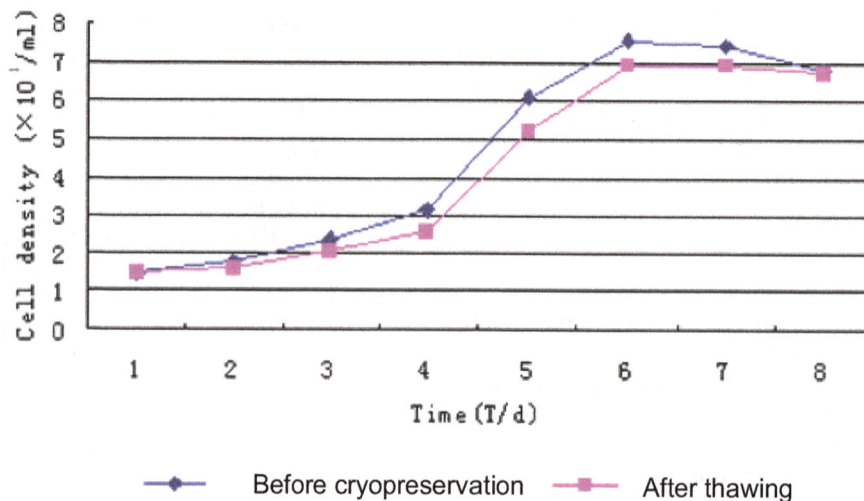

Figure 2. The growth curve of Xiaoshan chicken fibroblasts. The growth curve was typical "S" shape. Lag of around 48 h was observed after cells were seeded. Then, cells proliferated and entered the logarithmic phase. The PDT was approximately 47 h. From the fiveth day, cells entered the plateau phase.

at a density of about 3.0×10^6 cells/ml. The biological characteristics, especially the genetics related ones, may be altered by *in vitro* culture after many passages, so a minimal number of passages are recommended to protect them from degeneration.

Morphological observation indicated that there were both epithelial cells and fibroblasts in the primary cultures of the explanted tissues. Due to their different levels of tolerance to trypsinization, the fibroblasts detached from the flasks earlier when treated with trypsin and adhered again quickly after passage, whilst most epithelial cells were difficult to adhere, or only did so in an unstable manner and fell off when vibrated (Xue et al., 2001). For this reason, a purified fibroblast line could be obtained after 2 to 3 passages.

Isoenzyme polymorphisms and karyotyping together can effectively confirm the origin of a cell line and identify possible cross-contamination. The genetic stability of cell

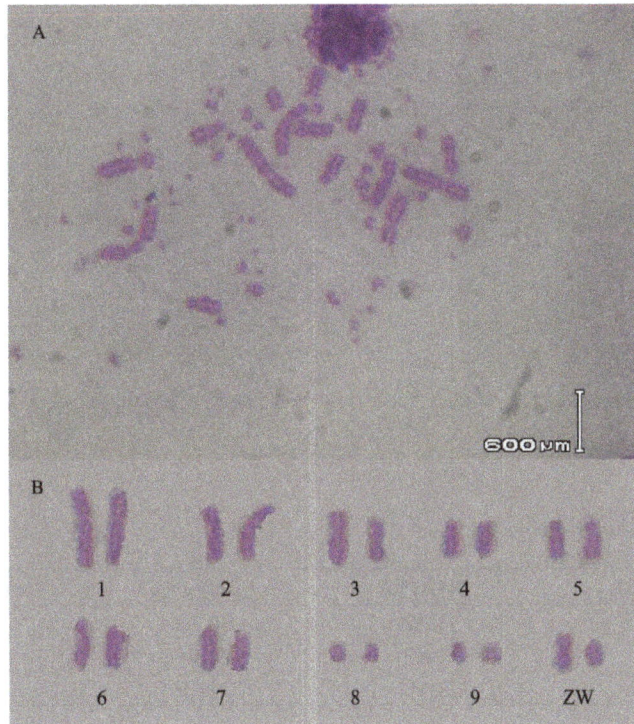

Figure 3. Chromosomes at metaphase (A) and karyotype (B) of the Xiaoshan chicken fibroblasts (♂).

Table 1. Chromosome parameters of Xiaoshan chicken (♂).

Choromosome No.	Relative length (%)	Centromere index (%)	Kinetochore type
1	23.25±0.12	39.43	SM
2	21.18±0.31	37.32	SM
3	19.75±0.35	31.27	M
4	15.78±0.58	25.98	SM
5	11.25±0.95	19.32	SM
6	11.17±0.19	14.62	SM
7	6.82±0.64	0	T
8	4.17±0.32	0	T
9	3.92±0.55	0	T
Z	13.28±0.53	33.71	M
W	6.91±0.31	0	T

Note: SM, Submetacentric chromosomes; T, telocentric chromosomes; M, metacentric chromosomes.

line is critical to preserve the genetic resources, the fibroblasts must maintain the same diploidy as cells *in vivo*. International standards of poultry karyotype were as 8 pairs of large chromosomes with sex chromosomes Z and W, and 30 pairs of microchromosomes (Ladjali-Mohammedi et al., 1999). Avian diploid chromosome number are greatly diversified, the majority of which ranged from 78 to 82. Chicken macrochromosome number is 7.8±0.9, fluctuating from 6 to 9, and the microchromosome number is 31.9±2.5, the range fluctuating from 24 to 35. In this study, results showed that the diploid chromosome number of Xiaoshan chicken was 78±0.9 in 100 cells, with 10 pairs of macrochromosomes and 29 pairs of microchromosomes. The proportion of 2n = 78 cells were 98% as detected in 100 cells. Most Xiaoshan chicken chromosomes were very small ones, which were easily lost in the preparation process and by the interference of dye, rendering the difficulty to count chromosome number and to observe the morphology. Therefore, time point and duration of

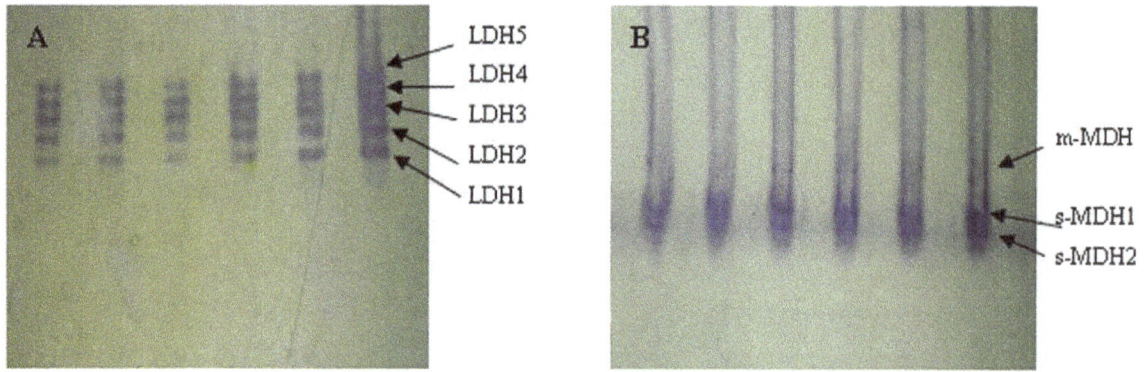

Figure 4. Isoenzyme patterns of LDH and MDH from three cell lines. 1 2 Xianju chicken; 3 4 Xiaoshan chicken; 5 6 Langshan chicken.

Table 2. Relative migration fronts of LDH.

Breeds	LDH1 (%)	LDH2 (%)	LDH3 (%)	LDH4 (%)	LDH5 (%)
Xianju chicken	38.52	33.82	30.14	25.67	19.18
Xiaoshan chicken	37.18	31.89	28.97	23.76	17.13
Langshan chicken	36.71	33.18	29.16	24.29	18.53

Table 3. Relative migration fonts of MDHs.

Breeds	m-MDH (%)	s-MDH1 (%)	s-MDH2 (%)
Xiaoshan chicken	9.13	14.85	23.87
Xianju chicken	9.58	15.13	21.53
Langshan chicken	8.96	14.57	22.86

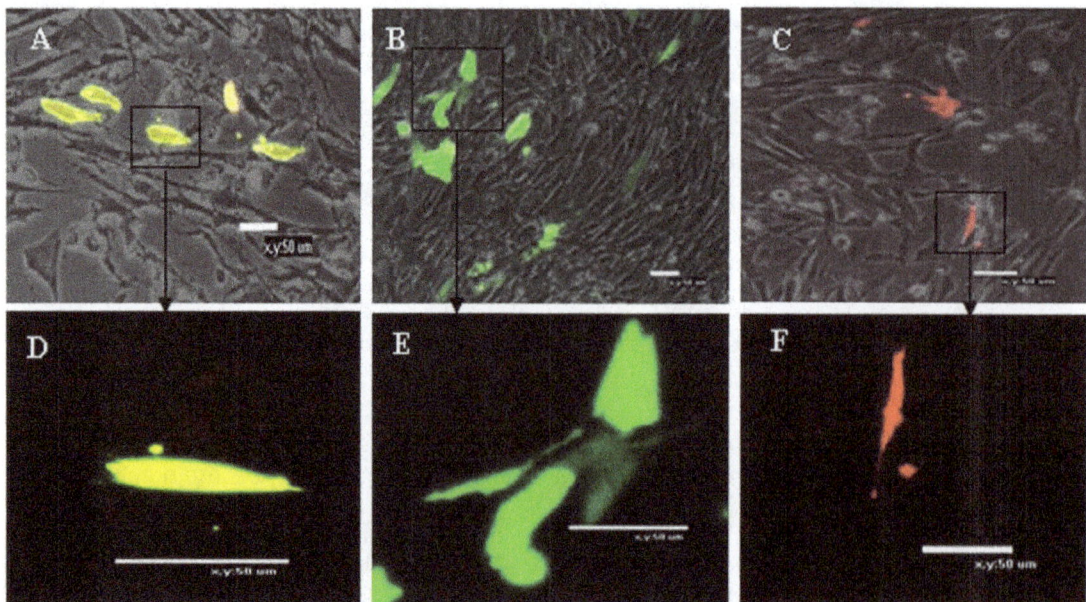

Figure 5. Expression of three fluorescent protein genes at 48 h after transfection. The photos were taken using a laser scanning confocal microscope (Nikon TE-2000-E, Japan) with the excitation wavelengths of 488 and 543 nm. a and d, b and e, c and f were the transfection results of pDsRed1-N1, pEGFP-N3 and pEYFP-N1 respectively. Bar = 50 um.

Table 4. Transfection efficiencies of three fluorescent protein genes.

Time (h)	Three fluorescent plasmids		
	pEGFP-N3 (%)	pEYFP-N1 (%)	pDsRed1-N1 (%)
24	14.18±0.31	16.26±0.72	14.74±0.32
48	21.41±0.55	20.43±0.36	20.16±0.57
72	18.63±0.27	19.21±0.31	18.78±0.19

colchicine administration should be precisely controlled in the experiments. The cells were incubated with 0.1 µg/ml colchine for 4 h when 70 to 80% confluent. Low-osmotic treatment is another major influential factor of karyotyping which aims to cell swelling and surface loosening of chromosome. So the low-osmotic duration must be tightly controlled within 40 min.

Isoenzyme polymorphisms exist in different species, different races, different individuals and even different tissues of the same species. Therefore, biochemical analysis of isozyme polymorphisms constitutes a standard method for the detection of cell line cross contamination in the current important biological resource centers throughout the world such as the American type culture collection (ATCC), European collection of animal cell cultures (ECACC) and Deutsche Sammlung von Mikroorganismen und Zellkulturen GmbH (DSMZ) (Drexler et al., 1999). It could be distinguished if the proportion of other cells would be more than 10% (Nims et al., 1998). Isoenzyme polymorphism was commonly selected to confirm the species origin of the cells, distinguish between normal and tumor cells, and act as biochemical index of animal classification. In almost all vertebrate tissues examined, the isoenzyme LDH has been shown to consist of five distinct bands. Using isoelectric focusing at a pH range of 3.0 to 9.0, a good and clear separation of all five LDH bands in chicken organs was achieved by Heinova et al. (1999). Moreover, m-MDH and s-MDH were both obtained during early and middle development (1 to 16 days) of chicken embryos and newly hatched chicks. In this study we detected the isoenzyme patterns of LDH and MDH, and improved the ATCC starch gel electrophoresis method. The LDH and MDH isoenzyme bands of Xiaoshan chicken fibroblasts were clear and distinct with the similar isoenzyme activities to that of the original tissues.

The three enhanced fluorescent protein genes in the present study are characterized with stable structure, effective and germ-line independent expression. They are characterized with brighter fluorescences, more efficient transcription and expression than lacZ, CAT and other common fluorescent markers in animal cells (Heim et al., 1995). High transfection efficiencies were obtained at 48 h after transfection. When the transfection efficiencies decreased, intensified fluorescences could still be observed even after 2 weeks, indicating that the exogenous genes could be replicated, transcribed, translated and subsequently modified in the fibroblasts.

The results provided solid theoretical and technical basis for structural genomics, functional genomics and transgenic researches concerning the Xiaoshan chicken in the future.

In summary, the Xiaoshan chicken fibroblast line was successfully established, containing biologically normal and genetically stable fibroblasts, all aspects met the cell line quality control standards of major international culture collection centers. Exogenous gene expression results supported their strong vitality and rationalized their applications in transgenic therapies and genetics. The precious genetic resource of Xiaoshan chicken were well preserved at cell level and will be used in the future as biological materials for genetics, biomedical sciences, cell and molecular biology, immunology and so on.

ACKNOWLEDGEMENTS

This research was supported by the Ministry of Agriculture of China for Transgenic Research Program (2011ZX08009-003-006, 2011ZX08012-002-06) and the central level, scientific research institutes for R & D special fund business (2011cj-9).

REFERENCES

Costa UM, Reischak D, da Silva J, Ravazzolo AP (2005). Establishment and partial characterization of an ovine synovial membrane cell line obtained by transformation with Simian Virus 40 T antigen. J. Virol. Meth., 128(2): 72-78.

Doyle A, Hay R, Kirsop BE (1990). Animal cells: living resources for biotechnology, Cambridge University Press, Cambridge, U K. pp: 81-100.

Drexler HG, Dirks WG, MacLeod RA (1999). False human hematopoietic cell lines: cross-contaminations and misinterpretations. Leukemia, 13(10): 1601-1607.

Guan W, Ma Y, Zhou X, Liu G, Liu X (2005). The establishment of fibroblast cell line and its biological characteristic research in Taihang black goat. Rev. China Agric. Sci. Technol., 7(6): 25-33.

Hay RI (1992). Cell line preservation and characterization. In: Freshney, R.I. (Ed.), Animal Cell Culture: A Practical Approach, 2nd ed. Oxford University Press, Oxford. pp: 104-135.

Heim R, Cubitt AB, Tsien RY (1995). Improved green fluorescence. Nature, 6516(373): 663-664.

Heinova D, Rosival I, Avidar Y, Bogin E (1999). Lactate dehydrogenase isoenzyme distribution and patterns in chicken organs. Res. Vet. Sci., 67(3): 309-312.

Hirofumi S, Kentaro Y, Kouichi H, Tsuyoshi F, Norihiro T, Norio N (2006). Efficient establishment of human embryonic stem cell lines and long-term maintenance with stable karyotype by enzymatic bulk passage. Biochem. Biophys. Res. Commun., 345(3): 926-932.

Ladjali-Mohammedi K, Bitgood JJ, Tixier-Boichard M, Ponce de Leon

FA (1999). International System for Standardized Avian Karyotypes (ISSAK): Standardized banded karyotypes of the domestic fowl (*Gallus domesticus*). Cytogenet Cell Genet., 86(3): 271-276.

Li L, Yue H, Ma J, Guan W, Ma Y (2009). Establishment and characterization of a fibroblast line from Simmental cattle. Cryobiology, 59(1): 63-68.

MacLeod RA, Dirks WG, Matsuo Y, Kaufmann M, Milch H, Drexler HG (1999). Widespread intraspecies cross-contamination of human tumor cell lines arising at source. In. J. Cancer., 83(4): 555-563.

Nims RW, Shoemaker AP, Bauternschub MA, Rec LJ, Harbell JW (1998). Sensitivity of isoenzyme analysis for the detection of interspecies cell line cross-contamination. *In Vitro*. Cell. Dev. Biol. Anim., 34(1): 35-39.

Qi Y, Tu Y, Yang D, Chen Q, Xiao J, Chen Y, Fu J, Xiao X, Zhou Z (2007). Cyclin A but not cyclin D1 is essential in c-myc-modulated cell cycle progression. J. Cell. Physiol., 210(1): 63-71.

Simpson RJ (2003). Proteins and Proteomics: A Laboratory Manual, Science Press, Beijing.

Sun Y, Lin C, Chou Y (2006). Establishment and characterization of a spontaneously immortalized porcine mammary epithelial cell line. Cell Biol. Int., 30(12): 970-976.

Weijun Guan (2002). The construction and identification of the cell bank of species of domestic animal on the brink of extinct. Rev. China Agric. Sci. Technol., 6(2): 66-67.

Werners AH, Bull S, Fink-Gremmels J, Bryant CE (2004). Generation and characterization of an equine macrophage cell line (e-CAS cells) derived from equine bone marrow cells. Vet. Immunol. Immunopathol., 97(3): 65-76.

Wu CX (1999). The theory and technology of the conservation of animal genetic resources-the species foundation of animal agricultural continuing development in 21 century. J. Yunnan Univ., 21(3): 7-10.

Wu HM, Guan WJ, Li H, Ma YH (2008). Establishment and characteristics of white ear lobe chicken embryo fibroblast line and expression of six fluorescent proteins in the cells. Cell Biol Int., 32(12): 1478-1485.

Xue QS (2001). The principle and technique of *in vitro* culture. Science Press. Beijing, pp. 432-444.

Zhou XM, Ma YH, Guan WJ, Zhao DM (2004). Establishment and identification of Debao pony ear marginal tissue fibroblast cell line. Asian-Austr. J. Anim. Sci., 17(5): 1338-1343.

23

Structure and population dynamics of myxobolus infections in wild and cultured *Oreochromis niloticus* Linnaeus, 1758 in the Noun division (West-Cameroon)

Elysée NCHOUTPOUEN*, Guy Benoît LEKEUFACK FOLEFACK and Abraham FOMENA

Laboratory of General Biology, Department of Animal Biology and Physiology, Faculty of Science,
University of Yaounde I, P. O. Box 812, Yaounde, Cameroon.

Myxosporidian parasites of *Oreochromis niloticus* Linnaeus, 1758 from the Noun River at Kouoptamo and the Foumban fish ponds in west Cameroon, were investigated from May 2008 to June 2009. Out of 537 Tilapia (267 cultivated and 270 wild) examined, 64.8% (n=173) specimens from the fish farming and 61.1% (n=165) from the Noun River harbored Myxosporean parasites. A total of ten parasite species were found. *Myxobolus kainjiae*, *Myxobolus sarigi* were scarce in both study sites; *Myxobolus Tilapiae*, *Myxobolus equatorialis* scarce in Foumban and Kouoptamo, respectively. *M. agolus*, *M. brachysporus*, *M. camerounensis*, *M. equatorialis*, *Myxobolus Heterosporus*, *Myxobolus israelensis* were secondary in the two sites. *M. Tilapiae*, *M. equatorialis* appeared secondary in the Noun River and the fish ponds respectively. Myxosporean spores were most encountered in the kidney (61.3 and 49.0%, respectively in cultured and wild fish) and the spleen (50.5% in Foumban and 47.5% in Kouoptamo) but no host sex preference was found. In the Foumban fish farm site, high significant infection rate was observed for *M. tilapiae*, *M. camerounensis* and *M. israelensis* during the rainy season, while in the Noun River, no significant seasonal effect was found. Older hosts were significantly most infected at the fish ponds while youngs Tilapia were most commonly infected in the River.

Key words: Myxosporean, *Oreochromis niloticus*, prevalence, fish-farm, Noun River, Cameroon.

INTRODUCTION

Myxosporean (Myxozoa: Myxosporea) are primarily fish parasite (Fomena et al., 2010; Eiras et al., 2010). With their pathogenic potentials, they can affect growth (Longshaw et al., 2010), reproduction (Obiekezie and Okaeme, 1990) and involve epizooties being able to cause the death of the host (Gbankoto et al., 2001; Feist anand Longshaw, 2005). Economic losses caused by these parasites in aquaculture have been well documented (Barassa et al., 2003; Lom and Dykovà, 2006). According to FAO (2008), fish represents nearly 50% of animal proteins intake of many countries in Africa. In addition, the economic interest of *Oreochromis niloticus*

and its generalized use in the development of fish farming projects in Africa, make this species one of most important. In natural environment, parasitism is frequent and the parasitic diseases are expressed only when the conditions of the environment allow the proliferation of the parasites (Odewage and Van As, 1987; Martins et al., 1999). In fish farming medium on the other hand, hosts containment increase the parasitic load, but also maintain the development of parasite life cycle (Hedrick, 1998; Abakar et al., 2007; Milanin et al., 2010). Under natural environment or during farming (Feist and Longshaw, 2006; Eiras et al., 2008), the Genus *Myxobolus* Bütschli, 1882 is the largest group among Myxosporean with approximately 790 valid species (Eiras et al., 2005; Lom and Dykovà, 2006; Umur et al., 2010). Many of these species are potential pathogens of fish. Fomena and

*Corresponding author. E-mail: enchoutpouen2002@yahoo.fr.

Bouix (1997) counted ten species of Myxosporean of the genus *Myxobolus* infesting various organs of *Oreochromis niloticus* and Abakar et al. (2007) identified 11 species of the same genus on *Oreochromis niloticus* and *Sarotherodon galilaeus* in Chad. These findings corroborate the idea of Combes (1995) who revealed that pathogenic effect is scarcely due to only one parasitic species.

In Cameroon, studies on Myxosporean fishes parasites are essentially descriptive (Fomena et al., 2008, 2010). Data provided by few authors such as Fomena (1995) in fish ponds, Tombi and Bilong Bilong (2004) and Lekeufack Folefack (2010) in natural environment; highlight aspect of the structure and the population dynamics of this group of parasites.

In this work, we study the structure and population dynamics of Myxosporean that infest *O. niloticus* Linneaus, 1758 under breeding situations (Fish ponds at Foumban) and in nature (River Noun at Kouoptamo) in the western region of Cameroon. The objective of this study is to determine the occurrence of Myxosporean parasites of *O. niloticus* in the two different biotopes. We also investigated the factors (seasonal variation, size and sex of the host) which can influence the parasitic prevalence.

MATERIALS AND METHODS

From May, 2008 to June, 2009, 537 fishes among which 267 taken in the fish ponds of the Zootechnical and Veterinary Training centre of Foumban and 270 captured in the river Noun (tributary of Mbam river) at Kouoptamo were examined. The head quarter of the survey region Foumban (5°43' 54.5"N; 10°54' 09.8"E) and the sampled locality Kouoptamo (5°39' 17" N, 10°37' 1" E) belong administratively to the west region of Cameroon (Figure 1). Its climate belongs to the tropical mountain subset characterized by two seasons: a short dry season from November to February and a long rainy season from March to October. The temperatures in this region is definitely lower than in other parts of the country with an annual average varying between 19.8 and 22°C. We can note an oceanic influence resulting in important precipitations in the locality. The mean annual rainfall varies between 1313.7 and 1988.6 mm. The sub highland forest is often degraded by coffee plantations and other food crops (Olivry, 1986). Monthly, Fish were captured during the day and night, using a fish net or fishing canes. Immediately after harvesting, the captured fish were stored in a formalin solution (10%) until further examination. These fishes were identified according to Stiassny et al. (2007). At the laboratory, each fish was measured to the closest millimeter using a slide caliper of the brand Stainless Hardened. The sex was determined after dissection and examination of the gonads based on the work of Obiekezie and Okaeme (1990)

The external (eyes, skin, operculum, fins, scales) and internal organs (gills, liver, intestine, stomach, kidneys, spleen, gall bladder, urinary bladder, ovaries) were first examined macroscopically with the naked eye and then with a Olympus Bo 61 stereoscopic microscope to search for cysts. The smears of the kidneys, spleen, liver and gonad were mounted and examined with objective 100X of the light microscope to search for spores. Cysts found were crushed between slide and cover glass and their content identified with the objective 100X of the microscope Wild M-20.

The various parasitic species were identified according to Lom and Arthur (1989). The sample of studied hosts was grouped by classes of size based on the modified formula of Yule (Mouchiroud, 2002). Prevalence (P) is calculated as being the number of individuals of a host species infested by a parasite species divided by the number of hosts examined for that parasite species: it is often expresses as a percentage (Margolis et al., 1982). Analysis of the status of each parasitic species was made according to Valtomen et al. (1997). Thus, the species are qualified as frequent (or common or principal) if $P > 50\%$. Less frequent (or secondary or intermediate) if $10\% \leq P \leq 50\%$, and scarce (or satellite) if $P < 10\%$. Comparison of parasitic prevalence of the various parasite species was made using the χ^2 test. The security level retained in our analysis is 95%, that is, error probability < 0.05.

RESULTS

Population structure of *Oreochromis niloticus*

The size (LS) of *O. niloticus* ($n_1=270$ and $n_2=267$) sampled respectively in Kouoptamo and Foumban varied from 20 to 226 mm in the natural environment, and from 35 to 175 mm in fish pond medium. These fishes were grouped in 3 classes based on size of amplitude 50 mm (Figure 2). The modal class is (70-120) in the two study sites (Foumban and Kouoptamo). Approximately 92.2% (n=249) and 64.8% (n=173) individuals examined respectively in natural environment and in breeding situation belonged to class (70, 120). The sex ratio is skewed toward females (1.03) in fish pond and toward males (0.82) in river Noun.

Species richness and status of parasitic species

Except *Myxobolus nounensis*, which was collected only among hosts captured in the Noun River, the fauna of Myxosporeans collected is identical in fish ponds and in Noun river (Table 1). Therefore, ten species of Myxosporeans have been identified in natural environment while nine species were recorded in breeding fish ponds (Figure 3).

In the two biotopes, the number of parasitic species carried by each individual host varied from 1 to 6 (Figure 4). 37% (that is, 99/267 individuals hosts) at Foumban carried 4 to 6 species of parasites, whereas 49% (that is, 132/270 individuals) of the population of hosts examined in the natural environment were infested by 1 to 3 species of parasites. 38.8 (105/270) and 35.2% (94/267) individuals hosts examined respectively in Kouoptamo and Foumban were free of Myxosporidian.

In the two biotopes, *Myxobolus camerounensis*, *Myxobolus agolus*, *Myxobolus brachysporus*, *Myxobolus israelensis* and *Myxobolus heterosporus* are secondary (10 ≤ P ≤ 50%). However, *Myxobolus Tilapiae* is scarce (P < 10%) in Foumban and secondary in Kouoptamo, whereas *Myxobolus equatorialis* is secondary in Foumban and rare in Kouoptamo. *M. sarigi* and *M kainjiae* are rare in the two study sites. On the other hand,

Structure and population dynamics of myxobolus infections in wild and cultured Oreochromis niloticus Linnaeus, 1758...

141

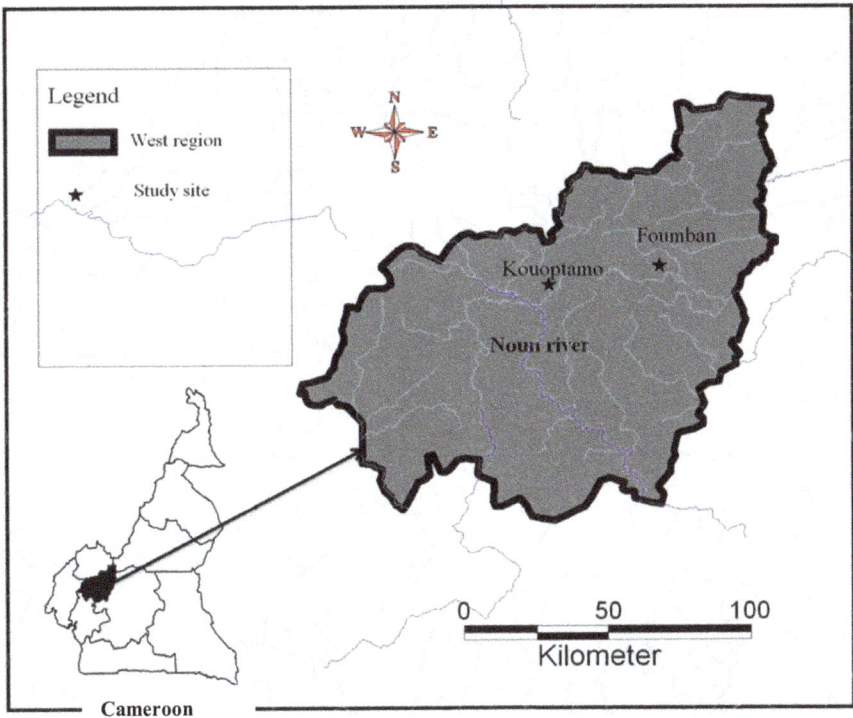

Figure 1. Cameroon map showing the study area.

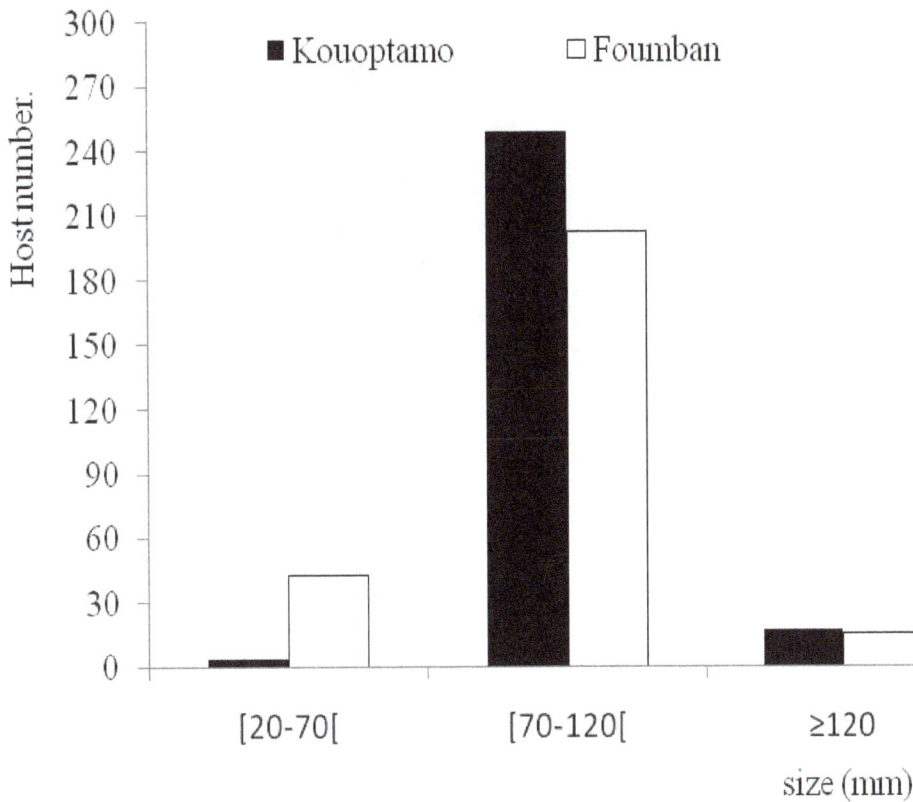

Figure 2. Host distribution (*O. niloticus*) as a function of the host size class at Kouoptamo and Foumban.

Table 1. Comparison of the Myxosporean infection rate (%) in *O. niloticus* at Foumban and Kouoptamo.

Parasite species	Locality			
	Foumban	Kouoptamo	x^2	P
M. agolus	31.8 (85) **	12.2(33) **	30.1	0.001
M. brachysporus	48.3(129) **	49.3(133) **	0.105	0.827
M. camerounensis	19.5(52) **	16.0(16) **	22.8	0.001
M. heterosporus	40.8(109) **	15.2(41) **	43.83	0.001
M. israelensis	39.0(104) **	16.3(44) **	34.15	0.001
M. equatorialis	15.0(40) **	4.4(12) *	17.04	0.001
M. tilapiae	8.6(23) *	16.7(45) **	7.87	0.005
M. kainjiae	1.9(5) *	1.5(4) *	1.25	0.724
M. sarigi	2.3(6) *	3.4(9) *	0.58	0.445
M. nounensis	0	14.4(39) **	-	-

The rate infections are followed in brackets by the number of hosts species harboring at least one parasitic species of the population examined. **, secondary species; *, scarce species. statistical analyses is not doing concerning *M. nounensis*.

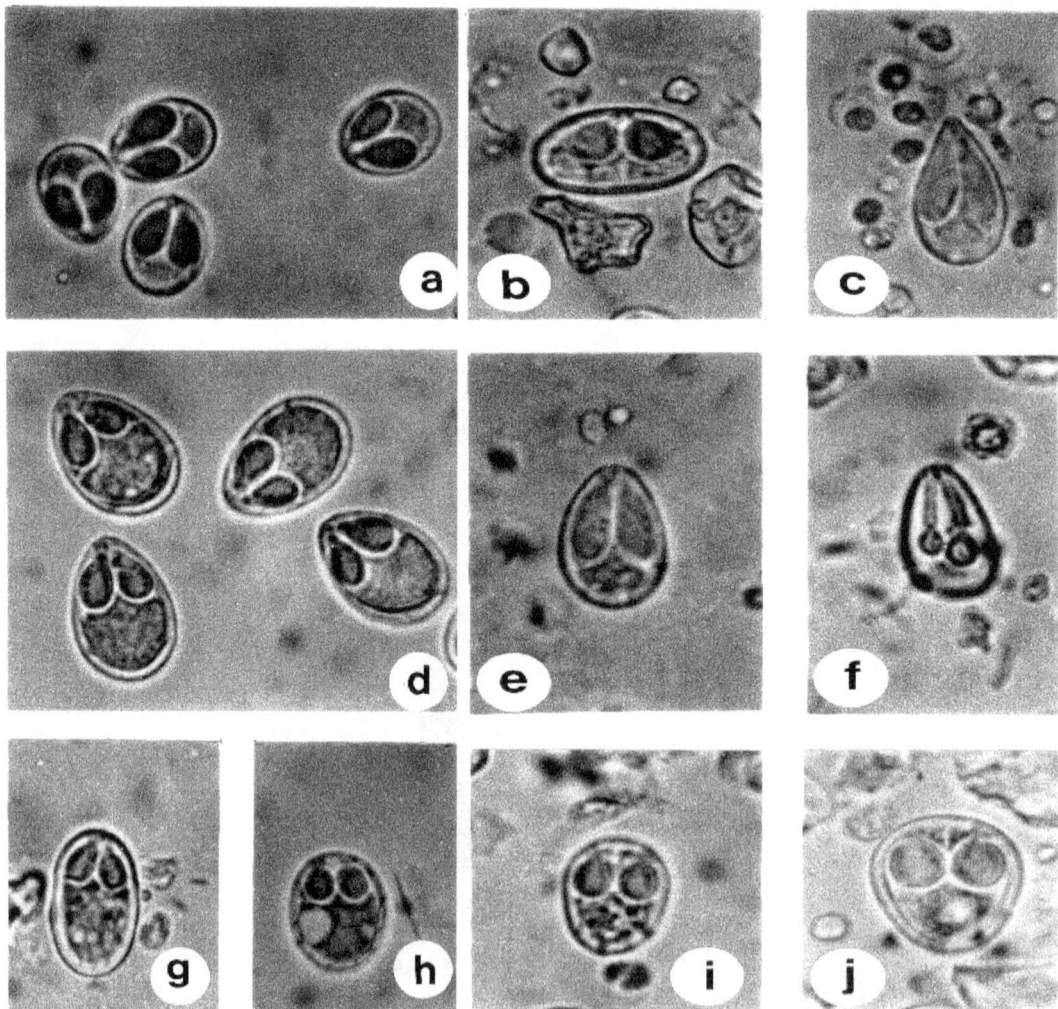

Figure 3. Spores of different species of Myxobolus studied. a: *Myxobolus agolus* (X 1500); b: *M. brachysporus* (X 1500); c: *M. heterosporus* (X 1500); d: *M. camerounensis* (X1500); e: *M. israelensis* (X 1500); f: *M. equatorialis* (X 1500); g: *M. tilapiae* (X 1500); h: *M. kainjiae* (X 1500); i: *M. sarigi* (X 1500); J: *M. nounensis* (X 1500).

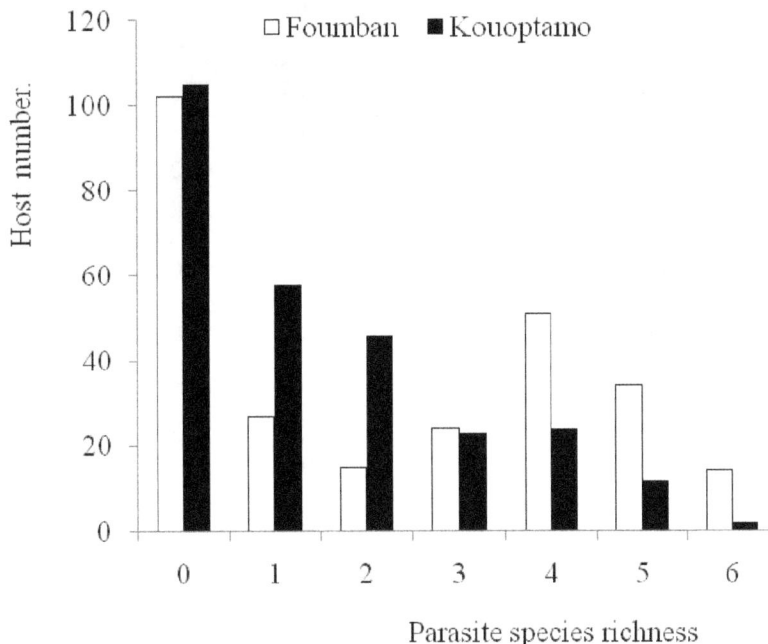

Figure 4. Host distribution (*O. niloticus*) as a function of the Myxosporean species richness.

Table 2. Percentage of parasitize organs by Myxosporean in wild and cultured *O. niloticus* at Foumban/kouoptamo; *M. Myxobolus.*

Parasite species	Parasitize organs									
	Gill	Liver	Gonad	Operculum	Skin	Spleen	Kidney	eyes	x^2 Value	P Value
M. agolus	1.8/0	2.7/2.7	0/0	0/11	0/0	22.2/8.3	34.6/10.8	0/0	x^2= 46.7	P< 0.001
M. brachysporus	0/0	3.1/2.7	0/0	0/0	0/0	43.8/40.0	51.9/40.9	0/0	x^2= 2.38	P= 0.12
M. camerounensis	23.7/4.0	0/0	0/0	1.5/3.3	11.1/0	0.4/0.4	0.9/0.8	26,9/0	x^2=32.36	P< 0.001
M equatoriais	0/0	0/0	0/0	0/0	0/0	5,8/4.2	10.0/3.1	0/0	x^2=7.49	P=0.06
M. heterosprus	0/0	2.2/1.6	0/0	0/0	0/0	29.2/12.8	39.0/15.4	0/0	x^2=40.0	P< 0.001
M. israelensis	0/0	3.6/1.2	0/0	0/0	0/0	32.9/13.9	39.4/14.7	0/0	x^2=51.4	P< 0.001
M. kainjiae	0/0	0/0	2.5/1.3	0/0	0/0	0/0	0/0	0/0	x^2=1.22	P= 0.26
M. sarigi	0/0	0/0	0/0	0/0	0/0	0.8/1.9	1.7/1.9	0/0	x^2= 1.7	P= 0.18
M. tilapiae	0/0	0/0	0/0	0/0	0/0	2.1/9.4	4.8/11.6	0/0	x^2=21.0	P< 0.001

M. nounensis is secondary and is present only in the River Noun.

Comparison of infestations rates of *Oreochromis niloticus* in the two study sites.

Difference between the prevalence of *M. brachysporus* (x^2=0.105; P=0.82), *M. sarigi* (x^2=0.58; P=0.44) and *M. kainjiae* (x^2=1.25; P=0.72) in the natural environments and the fish ponds are statistically not significant. *M. agolus, M. camerounensis, M. equatorialis, M. heterosporus M. tilapiae,* and *M. israelensis* have

statistically different rates of infestation (P < 0.001) between the two study sites. *M. nounensis* was not found in Foumban (Table 1).

Parasitism according to the infected organs

The various species of Myxosporean infest the same organs in the two biotopes except *M. camerounensis* (collected on the skin and eyes) and *M. agolus* found in the gills of *O. niloticus* in fish ponds (Table 2). The highest prevalence was observed in the kidneys (61.3 and 49.0%, respectively in fish ponds and natural

Table 3. Infection rate (%) of Myxosporean parasite in *O. niloticus* as a function of the size class at Foumban and Kouoptamo.

Parasites species	Localities	Size class			χ^2	P
		(20-70)	(70-120)	≥ 120		
M. agolus	Foumban	30.6	25.0	71.4	29.53	0.001
	Kouoptamo	50.0	11.2	13.3	5.6	0.061
M. brachysporus	Foumban	66.7	40.8	714	16.75	0.001
	Kouoptamo	100	51.0	40.0	6.4	0.04
M. camerounensis	Foumban	22.2	17.3	28.6	19.33	0.001
	Kouoptamo	0.0	7.3	1.7	2.88	0.24
M. equatorialis	Foumban	11.1	14.3	22.0	2.2	0.33
	Kouoptamo	0.0	2.9	10.0	5.6	0.56
M. heterosporus	Foumban	2.8	0.5	11.4	16.11	0.001
	Kouoptamo	50.0	16.0	15.0	3.4	0.18
M. israelensis	Foumban	69,9	31.1	57.1	19.33	0.001
	Kouoptamo	50.0	16.0	15.0	3.4	0.18
M. kainjiae	Foumban	0.0	2.6	0.0	1.84	0.34
	Kouoptamo	25.0	3.9	0.0	8.09	0.18
M. sarigi	Foumban	2.8	0.5	11.4	16.16	0.001
	Kouoptamo	25.0	3.9	0.0	8.09	0.18
M. tilapiae	Foumban	2.8	9.7	8.6	1.85	0.4
	Kouoptamo	100	13.6	21.7	22.5	0.001

M, Myxobolus; ddl= 2.

environment) and in the spleen (50.5% in Foumban and 47.5% in Kouoptamo). Variation of the rates of infestation according to the infested organs (gills, skin, spleen, gonad, operculum, kidney, eyes and liver) shows a statistically significant difference in Foumban (χ^2=274.20; P< 0.01) and in Kouoptamo (χ^2=427.4; P<0.001). *M. camerounensis* with 4 organs (gills, operculum, kidneys and spleen) parasitized in natural environment and 6 organs (gills, operculum, kidneys, spleen, skin and eyes) infested in fish ponds is the species which presents the largest spectrum of target organs in the two biotopes. The parasitic rates of infestation are statistically higher (χ^2=8.48; P=004) in the organs of *O. niloticus* in fish ponds than in the natural milieu. *M. agolus* (χ^2= 46.7; P<0.001); *M. heterosporus* (χ^2=40.0; P<0.001) and *M. israelensis* (χ^2=51.4; P<0.001) presents the highest prevalence in the kidneys and spleen of *O. niloticus* in fish pond than in natural environment. On the other hand, *M. tilapiae* show a higher percentage of infestation (χ^2= 21.0; P<0.01) in natural environment than in fish pond in

these same organs. The gills of *O. niloticus* are parasitized by *M. camerounensis* in Foumban (χ^2=28.4; P<0.01) than in Kouoptamo. In the two situations, the rare species *M. kainjiae*, infests only the gonads whereas the majority of species following the example of *M. brachysporus*, *M. israelensis*, *M. heterosporus*, *M. tilapiae*, *M. sarigi* and *M. equatorialis* infect two organs (kidneys and spleen).

Parasitism according to the size of the host

The parasitic infection rates are statistically not significant (χ^2=3.38; ddl= 2; P=0.18) in the various size of classes in natural environment (Table 3). In breeding situation on the other hand, the bigger fish (LS≥ 120mm) are generally infested (χ^2=5.84 ddl=2; P=0.054) than small hosts and those of intermediate size. The comparison of the various sets of class between the two sites shows that the parasitic rates of infection are significantly higher

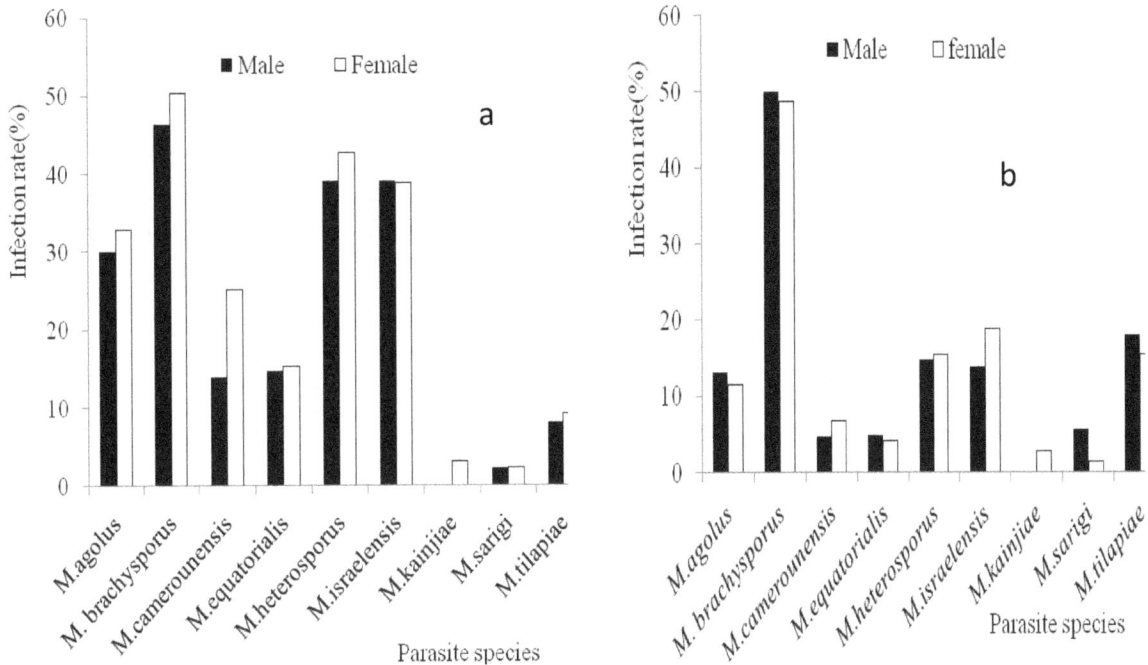

Figure 5. Parasitism (%) as a function of the sex of *O. niloticus* at Foumban (a) and Kouoptamo (b).

($P<0.001$) among the bigger fish in fish pond than in the natural environment. *M. agolus* (χ^2=32.9; *P<0.001*); *M. brachysporus* (χ^2=8.7; P=0.003); *M. heterosporus* (χ^2= 27.9; P<0.001) *M. israelensis* (χ^2=18.5; *P<0.01*) and *M. sarigi* (χ^2=7.15; P=0.007) infest mostly the bigger fish in fish ponds than in the natural environment. *M. agolus* (χ^2=5.6; P=0.06); *M brachysporus* (χ^2=6.4; P=0.04), *M. israelensis* (χ^2=16. 06; *P<0.001*) and *M. sarigi* (χ^2=22.5; *P<0.001*) infest preferentially young fishes in the River noun than in fish ponds.

Parasitism according to the sex of the host

The various species of Myxosporean collected parasitize indifferently the males and females in fish pond (χ^2= =2.461; P= 0.117) than in natural situation (χ^2= 0.508; P= 0.476) (Figure 5). The hosts of the two sexes are statistically (*P<0.001*) more parasitized in the fish ponds than in River Noun by *M. camerounensis*, *M. israelensis*, *M. brachysporus*, *M. agolus* and *M. equatorialis*. In addition, *M. brachysporus, M. sarigi* and *M. kainjiae* have statistically comparable rates of infestation between the males and the females of *O. niloticus* in the two study sites.

Parasitism according to seasons

In the fish ponds basins of Foumban, the rates of infestation of the various parasitic species are generally

higher (χ^2=5.84; P=0.16) in the rainy season than in the dry season (Figure 6). In natural environment (in Kouoptamo), the influence of the season on parasitism is not significant (χ^2=0.27; P=0.604). In Foumban, *M. tilapiae* presents a higher percentage of infestation (χ^2=4.0; P=0.27) during the dry season, whereas the rates of infestation of *M. camerounensis* (χ^2=6.12; P=0.013) and *M. israelensis* (χ^2=4.91; P=0.27) are higher in the rainy season. The following species, *M. agolus, M. heterosporus, M. brachysporus, M. sarigi, M. kainjiae* and *M. equatorialis* present statistically identical prevalence in both seasons. In Kouoptamo, the parasitic prevalence is statistically comparable in the two seasons, except for *M. israelensis* (χ^2=5.71; P=0.017) whose rates of infestation are high during the rainy season. A comparison between the study sites shows that the rate of infestation of *M. equatorialis* is higher in the fish pond basin than in River Noun during the rainy season (χ^2=8.09; P=0.004) as well as in the dry season (χ^2=11.13; *P<0.01*). On the contrary, *M. tilapiae* in the natural milieu parasitizes more fish than in breeding situation during the rainy season (χ^2=7.86; P=0.005) as well as in the dry season (χ^2=4.69; P=0.49).

DISCUSSION

This work carried out in a natural environment (River Noun) and in breeding situation (Fish pond basin), show in both cases a polyparasitism of fish by Myxosporean. The parasite species *M. nounensis* is present only in River Noun where it was originally described

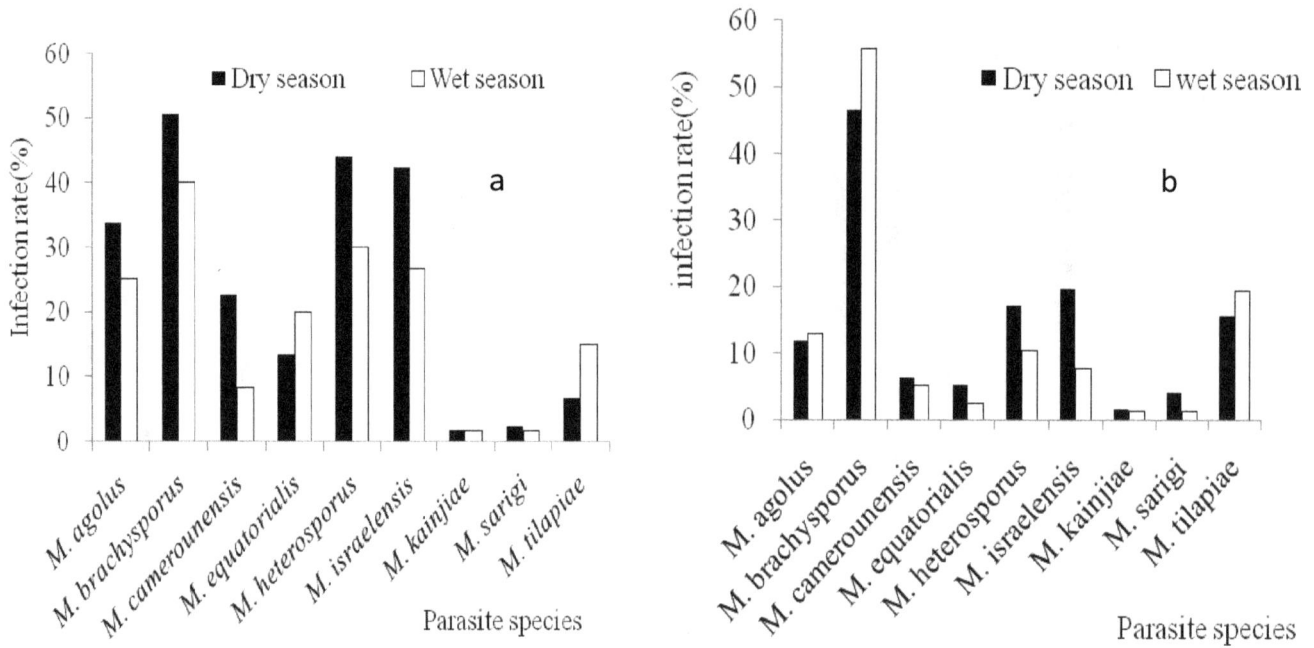

Figure 6. Parasitism (%) as a function of the season in *O. niloticus* at Foumban (a) and Kouoptamo (b).

(Fomena and Bouix, 2000). All other parasite species: *M. camerounensis*, *M. agolus*, *M. brachysporus*, *M. israelensis*, *M. heterosporus*, *M. kainjiae*, *M. sarigi*, *M. equatorialis* and *M. tilapiae* are common in both biotopes. The presence of *M. nounensis* only in River Noun could be explained by the fact that this medium may provide better eco-climatic conditions for the development of this parasite and could facilitate the contact between the infesting stages of this Myxosporean and the host fish (El-Tantawi, 1989). The various species of parasites recorded in Foumban and Kouoptamo were already documented in the same host, in the same organs by Obiekezie and Okaeme (1990) in Nigeria, Fomena and Bouix (1997) in Cameroon and Abakar et al. (2007) in Chad. The identified parasites present to a certain degree the same status in the two sites.

It results from our work that no parasitic species appeared frequent (Prevalence > 50%), whereas studies undertaken in Israel (Landsberg, 1985), in Nigeria (Obiekezie and Okaeme, 1990) and in Chad (Abakar, 2006) show that *M. agolus*, *M. brachysporus* and *M. heterosporus* are frequent in the Cichlidae *O. niloticus* and *S. galilaeus*. Poulin (2006) thinks that parameters (prevalence, intensity, abundance) traditionally used to qualify populations of parasites or the severity of the parasitic infection is prone to variations. For a given species of parasite, the proportion of the infested hosts is not fixed through its geographical distribution range. The status of parasitic species varies according to environmental conditions (El-Tantawi, 1989) and hosts species (Brummer-Korvenkontio et al., 1991).

The rates of infection of some collected species of Myxosporidian are higher in the fish ponds than in the River Noun. Our results corroborate those of Alvarez Pellitero and Sitjà Bobadilla (1989) and Sitjà Bobadilla and Alvarez Pellitero (1990) who in their work, showed that the rate of infestation of *Ichthophomus* sp and *Ceratomyxa* sp are higher in a breeding medium than in a natural environment. On the other hand, Sitjà Bobadilla and Alvarez Pellitero (1993) show that the rate of infestation of *Sphaerospora dicentrarchi* is higher in fish resulting from the natural environment as compared to the hosts from fish pond. These authors explain this situation by the age of fishes (They worked on old hosts) and the fact that the infesting stages of this parasite would be more viable in natural environment than in fish pond medium. Euzet and Pariselle (1996) think that, the pathogenic effect of the parasites under natural conditions is reduced, consequence of the balance established during the evolution in the host-parasite system. Indeed in the fish pond medium, the confinement of hosts, the presence of the muddy vase, the weak oxygenation which is at the origin of an increased sensitivity to the parasites attacker and the low depth of the basins would favor the transmission of the parasites (Obiekezie and Okaeme 1990; Fomena, 1995; Barassa et al., 2003; Tombi and Bilong Bilong, 2004).

In natural environment as in breeding situation, the various parasitic species infest indifferently the males and females. Our results agree with those of Fomena (1995) which did not note any influence of sex on the parasitism of *O. niloticus* in the various species of Myxosporean in fish ponds station of Melen in Yaoundé. Based on the same predictions, Özer (2003) reveals that there is no difference of infestation with respect to the sex in *Gasterosteus aculeatus* by *Sphaerospora elegans* and *Myxobilatus gasterostei* in Turkey. However, authors such as Viozzi and Flores (2003); Milanin et al. (2010) think that concerning Myxosporidiosis, the prevalence is independent of the sex of the host. On the other hand, Tombi and Bilong Bilong (2004) revealed that *M. njinei* parasite of *Barbus martorelli* (Cyprinidae) infests preferentially females than males. According to Sakiti (1997) and Gbankoto et al. (2003), parasitic infections often present higher prevalence in males than in females. According to Poulin (1996), the elevated level of testosterone secreted by the males can involve under certain conditions an immune depression making these individuals more vulnerable to parasitism. A good knowledge of the biology of parasites and their various hosts would make it possible to better explain the parasitism of Myxosporean with respect to sex.

In nearly all cases, parasitic species found in *O. niloticus* (Cichlidae) in natural environment and in breeding are present in their hosts during all the study period. At Foumban, the rates of parasitic infestation are higher in the rainy season and generally low in the dry season. On the other hand, no statistically appreciable fluctuation of parasitism is observed for the wild populations of hosts. Our results are close to those of Sitjà Bobadilla and Alvarez Pellitero (1993) who showed that in Spain, the rate of infestation of *S. dicentrarchi*, parasite of *Dicentrarchus labrax* varies with seasons in the fish pond basins, whereas in natural environment these authors do not note a difference between the percentages of infestation according to seasons. In this work, seasons do not seem to have an influence on Myxosporean in the natural environment. This remark corroborates many observations already made by authors such as Fomena (1995) on *M. camerounensis*, parasite of *O. niloticus*, in Cameroon; Gbankoto et al. (2001) on *Myxobolus* sp and *M. zillii*, gills parasites of *Tilapia zillii* and *Sarotherodon melanotheron melanotheron* in Benin and Milanin et al. (2010) on *M. oliveirai*, parasite of *Brycon hilarii* (Characidae) in Brazil. However, our results are contrary with those of Gbankoto et al. (2003) who note a seasonal difference of the infestation of *Tilapia zillii* by *M. heterosporus* in Benin.

During wet season, the higher water temperature and the presence of muddy vase are the major factors affecting the prevalence of both myxosporean and their definitive host (Tubifex tubifex). According to Özer et al. (2002), mud substrat allow rapid growth and multiplication of Oligochaete. Therefore, during wet season, annelids Oligochaete are very abundant and their infecting stages (actinospores) multiply rapidly, a situation which is favorable to fish infection (Abakar et al., 2007).

The spectrum of target organs of the various studied species of Myxosporean varies from 1 to 6 in *O. niloticus* in the two mediums. Abakar (2006) showed that the spectrum of organs colonized by *O. niloticus* and *S. galilaeus* in Chad varies from 1 to 5. *M. camerounensis* appeared as the species having a broad spectrum of organs colonized in this work. We found *M. camerounensis* spores or cyst in six organs in fish pond medium and in four organs in the natural environment. Abakar et al. (2007) found this species infesting six different organs in *O. niloticus* and *S. galilaeus*. According to Obiekezie and Okaeme (1990), Myxosporean of *Tilapia* have a general distribution in tissues. In addition, Alvarez Pellitero and Sitjà Bobadilla (1989) think that spores of the various species of Myxosporean can disperse in the kidneys and other internal organs via blood, causing severe infections. The kidneys and the spleen are organs that are most frequently infested. These results are close to those of Fomena (1995) and Abakar et al. (2007). The kidneys and the spleen may be the site of predilection for certain species of Myxosporean. These organs according to Fomena (1995) constitute the sites of initiation of the developmental cycle for many species of Myxosporean. In addition, Alvarez Pellitero and Sitjà Bobadilla (1989) think that the capacity of Myxosporean to locate itself in a given tissue or cavities of fish could have an influence on the type of damage which these parasites cause to their hosts. The gills constitute one of the organs of predilection in Myxosporean (Lekeufack Folefack, 2010) mean while, the massive infestation of this organ by *M. camerounensis* in the fish pond basin in Foumban, could be at the origin of the reduction in the respiratory capacity of the host. In the same regard, Gbankoto et al. (2001) reveal that this situation would affect its reproductive success negatively.

The rate of infestation of almost all disease-causing agents does not vary statistically between the various sets of classes. The percentage of infestation of the various parasites species is higher in breeding condition than in natural environment in old fish. In natural environment on the other hand, younger hosts are more threatened by certain parasitic species. According to Tombi and Bilong Bilong (2004), the parasitic load is high among hosts of small size host compared to those of big size. On the other hand, Sitjà Bobadilla and Alvarez Pellitero (1993) revealed a progressive increase in the infection of *D. labrax* by *S. dicentrarchi* with the age of the host in agreement with a weak pathogenicity of the parasite. These authors explain this situation by the effect of parasites accumulation in organs with age. In the fish ponds at Foumban, the young hosts would succumb more quickly to the infections because of confinement. According to Lom and Dykovà (1992), young fish are more susceptible to Myxosporean and certain infections decrease with host age.

This study had showed a polyparasitism of *O. niloticus* both in the Noun River and the fish ponds ecosystem in Cameroon. In general, fishes are more infected by Myxosporean when cultured than in Wild environment. This study raised the necessity to pay more attention concerning Myxosporean parasite of fish in a fish farm station because of the hosts confinement. In addition, further investigation for these Myxosporean species is necessary to determine the real effect of these parasites on the health of the hosts.

ACKNOWLEDGEMENTS

We appreciate the technicians of the Foumban farm station for their help during the sampling data. Special thank to Dr. Kekeunou and Dr. Ndassa for their useful comments.

REFERENCES

Abakar O (2006). Les Myxospories (Myxozoa : Myxosporea) parasites des poissons d'eau douce du Tchad : faunistique et biologie des espèces inféodées à *Oreochromis niloticus* (Linné, 1758) et *Sarotherodon galilaeus* (Linné, 1758) Cichlidae. Thèse de Doctorat/PhD, Université de Yaoundé I, 2006, p. 163.

Abakar O, Bilong CF, Njine T, Fomena A (2007). Structure and dynamics of myxosporean parasites component communities in two fresh water Cichlids in the Chari River (Republic of Chad). Pak. J. Biol. Sci., 10: 692-700.

Alvarez P, Bobadilla SA (1989). On the influence of culture and stress condition on *Ceratomyxa spp.* (Myxozoa) infections in sea bass (*Dicentrarchus labrax*) from the Spanish Mediterranean area. Proceedings of the IVt1 EAFP Conference on Fish and Shellfish Diseases, Santiago de Compostela, Spain, p. 27.

Barassa B, Adriano EA, Arana S, Cordeiro NS (2003). *Henneguya curvata* sp. n. (Myxosporea; Myxobolidae) parasitizing the gills of *Serrasalmus spilopleura* (Characidae: Serrasalmidae), South American fresh water fish. Folia Parasitologica, 50: 151-153.

Brummer-Korvenkontio H, Valtomen T, Pugachev ON (1991). Myxosporea parasites in roach, *Rutilus rutilus* (Linnaeus) from four lakes in Central Finland. J. Fish Biol., 38: 573-586.

Combes C (1995). Interactions durables. Ecologie et évolution du parasitisme. Paris, France, Masson (Collection écologie, 26: 524.

Eiras JC, Molnár K, Lu YS (2005). Synopsis of the species of *Myxobolus* Bütschli, 1882 (Myxozoa, Myxosporea : Myxobolidae). System. Parasitol., 52: 43-54.

Eiras JC, Monteiro CM, Brasil-Sato MC (2010). *Myxobolus franciscoi sp.* Nov. (Myxozoa: Myxosporea), a parasite of *Prochilodus argenteus* (Actinopterygii: Prochilodontidae) from the Upper Sao Francisco River, Brazil, with a revision of *Myxobolus spp.* from South America. Zoologia, 27(1): 131-137.

Eiras JC, Takemoto RM, Pavanelli GC (2008). *Henneguya caudicula* n. sp. (Myxozoa, Myxobolidae) a parasite of *Leporinus lacustris* (Osteichthyes, Anostomidae) from the high Paraná River, Brazil, with a revision of *Henneguya* spp infecting South American fish. Acta Protozool., 47: 149-154.

El-Tantawi SAM (1989). Myxosporidian parasites of fishes in lakes Dgal Wielki and Warniak (Mazurian Lakeland, Poland). I. Survey of parasites. Acta Parasitol. Polonica, 34(3): 203-219.

El-Tantawi SAM (1989). Myxosporidian parasites of fishes in lakes Dgal Wielki and Warniak (Mazurian Lakeland, Poland). II. Infection of fishes. Acta parasitol. Polonica, 34(3): 221-233.

Euzet L, Pariselle A (1996). Le parasitisme des poissons Siluroidei: un danger pour l'aquaculture? Aquatic Living Resour., 9: 145-151.

FAO (2008) Comité des pêches continentales et d'aquaculture pour l'Afrique (CPCAA). Renforcer le CPCAA dans le cadre des initiatives de coopération régionale. Quinzième session, Lusaka Zambie, 9-11 décembre 2008.

Feist SW, Longshaw M (2006). Phylum Myxozoa. 230-297. In Pik Woo (Ed), fish diseases and disorders, protozoan and metazoan infestion 2nd Ed., CABI publishing, UK.

Feist SW, Lonshaw M (2005). Myxozoan diseases of fish and effects on host population Acta Zool. Sin., 51(4): 758-760.

Fomena A (1995). Les Myxosporidioses et Microsporidioses des poissons d'eau douce du Sud-Cameroun: Etude faunistique, Ultra structure et Biologie. Thèse de Doctorat d'Etat, Université de Yaoundé I, p. 397.

Fomena A, Bouix G (2000) *Henneguya mbakaouensis* sp. nov. *Myxobolus nounensis* sp. nov. And *M. hydrocyni* Kostoingué & Toguebaye, 1994, Myxosporea (Myxozoa) parasites of Centropomidae, Cichlidae and Characidae (Teleosts) of the Sanaga basin in Cameroon (Central Africa). Parasite, 7: 209-214.

Fomena A, Bouix, G (1997). Myxosporea (Protozoa : Myxozoa) of freshwaterfishes in Africa : Keys to genera and species. System. Parasitol., 37: 161-178.

Fomena A, Lekeufack Folefack GB, Bouix G (2008). Three new species of Henneguya (Myxozoa: Myxosporea), parasites of fresh water fishes in Cameroon (Central Africa). J. Afrotrop. Zool., 4: 93-103.

Fomena A, Lekeufack Folefack GB, Bouix G (2010). Deux espèces nouvelles de Myxidium (myxosporea: Myxidiidae) parasites de poisons d'eau douce du Cameroun. Parasite, 17: 9-16.

Gbankoto A, Pampoulie C, Marques A, Sakiti GN (2001). Occurrence of Myxosporean parasites in the gills of tilapia species from Lake Nokoue (Benin, West Africa): effect of host size and sex, and seasonal patterns of infection. Dis. Aqua. Organ., 44: 217-222.

Gbankoto A, Pampoulie C, Marques A, Sakiti GN, Dramane KL (2003). Infection patterns of *Myxobolus heterospora* in two tilapia species (Teleostei: Cichlidae) and its potential effects Dis. Aquat. Organ, 55: 125-131.

Hedrick RP (1998). Relationship of the host pathogen, and environment: implication for diseases of cultured and wild fish populations. J. Aqua. Anim. Health, 10: 107-111.

Landsberg JH (1985). Myxosporean infections in cultured Tilapias in Israel. J. Protozool., 32: 194-201.

Lekeufack Folefack GB (2010). Faunistique et biologie des Myxosporidies (Myxozoa: Myxosporea) parasites de quelques Téléostéens dans la rivière Sangé (sous affluent du Wouri) au Cameroun. *Thèse de Doctorat PhD*. Université de Yaoundé I, p. 181.

Lom J, Arthur JR (1989). A guideline for the preparation of species descriptions in Myxosporea, J. Fish Dis., 12: 151-156.

Lom J, Dykova I (1992). Protozoan parasites of fishes. Dev. Aquacult. Fish. Sci., 26: 1-315.

Lom J, Dyková I (2006). Myxozoan genera: definition and notes on taxonomy, life-cycle terminology and pathogenic species. Folia Parasitol., 53: 1-36.

Longshaw M, Freak PA, Nunn AD, Cowx IG, Feist SW (2010). The influence of parasitism on fish population success. Fish. Manage. Ecol., 17: 246-434.

Margolis L, Esh GW, Holmes JC, Kuris AM, Schad GA (1982). The use of ecological terms in parasitology (Report of an hoc committee of American Society of Parasitologists). J. Parasitol., 68(1): 131-133.

Martins ML, Souza VN De, Moraes JR, Moraes FR De (1999). Gill infection of *Leporinus macrocephalus* by *Henneguya leporinicola* n. sp. (Myxozoa: Myxobolidae). Description, histopathology and treatment Rev. Bras. Biol., 59(3).

Milanin T, Eiras JC, Arana S, Maia AAM, Alves AL, Silva MRM, Carriero MM, Ceccarelli, PS, Adriano E (2010). Phylogeny, ultrastructure, histopathology and parasite of *Brycon hilarii* (Characidae) in the Pantanal Wetland, Brazil. Mem Inst Oswaldo Cruz, Rio de Janeiro, 105(6): 762-769.

Mouchiroud D (2002). Chapitre 5: Statistique descriptive Mathématique: Outil pour la biologie. *DEUG sv1. UCBL.* Mathsv.univ-lyon1.fr/cours/pdf/stat/chapitre5.pdf., p. 19.

Obiekezie AI, Okaeme AN (1990). Myxosporea (Protozoa) infections of cultured Tilapias in Nigeria. J. Afr. Zool., 104: 77-91.

Oldewage WH, Van As JG (1987). Parasites and winter mortalities of Oreochromis mossambicus. S. Afr. J. Wldl. Res., 17(1).

Olivry JC (1986). Fleuves et rivières du Cameroun. O.R.S.T.O.M. (éd), p. 733.

Özer A (2003) Sphaerospora elegans Thelohan, 1892 and Myxobilatus gasterostei Davis, 1944 (Phylum: Myxozoa) infection in the Three-Spined Stickleback, Gasterosteus aculeatus L., 1758 in Turkey. Turk. J. Zool., 27: 163-169.

Özer A, Wootten R, Shinn AP (2002). Infection prevalence, seasonality and host specificity of actinosporean types (Myxozoa) in an Atlantic salmon fish farm located in Northeen Scotland, Folia Parasitol., 49: 263-268.

Poulin R (1996). Sexual inequalities in Helminth infections: a cost of being a male? Am. Naturalist, 147: 287-295.

Poulin R (2006). Variation in infection parameters among population within parasite species: Intrinsic properties versus local factors. Int. J. Parasitol., 36: 877-885.

Sakiti GN (1997). Myxosporidies et Microsporidies de poisons du Sud Benin: Faunistique, Ultrastructure, Biologie. Thèse de Doctorat D'Etat. Université du Benin, p. 296.

Sitja-Bobadilla A, Alvarez-Pellitero P (1993). Populations dynamics of Sphaerospora dicentrarchi Sitja-Bobadilla et Alvarez-Pelletero, 1992 and S. testicularis Sitja-Bobadilla et Alvarez-Pellitero, 1990 (Myxosporea : Bivalvulida) infections in wild and cultured Mediterranean sea bass (Dicentrarchus labrax L.) Parasitology, 106: 39-45.

Sitja-Bobadilla, Alvarez-Pellitero P (1990). Sphaerospora testicularis sp. Nov. (Myxosporea : sphaerosporidae) in wild and cultured sea bass, Dicentrarchus labrax (L.), from the Spanish Mediterranean area. J. Fish Dis., 13: 193-203.

Stiassny MIG, Teugels GG, Hopkins CD (2007). Poissons d'eaux douces et saumâtres de la basse Guinée, Ouest de l'Afrique Centrale. Collection faune et flore tropicales, IRD (éd), Paris I: 797.

Tombi J, Bilong Bilong CF (2004). Distribution of gill parasites of the freshwater fish Barbus martorelli Roman, 1971 (Teleostei : Cyprinidae) and tendency to inverse intensity evolution between Myxosporidia and Monogenea as a function of the host age. Revue d'élevage et de Medecine Vétérinaire des pays tropicaux, 57(1-2): 71-76.

Umur S, pekmerezci GZ, Beyhan YEB, Gurler AT, Acici M (2010). First record of Myxobolus muelleri (Myxosporea: Myxobolidae) in flathead grew mullet Mugil cephalus (Teleostei, Mugilidae) from Turkey Ankara univ Vet Fak Derg, 57: 205-207.

Valtomen ET, Holmes JC, Koskivaara M (1997). Eutrophisation, pollution and fragmentation: effects on parasite communities in roach (Rutilus rutilus) and perch (Perca fluviatilis) in four lakes in Central Finland. Can. J. Fish. Aquat. Sci., (54): 572-585.

Viozzi GP, Flores VR (2003). Myxidium biliare sp. n. (Myxozoa) from gall bladder of Galaxias maculates (Osmeriformes: Galaxiidae) in Patagonia (Argentina). Folia Parasitolol., 50: 190-194.

24

A comparative study on the morphology, anatomy, physiology and karyotype of the gadwall and red crested pochard that migrate to south Iraq marshes

Khitam Jassim Salih[1]* and Majeed Hussein Majeed Al-Sarry[2]

[1]Vertebrate Department, Marine Science Center, Basrah University, Iraq.
[2]Nursing College, Basrah University, Iraq.

Although the gadwall *Anas strepera* and red-crested pochard *Netta ruffina Netta* are reclassified back to the same family origin, most of their characters such as morphology, anatomy, and physiology were different. A total of twenty birds of the gadwall and red-crested pochard were included in this study. These birds were hunted in the south of Iraq marshes by the use of catch drag. The results revealed cardinal differences in the blood picture between *A. strepera* and *N. ruffina Netta* summarized in increase red blood cell count, white blood cell count, hemoglobin level, packed cell volume and differential white blood cell count in *N. ruffina Netta* compared to *A. strepera*. In addition, this study showed no differences in the karyotype of the two species.

Key words: Gadwall, red crested pochard, *Anas strepera*, *Netta ruffina Netta* dabbling ducks, south Iraq marshes.

INTRODUCTION

The aim of this study is to record some aspect data base about two different species of birds that emigrant to Iraq marshes. Various bird populations migrate long distances along a flyway. The most ordinary guide involves flying north in the spring to breed in the temperate or arctic summer and returning in the autumn to wintering grounds in warmer regions to the south. The gadwall *Anas strepera* is a common and widespread duck of the family Anatidae. This species was first described by Linnaeus in his system nature in 1758 under its current scientific name (http/www.Gadwall. Wikipedia, the free encyclopedia, 2011). The gadwall's closest relative within the genus *Anas* is the Falcated Duck, followed by the wigeons (Johnson and Sorenson, 1999). In addition, there are two subspecies although one is extinct, the nominate common gadwall *A. strepera strepera* and the Coues' Gadwall, extinct circa 1874 *A. strepera couesi*, that was located on Fanning Island (Clements, 2007). The diving ducks, commonly called pochards or scaups, are a category of duck which feed by diving under the surface of water. They are part of the varied and very large Anatidae family that includes ducks, geese and swans. The diving ducks are located in a distinct sub-family, Aythyinae. While they are morphologically close to the dabbling ducks, the red-crested pochard is a Palearctic species (Cramp and Simmons, 1977) of Sarmatic origin (Voous, 1960). Its breeding circulation extends just about between the latitudes 35° and 55° north, in continental, temperate and Mediterranean climatic regions, from the British Isles to China (Scott and Rose, 1996).

MATERIALS AND METHODS

A total of twenty bird of the gadwall *A. strepera* and the red-crested Pochard *N. ruffina Netta*, ten samples (five males and five females) each, were hunted in the south Iraq marshes by the use of catch drag. Afterwards, they were reared for one week to investigate the morphology, anatomy, genetic (Sugiyama, 1971) and physiology differences between them (Haen, 1995).

RESULTS AND DISCUSSION

Although the gadwall *A. strepera* and the red-crested

*Corresponding author. E-mail: khitam_36@yahoo.com.

Figure 1. The gadwall and red crested Pochard.

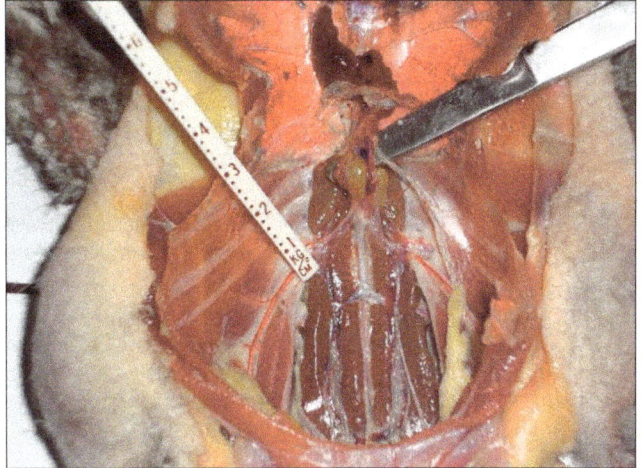

Figure 3. Kidney of the gadwall.

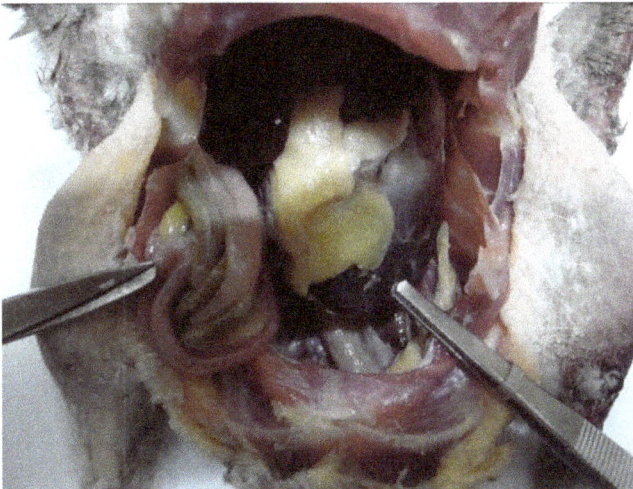

Figure 2. The viscera of the red crested Pochard.

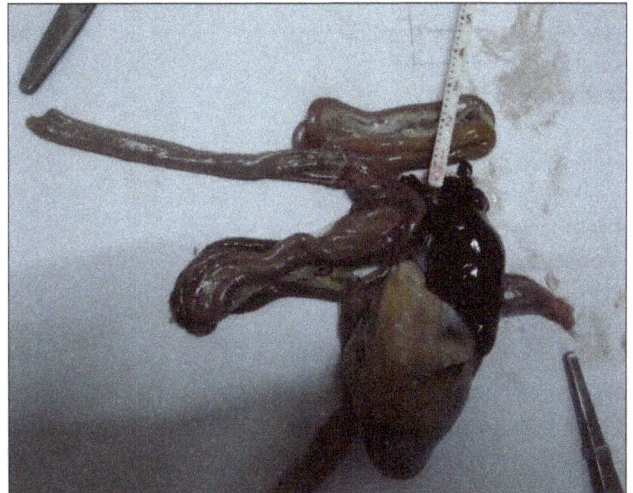

Figure 4. Viscera of the gadwall.

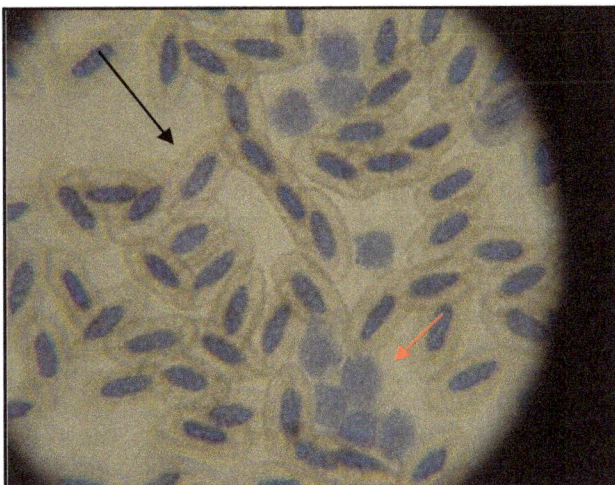

Figure 5. Erythrocytes and thrombocytes of the red-crested Pochard.

pochard *N. ruffina Netta* are placed back to the same family origin (Figure 1), the present study showed many differences between them in morphology and some physiological parameters enumerated in color of the beak, head, neck, chest and flank, wing and feet of the males and females of both species. Also, there are dissimilarities in the internal organ weight between them (Figures 2, 3 4, and Table 1). Moreover, the results revealed cardinal differences in the blood picture between the two species listed in the increase of red blood cell count, white blood cell count, hemoglobin level, packed cell volume and the differential white blood cell count. All these variations between the two species may be due to the behavior of each species concerned with the type of food and properties of diving under water. The red-crested Pochard, the number and level of blood parameters preceded those of the gadwall (Table 1, Figures 5, 6, 7, 8, 9, 10, 11, 12, 13 and 14).

Table 1. Shows the morphology, anatomy, and physiology comparison between the gadwall and red-crested pochard.

Parameters	Gadwall	Red-crested pochard
Head and neck	Male head and neck are brown spotted color while female has head and neck dark brown color	Male head and neck are bright henna color with short gold while female head is dark brown with grayish neck color
Beak	Male has black or brown greenish beak, while the female has orange beak	Male has carmine beak and female has grayish with red edges beak
Chest and flank	Male chest and flank are stripped with white and brown lines and flank stripped with grayish color. While female chest and flank are bright brown color with some dark brown spot	Male has dark chest and flank with some white feathers on the shoulders and the abdomen plumage dark brown black line striate, with white feathers in flank. The female has dark brown color with some dirty white spot chest
Wings	Both male and female have white wing bar	Both male and female have big white spot
Foot	Both male and female has dark orange	Male have orange foot and female has red foot
Red blood cell count million /cm³	2.46	3.08
White blood cell count thousands/cm³	22.50	24.30
Hemoglobin level (%)	48.30	41.05
Packed cell volume (%)	29.80	34.25
Lymphocytes (%)	25.88	26.15
Monocytes (%)	11.92	11.83
Hetrophils (%)	52.96	56.08
Eosinophils (%)	3.84	3.59
Basophils (%)	1.95	1.90
Liver weight (g)	23.13	28.50
Proventriculus (empty) weight (g)	10.60	13.20
Gizzard (empty) weight (g)	16.80	18.50
Gall bladder(fill) weight (g)	2.80	3.50
Kidney weight (g)	5.60	7.60
Alimentary canal length (cm)	108.50	115.10
Male length (cm)	48.70	55.50
Female length (cm)	45.80	48.90
Male weight (g)	910.70	980.20
Female weight (g)	830.50	910.50

The karyotype of *A. strepera* and *N. ruffina Netta* species is presented in Figures 16 and 18. It was extremely difficult to determine the chromosome number because of the large number of micro-chromosomes. The result of this study showed no differences in the karyotype of the two species. Of these chromosomes, the first pair appeared as metacentric, the second pair was subtelocentric and the pairs no. 3, 4, 5, 6, 7 and 8 were acrocentrics. The chromosome Z was identified as a submetacentric with a size larger than that of the fourth pair (Figure 15) and the chromosome W was small acrocentric and easily identified (Figure 17). The micro chromosomes were so numerous and often so small to a

degree that could not be morphologically identified, a fact that is also reported in other duck species (Belterman and Boer, 1984; Lucca and Rocha, 1985; Lucca and Waldrigues, 1985).

In individual organisms, the phenotype consequences from its genotype and the power from the ecosystem lead to individual variation. A considerable branch of the variation in phenotypes in a population is caused by the differences between their genotypes. The current evolutionary combination defines that evolution was the modified mean greater than the time in the genetic differences. The incidence of one exacting allele will vary, becoming more or less common relative to other forms of

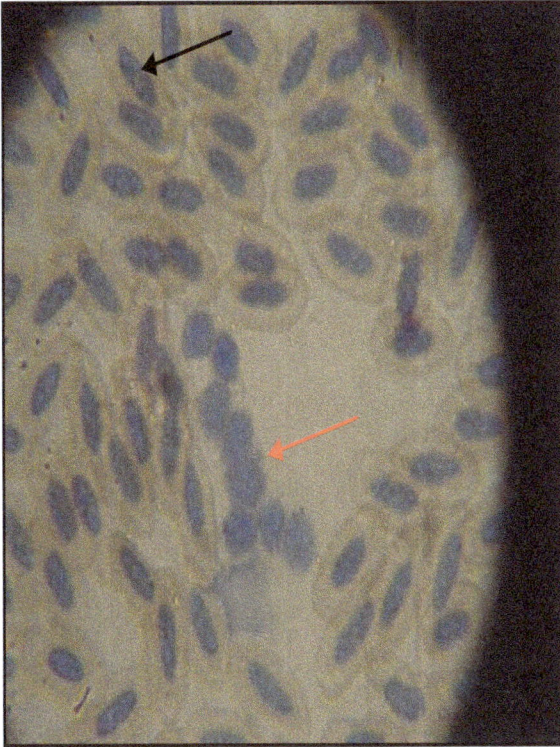

Figure 6. Erythrocytes and the gadwall.

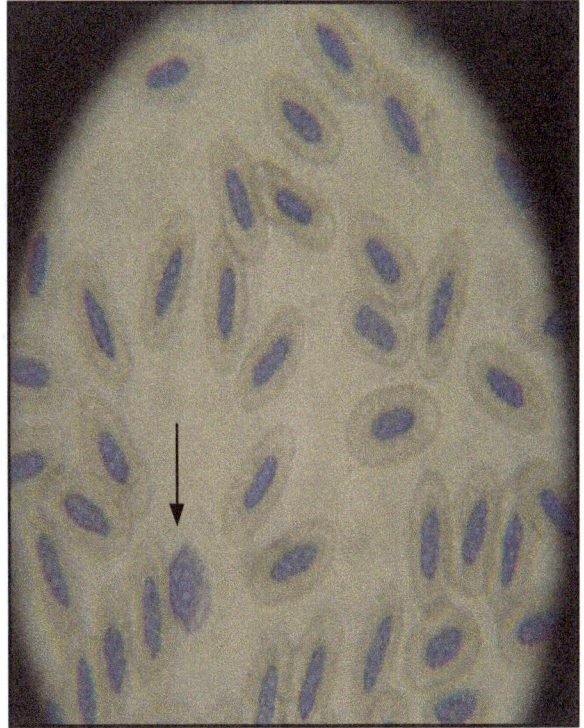

Figure 8. Lymphocyte of the gadwall.

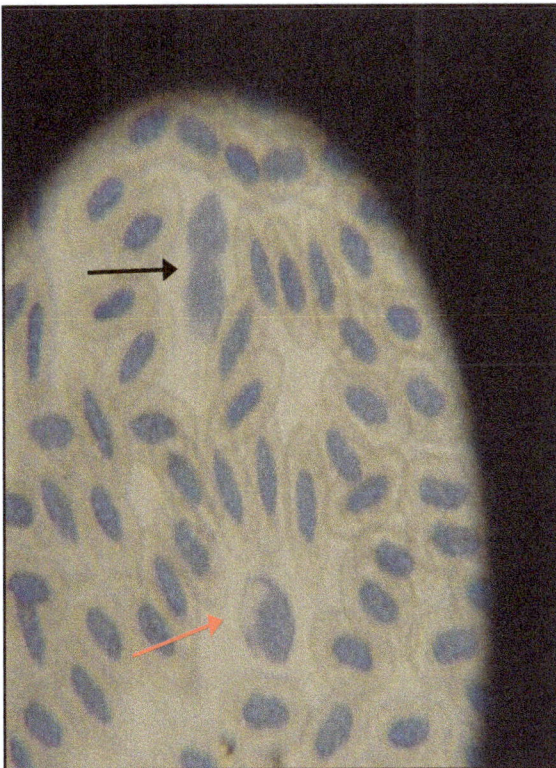

Figure 7. Lymphocyte and monocyts of the redcrested Pochard.

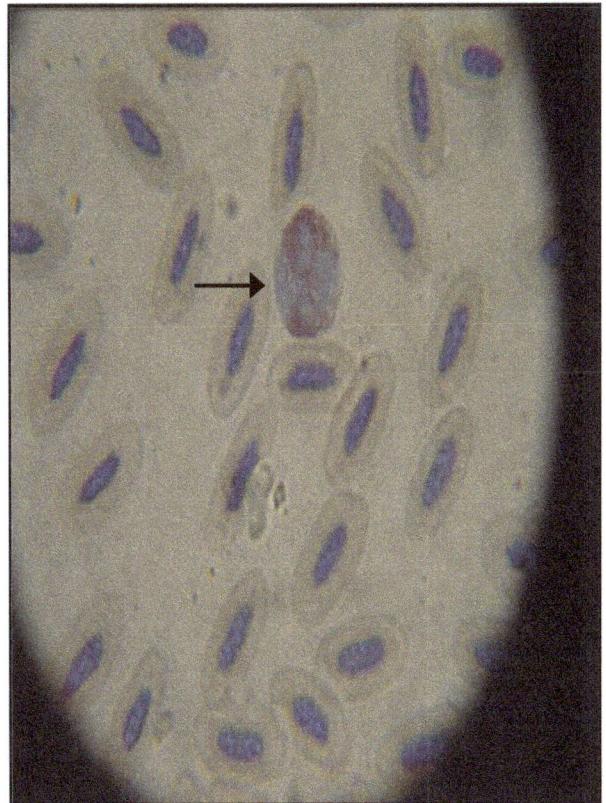

Figure 9. Hetrophil of the red-crested Pochard.

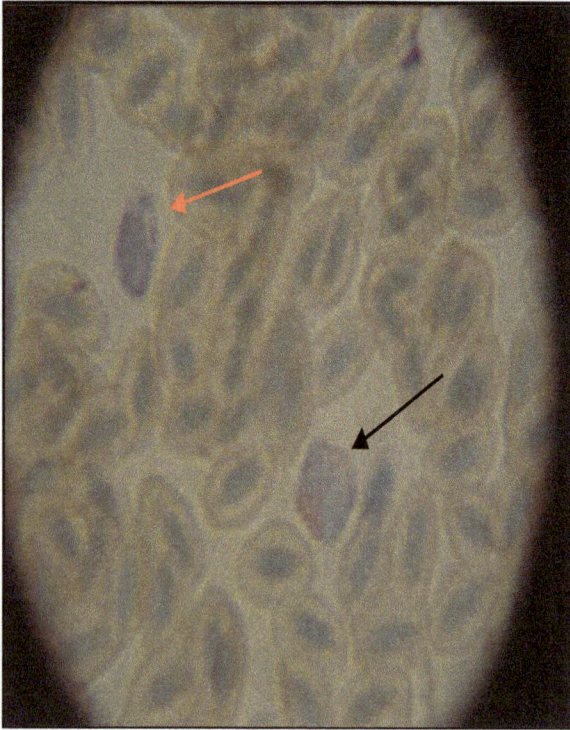

Figure 10. Hetrophil and monocytes of the gadwall.

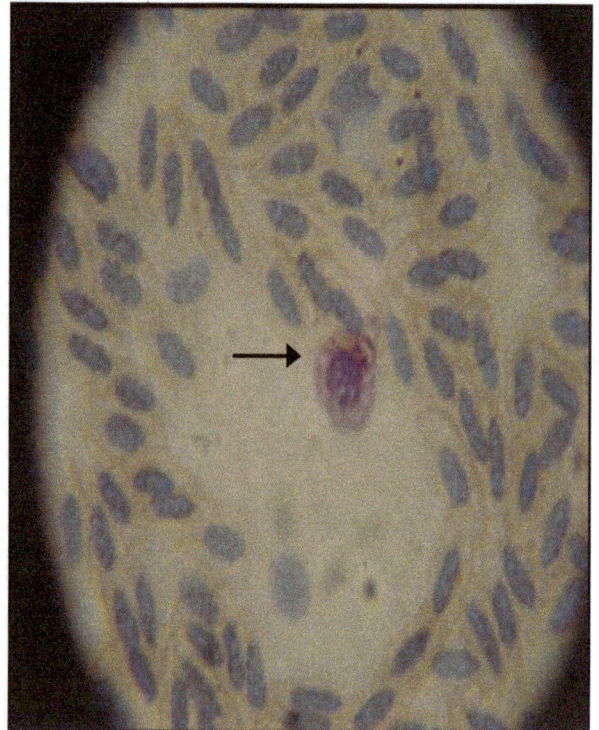

Figure 12. Basophil of the gadwall.

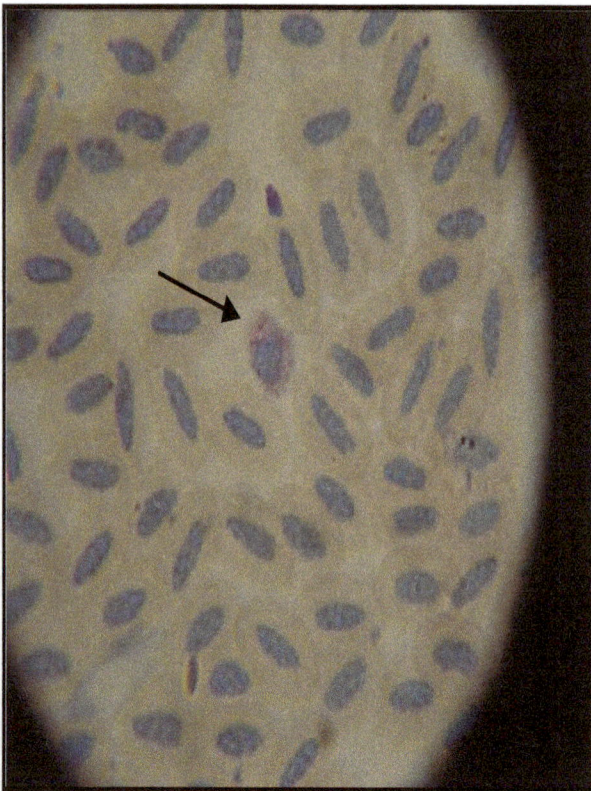

Figure 11. Basophil of the red-crested Pochard.

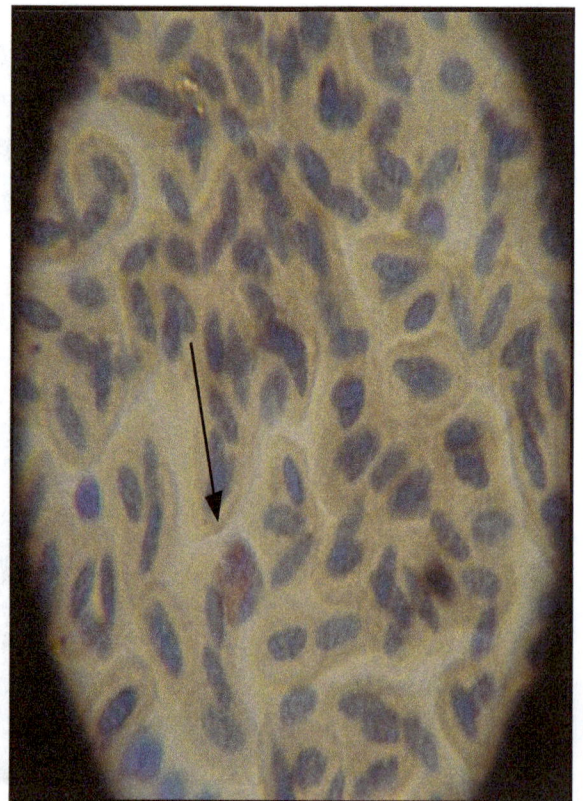

Figure 13. Eosinophil of the red-crested Pochard.

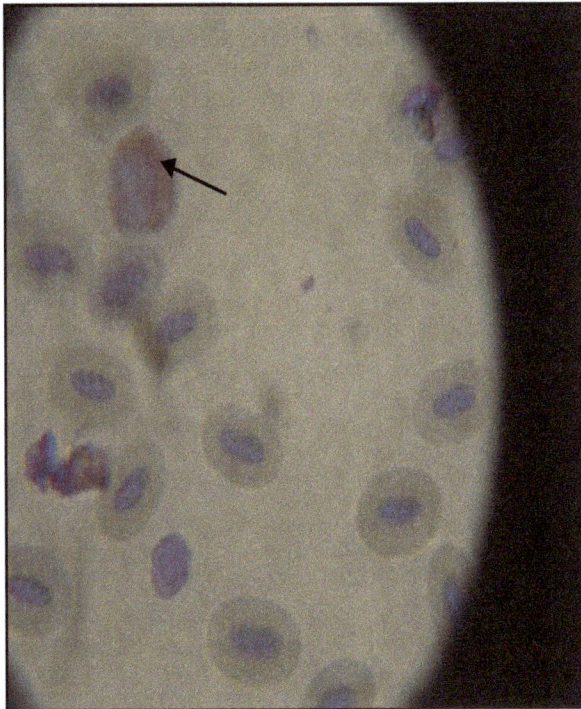

Figure 14. Eosinophil of the gadwall.

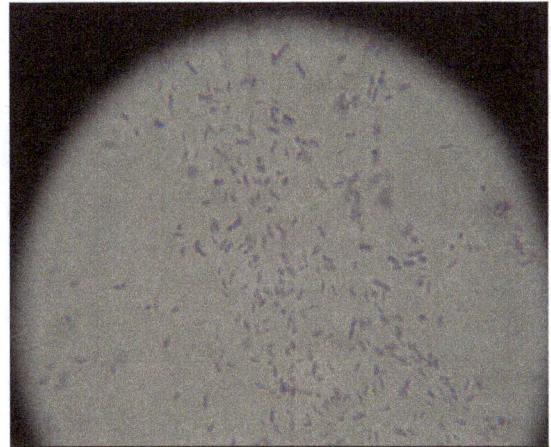

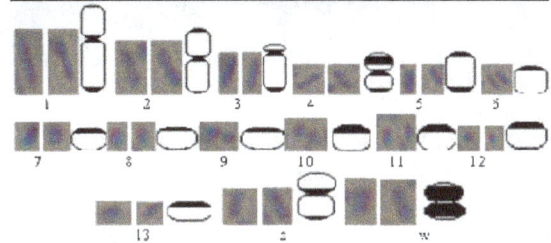

Figure 16. Ideogram and karyogram of the chromosomes of the gadwall (CBG banding).

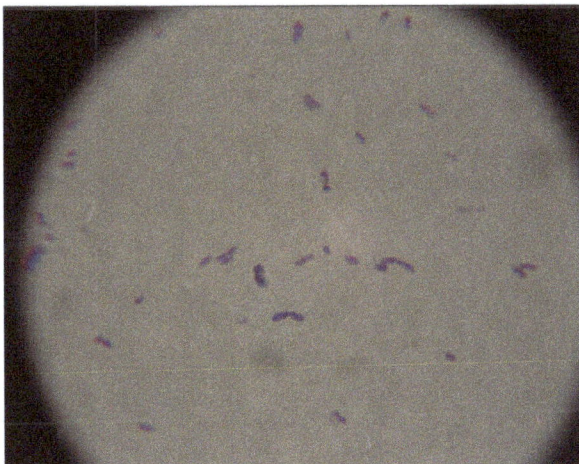

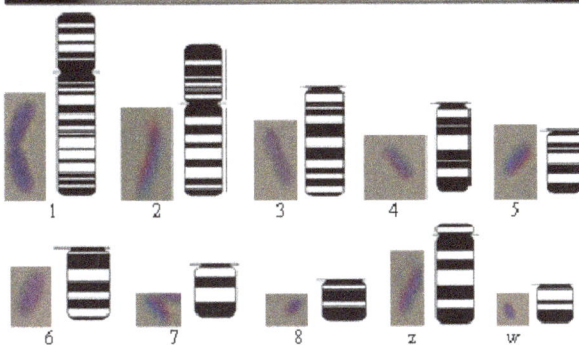

Figure 15. Ideogram and karyogram of the chromosomes of the gadwall (RBG banding).

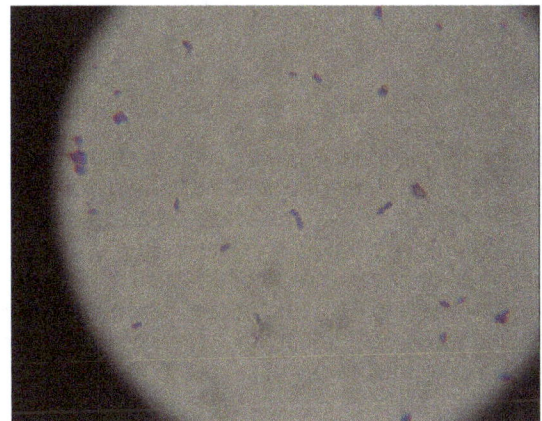

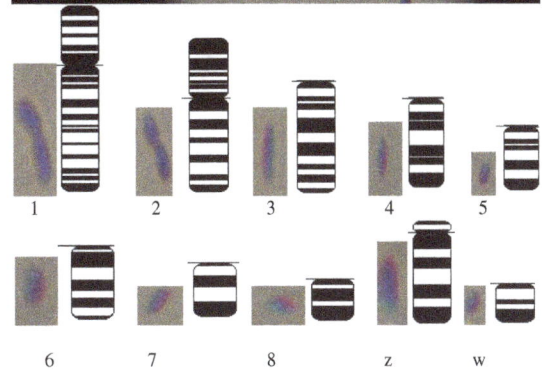

Figure 17. Ideogram and karyogram of the chromosomes of the red-crested Pochard (RBG banding).

also comes from exchanges of genes between different species; in spite of the stable forward of variation through these processes, most of the genome of a species is the same in all individuals of that species. However, even moderately small changes in genotype can lead to powerful changes in phenotype.

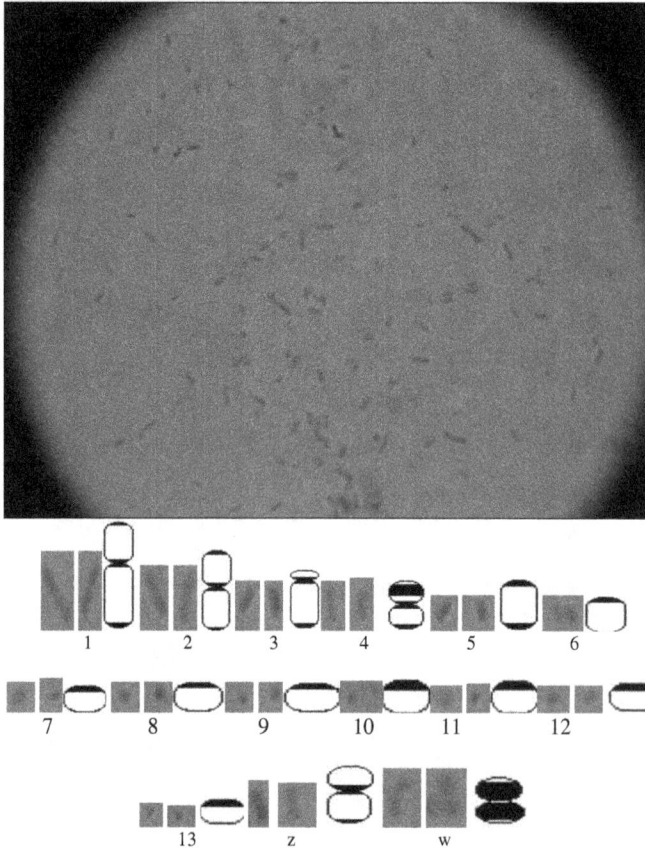

Figure 18. Ideogram and karyogram of the chromosomes of the red crested Pochard (CBG banding).

REFERENCES

Belterman RH, Boer LEM (1984). The karyological study of 55 species of birds, including karyotypes of 39 species new to cytology. Genetica, *65*: 39-82.

Clements J (2007). *The Clements Checklist of the Birds of the World.* Ithaca: Cornell University Press.

Cramp S, Simmons KE (1977). The Birds of the Western Palearctic. Vol. 1. Oxford Univ. Press.

Haen PJ (1995) Principles of hematology. Wm. C. Brown Publishers Chicago. USA.

Johnson KP, Sorenson MD (1999). Phylogeny and biogeography of dabbling ducks (genus Anas): a comparison of molecular and morphological evidence. Auk 116 (3): 792-805. Current Biol., 17(8): R283-R286.

Lucca EJ, Rocha GT (1985). Chromosomal polymorphism in *Zonotrichiacapensis* (Passeriformes-Aves). Rev. Bras. Genet. VIII: 71-78.

Lucca EJ, Waldrigues A (1985). The karyotypes of nine species of Passeriformes. Egypt. J. Genet. Cytol. 14: 41-50.

Scott DA, Rose PM (1996). Atlas of Anatidae Populations in Africa and Western Eurasia. Wetlands International Pub. n° 41, Wetlands International, Wagenigen, The Netherlan.

Sugiyama T (1971). Specific vulnerability of the largest telocentric chromosome of rat bone marrow cell to 7.12 dimethyle benz[á]anthrance. Nalt. Cancer inst., 47: 1264.

Voous KH (1960). Atlas of European Birds. Nelson, Edinburgh.

that gene. Evolutionary power proceeds by forceful of these changes in allele incidence in one direction or any more. Differences disappear when an allele reaches the end of fascination, when it either disappears from the population or replaces the inherited allele completely.

Variation comes from mutations in genetic material, migration between populations (gene flow), and the reshuffling of genes through sexual breeding. Variation

25

Photoperiod as a factor for studying fluctuations of seminal traits during breeding and non-breeding seasons

M. M. Pourseif and G. H. Moghaddam*

Department of Animal Science, Faculty of Agriculture, University of Tabriz, Tabriz, Iran.

The main purpose of this study was to evaluate the influence of photoperiod on the seminal traits of crossbreed wool-producing rams throughout one year period. For the effect of photoperiod (PTP), two periods were considered: Decreasing daylight length (summer and autumn) and the other, increasing daylight length. For this study, 5 Baluchi × Moghani (BL × MG) and 5 Arkharmerino × Moghani (AM × MG) rams were used. Semen collection started from first of October 2010 to end of September 2011. After a training period of 2 weeks, semen ejaculates were evaluated for volume, total sperm/ejaculate (TSE), concentration (SC), color, wave motion (WM), percentage of progressive motility (PM), percentage of live sperm (LS) and abnormal sperm (SAB), pH, methylene blue reduction time (MBRT) and semen index (SI). Analysis of the year long data showed that semen with the best quality was collected in September to November (P < 0.05). Significant seasonal variations of semen traits were observed for all seminal traits except for PM, LS and MBRT. Yet, no statistical differences were found between the crosses (P > 0.05). Although, there were significant seasonal changes in seminal traits of the crosses, the fresh semen showed adequate quality to be used for artificial or natural insemination throughout the year. PTP was found to influence semen production in two genetic groups at 38°02' N, 46°27' E and an altitude of 1567 m above sea level of Iran. However, these effects were not detrimental to the use of rams for breeding purposes throughout the year.

Key words: Genetic group, crossbreed ram, photoperiod, seasonal variation, spermatozoa.

INTRODUCTION

Sheep production is a traditional economic activity that is mostly used for meat and milk production in Iran. One of the most important factors for economical development of sheep industry is lambing throughout the year. Seasonal breeding is a limiting factor in this species. Therefore, it was determined to evaluate the quality of semen in breeding and non-breeding seasons before using artificial insemination (AI) in sheep. Unlike most domestic livestock species, sheep are widely known for their marked seasonality of breeding activity linked to annual cycle of daily photoperiod (PTP) (Rosa and Bryant, 2003). The annual cycle of daily PTP has been identified as the major determinant for this phenomenon in sheep (Rosa and Bryant, 2003). Understanding the fertility quality in non-breeding season will be helpful in developing sheep industry. In contrast to ewes and most horse mares, that become anovulatory outside the breeding season, stallions and rams are not azoospermic during the non-breeding season despite a significant reduction in sperm production or quality (Aurich et al., 1996). Also, overall physiological and behavioral sexual variations are also less pronounced in rams than ewe (Rosa and Bryant, 2003). Therefore, yearlong comparative studies comprising breeding and non-breeding seasons in rams will be useful for understanding their reproductive physiology. As a result of the revolution in assisted reproductive technologies in domestic animals

---
*Corresponding author. E-mail: ghmoghaddam@tabrizu.ac.ir.

Table 1. Climatic data during the experiment (October 2010 until September 2011) at Khalat Poshan Research Center, University of Tabriz.

Month	Air temperature (°C)		Relative humidity (%)		Average
	Minimum	Maximum	Minimum	Maximum	Day length (h)
October	7.6	25.1	26.9	77.5	11.3
November	0.23	16.7	32.5	71.1	10.2
December	-4.08	12.4	34.7	67.9	9.6
January	-7.93	3.65	54.26	84.06	9.9
February	-7.85	4.2	51.33	85.1	10.9
March	-2.32	8.51	48.5	81.75	12
April	2.64	16.06	25.03	67	13.3
May	6.83	19.45	36.16	80.93	14.3
June	11.51	28.03	23	78.54	14.8
July	15.75	32.61	22.8	57.87	14.6
August	16.83	33.61	15.29	56.51	13.7
September	12.16	28.22	16.74	74.16	12.5

in Iran, a growing interest and necessity demands more information concerning the reproductive physiology of farm animals (Talebi et al., 2009). The breeding season starts in most ovine breeds during summer or early autumn (Chemineau et al., 1992) and its length varies largely among breeds but in general it ends during the winter (Hafez, 1952). Many other factors affect the semen characteristics, including nutrition, social environment, the presence of females, geographical location, age, testicle and body conformation, libido and management system, as reported in many studies (Mandiki et al., 1998; Al-Ghalban et al., 2004; Zamiri and Khodaei, 2005; Zarazaga et al., 2005), but the PTP and the breed are primary factors regulating the seasonal reproduction. Therefore, they became preference for many researchers (Simplicio et al., 1982; Ibrahim, 1997; Kafi et al., 2004; Barkawi et al., 2006; Talebi et al., 2009; Zamiri et al., 2010).

The Arkhar-Merino is a breed of sheep obtained by crossbreeding between wild Arkhar rams with ewes of the Novocaucasian Merino, Précoce and Rambouillet breeds (Ernst and Dmitriev, 2007). These two genetic groups are developed targeting the improvement of local breeds (Baluchi and Moghani) for wool traits. There is no published information on the seasonal variation of seminal traits of Baluchi × Moghani and Arkharmerino × Moghani genetic groups. In addition to gathering information on reference values for semen characteristics, evaluation of the effects of PTP on ram semen characteristics at this latitude was another pursuit for this study, along with the evaluation of putative differences on semen traits between the two genetic groups, to identify the most suitable line for breeding purposes. In intensive management systems, a significant number of ewes are inseminated in non-breeding season (Colas et al., 1988, 1990).

Therefore, detection of semen characteristic of the crosses in non-breeding season is necessary. Although, information is available on the physicochemical parameters of semen of the crosses and on the part of their reproductive traits (Moghaddam et al., 2012; Asadpour et al., 2012; 2012a), however, this yearlong research is the first report on the seasonal variation of semen characteristics of these crossbreed rams reared at the northwest of Iran.

MATERIALS AND METHODS

Location

This trial was performed at the Sheep Breeding Research Center, Tabriz (38°02' N, 46°27' E and an altitude of 1567 m above sea level), Iran. This experiment was carried out from October 2010 to September 2011.

Animals

Ten crossbreed and fertile rams consisting of 5 BL × MG and 5 AM × MG aged 3 to 6 years and with a live weight of 74 to 88 kg were used in this study. The animals were maintained under natural PTP and equal levels of nutrition per day; 20% concentrate (75% barley, 25% corn, soya, bran) and 80% alfalfa hay. The rams were initially trained (beginning of September) for 15 days in September to ejaculate semen by using artificial vagina (AV) in the mating pen (210 cm in length, 60 cm in width and 120 cm in height). Training and semen sampling was performed via an anoestrous teaser with quiet temperament. The rams were separated of the herd and housed in a large cover shelter with an open precinct for walking freely. All rams were sent to drink fresh water twice or three times a day. The other general management was checked during the study. The climate conditions of the research center were recorded during the experiment (Table 1).

Semen collection

Randomly, all the rams were divided into the two groups. Each group included 5 rams of different genetic groups. Semen collection

Table 2. Semen characteristics in BL × MG and AM × MG genetic group over the year.

Genetic Group		SV (ml)	WM (0-5)	PM (%)	SC (0-5)	TSE (×10⁹)	Conc (×10⁹)	SL (%)	SAB (%)	SI (×10⁹)	pH	MBRT (s)
BL×MG	N	334	334	334	334	334	334	334	332	334	333	331
	Mean	0.86	3.72	67.42	3.73	3.55	3.77	68.91	13.02	14769	6.54	119.39
	S.E.	0.08	0.09	1.70	0.16	0.33	0.17	1.60	0.76	1823.50	0.09	3.10
	Min	0.45	2.00	40.00	2.00	0.916	1.95	45.00	4.00	722.40	5.70	65.00
	Max	1.40	5.00	85.00	5.00	18.90	5.68	90.00	26.00	43834.50	8.20	230.00
AM×MG	N	334	334	334	334	334	334	334	330	334	334	331
	Mean	1.02	3.93	72.79	3.55	4.495	3.516	74.57	10.95	19472.29	6.57	111.70
	S.E.	0.08	0.09	1.59	0.16	0.327	0.182	1.57	0.71	1767.06	0.09	3.17
	Min	0.45	2.00	45.00	1.00	0.985	0.960	50.00	3.00	1395.2	5.90	45.00
	Max	1.90	5.00	90.00	5.00	31.55	5.81	94.00	21.00	104058	7.80	220.00

SV: Semen volume, WM: wave motion, PM: progressive motility, SC: semen color, TSE: total spermatozoa per ejaculate, Conc: spermatozoa concentration, SL: percentage of live spermatozoa, SAB: percentage of abnormal spermatozoa, SI: semen index. MBRT: methylene blue reduction time.

was done for every 2 days from 5 rams. Ejaculation intervals of each ram were five days throughout the study. Short form AV was used for semen collection. Collecting glass of AV was warmed at 37°C before the operation and was maintained at this temperature until processed. Immediately, the fresh samples were transferred to the laboratory (avoiding sunlight) and were surveyed.

Semen appraising

Seminal traits of the fresh semen were evaluated according to the procedure adopted by Evans and Maxwell (1987). Volume of ejaculates was calculated in a conical tube graduated at 0.1 ml intervals. Semen pH was surveyed with two methods, a) pen form pH-meter (with 0.1 grades, model 8685, AZ Instrument, Taiwan), b) indicator pH-meter strips (Merck, made in Germany). SC was determined by use of a Thoma slide (haemocytometer method). The fresh semen was diluted using 0.1 M sodium citrate dehydrate 2.9% (pH = 6.7- 6.9) plus one drop of formalin (1: 400) at 400× magnification under a microscope. TSE was then calculated (volume × density). WM was evaluated according to the stuyd of Evans and Maxwell (1987). The assessment of the PM was a visual scale from 0 to 100% on basis of suspended droplet slide and on a heated (37°C) stage using phase-contrast optics (400×). The evaluation was done in increments of 5 to 10 percentage points for viewing individual spermatozoa with more lucidity and estimating PM. For SAB and LS, semen was stained with eosin-nigrosin stain and examined microscopically (400×). From several parts of the slide, about 300 spermatozoa were evaluated for mortality and 200 for SAB percentages. Metabolic activity of spermatozoa was measured by MBRT method based on color change from blue to colorless at 37°C. In a thin and transparent tube (1 mm diameter), 0.2 ml semen was added to 0.2 ml of methylene blue and time for color change was recorded. SI (volume × SC/ml × LS% × PM%) was calculated, as an indicator for appraising semen quality.

Statistical analysis

All statistical analysis were performed using the MIXED Procedure of Statistical Analysis System (SAS, 1996) and outliers deleted for volume, concentration, abnormality and MBRT traits. Values were considered to be statistically significant at P ≤ 0.05. For volume, SC, abnormality and MBRT traits, the outlier data was deleted. Means values were compared with Tukey test. Pearson correlation coefficient was calculated to evaluate the relationship between quality and quantity of semen attributes.

RESULTS

The quantity traits of semen (semen volume, color, SC and TSE; Table 2) were significantly influenced by season of the year. In AM × MG and BL × MG genetic groups, minimum and maximum values of semen volume were recorded at spring and autumn, respectively (P < 0.01). Frequently, semen volume increased from end of June and received the highest values at the October, and again decreased gradually at the end of October. This falling process continued during autumn and winter except for BL × MG that had a sudden increase in April (Figure 1). The highest mean vales of LS were recorded in December (in BL × MG) and September (in MR × MG). In AM × MG

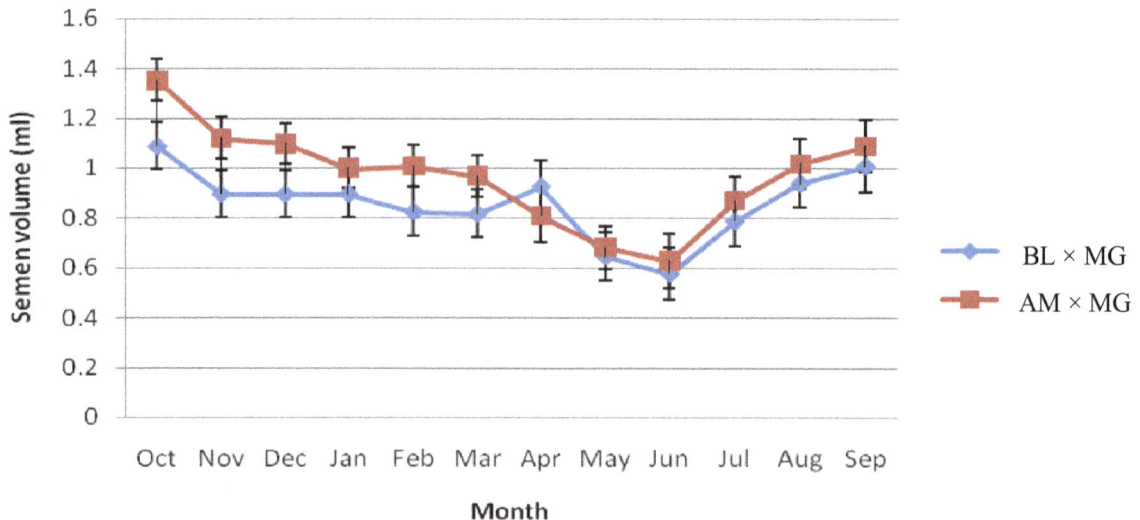

Figure 1. Monthly variations of semen volume in the two genetic groups throughout the year.

rams the significant differences was observed between the spring (the part of non-breeding season) and the other seasons (P < 0.01). But BL × MG rams had a significant difference between non-breeding season (spring and winter) and breeding season (P < 0.01). In the two genetic groups SC, TSE and semen color were highest in winter and the lowest was in spring. In the crosses, spring was the peak for semen pH. In BL × MG rams, mean values of semen pH increased concurrent with the spring. The most SI was observed during autumn (in both genetic group observed at October) and the lowest mean values was in spring (especially June). In AM × MG rams, there was significant difference in WM between breeding season (summer and autumn) and non-breeding season (Table 3). But most mean values in BL × MG rams were recorded in summer and the fewest were in winter (P < 0.01).

The results in the AM × MG and BL × MG rams demonstrated that individual progressive motility of spermatozoa was higher in breeding seasons (autumn and summer). In AM × MG genetic group, greatest value synchronized with the October (75.09 ± 1.53) and the lowest was in January (69.78 ± 1.58). In GH × BL genetic group, the highest levels were recorded in September (71.1 ± 1.59) and lowest in June (62.43 ± 1.70). The highest and lowest percentage of live spermatozoa in BL × MG genetic group were recorded in September (71.56 ± 1.56) and June (64.66 ± 1.65), respectively. But the highest and lowest LS in AM × MG rams were recorded in October (75.59 ± 1.54) and February (71.43 ± 1.51), respectively (Figure 2). In our study, the tail abnormalities were the most spermatozoa abnormality that occurred. In spite of these facts, semen quality from the viewpoint of sperm normality improved significantly during autumn in AM × MG and summer in BL × MG genetic groups. The season and genetic group did not influence the rate of

metabolic activity and WM (P > 0.05). The BL × MG genetic group showed the lowest MBRT in autumn and AM × MG group in summer. Correlation coefficients between various semen characteristics (Table 4) exhibited a near relationship between LS and motility traits (r = 0.90, P < 0.01) and sperm density and semen color (r = 0.30, P < 0.01). Semen volume showed a positive correlation with SC, color and TSE (r = 0.21, 0.24 and 0.39, respectively). The MBRT decreased over time and correlated with all semen traits (P < 0.01). SAB was correlated with all the semen quantity traits, except for TSE. The percentage of abnormal spermatozoa was significantly correlated with WM (r = -0.69, P < 0.01), PM (r = -0.88, P < 0.01) and LS (r = -0.92, P < 0.01). WM and individually progressive motility of semen samples showed a significant correlation with semen density (r = 0.19 and r = 0.33, respectively) and semen pH (r = -0.38 and r = -0.39, respectively). Moreover, semen pH showed high negative correlation with SC (r = -0.6, P < 0.01).

DISCUSSION

As expected, the summer and autumn with decreasing daylight length (breeding season) and winter and spring seasons with increasing daylight length (non-breeding season), affected the seminal indices of the crossbreed rams. The effect of season and/or PTP on seminal traits has been previously reported in different breeds of rams (Karagiannidis et al., 2000; Kafi et al., 2004; Zamiri and Khodaei, 2005; Deldar Tajangookeh et al., 2007; Zamiri et al., 2010) and also in other seasonal breeding animals such as buck (Barkawi et al., 2006; Karagiannidis et al., 1999) and stallion (Janett et al., 2003). Among quality traits, a significant effect of season was recorded on SAB, semen pH (P < 0.05) and SI (P < 0.01). PM, LS and

Table 3. Seasonal variations in semen characteristics (mean ± S.E.) of Baluchi × Moghani (BL × MG) and Arkharmerino × Moghani (AM × MG) rams.

Semen characteristics	Season	BL × MG (Mean±S.E.)	AM × MG (Mean±S.E.)
Total sperm/ejaculate (×10^9)	Spring	2.430±0.430^b	2.326±0.463^c
	Summer	3.210±0.298^a	3.730±0.290^b
	Autumn	3.636±0.316^a	4.336±0.285^b
	Winter	4.105±0.306^a	5.853±0.306^a
	Mean	3.345±0.337	4.061±0.336
Sperm concentration (×10^9)	Spring	3.443±0.195^b	3.315±0.199^b
	Summer	3.625±0.177^{ab}	3.555±0.185^{ab}
	Autumn	3.796±0.176^a	3.482±0.176^{ab}
	Winter	3.952±0.180^a	3.676±0.179^a
	Mean	3.704±0.182	3.507±0.184
Semen volume (ml)	Spring	0.69±0.09^c	0.70±0.09^b
	Summer	0.90±0.08^a	1.02±0.08^a
	Autumn	0.96±0.09^a	1.15±0.09^a
	Winter	0.84±0.09^b	1.06±0.08^a
	Mean	0.84±0.09	0.98±0.08
Semen color (0-5)	Spring	3.351±0.171^c	3.326±0.177^b
	Summer	3.512±0.156^{bc}	3.419±0.173^{ab}
	Autumn	3.712±0.159^{ab}	3.566±0.158^{ab}
	Winter	3.979±0.161^a	3.671±0.161^a
	Mean	3.638±0.162	3.495±0.167
Wave motion (0-5)	Spring	3.60±0.10^{bc}	3.81±0.09^b
	Summer	3.92±0.09^a	4.00±0.09^a
	Autumn	3.72±0.09^{ab}	4.05±0.11^a
	Winter	3.51±0.11^c	3.75±0.09^b
	Mean	3.68±0.09	3.90±0.09
Progressive motility (%)	Spring	65.37±1.61^b	72.00±1.75^a
	Summer	70.51±1.54^a	74.67±1.61^a
	Autumn	69.89±1.50^a	74.89±1.50^a
	Winter	64.40±1.73^b	71.00±1.54^a
	Mean	67.54±1.70	73.14±1.59
Live spermatozoa (%)	Spring	65.91±1.68^c	73.33±1.67^a
	Summer	70.37±1.66^{ab}	75.73±1.66^a
	Autumn	70.96±1.51^a	75.26±1.51^a
	Winter	66.54±1.54^{bc}	72.28±1.51^a
	Mean	68.44±1.60	74.16±1.57
Abnormal sperm (%)	Spring	12.71±0.76^{ab}	11.47±0.75^a
	Summer	11.86±0.70^b	9.67±0.67^b
	Autumn	12.65±0.74^{ab}	9.11±0.73^b
	Winter	14.04±0.75^a	10.84±0.77^a
	Mean	12.81±0.74	10.27±0.72
Semen index (×109)	Spring	11346±1859.61b	12123±1854.65b
	Summer	16511±1881.39a	19852±1769.43a
	Autumn	17249±1813.56a	22645±1827.49a
	Winter	14023±1797.32ab	20071±1769.46a
	Mean	14782±1837.97	18672±1805.25

Table 3. Continued.

Semen pH	Spring	6.88±0.10[a]	6.69±0.11[a]
	Summer	6.42±0.09[b]	6.60±0.10[a]
	Autumn	6.37±0.11[b]	6.57±0.09[ab]
	Winter	6.51±0.10[b]	6.51±0.09[b]
	Mean	6.54±0.10	6.59±0.09
MBRT (sec)	Spring	120.22±3.61[a]	111.18±3.70[a]
	Summer	110.74±3.77[a]	106.27±3.11[a]
	Autumn	116.69±3.81[a]	108.33±3.20[a]
	Winter	123.47±3.17[a]	116.08±3.25[a]
	Mean	117.78±3.59	110.46±3.31

[a, b, c] Means in the column of each parameter with different superscripts differ significantly ($P < 0.05$). Means within each column within each factor having the same letter does not differ significantly from each other ($P < 0.05$).

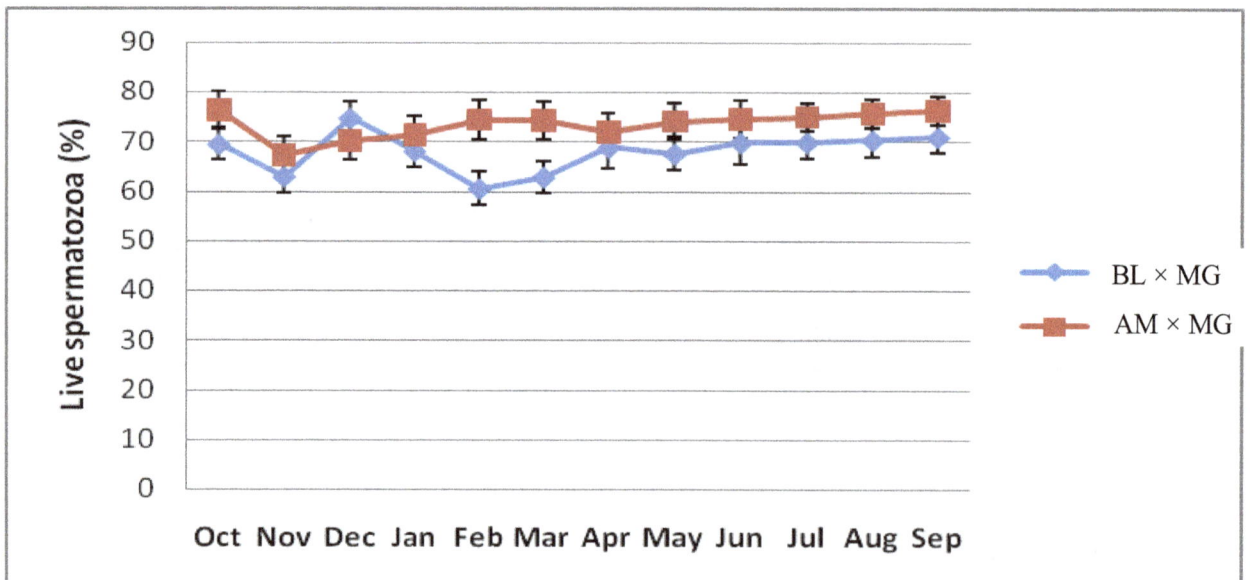

Figure 2. Monthly variations of live spermatozoa in two the genetic groups throughout the year.

Table 4. Correlation coefficients between various seminal traits of the rams.

Parameter	MBRT	pH	SI	SAB	SL	Conc	TSE	Color	PM	WM
SV	-0.21**	-0.20**	0.71**	-0.10	0.13*	0.21**	0.39*	0.24**	0.15**	0.21**
WM	-0.76**	-0.38**	0.50**	-0.69**	0.74**	0.19**	0.19*	0.37**	0.78**	
PM	-0.84**	-0.39**	0.59**	-0.88**	0.90**	0.33**	0.11*	0.34**		
Color	-0.67**	-0.56**	0.57**	-0.31*	0.30**	0.92**	0.34**			
TSE	-0.22**	-0.17*	0.41**	-0.05	0.09	0.29**				
Conc	-0.67**	-0.60**	0.61**	-0.28*	0.30**					
SL	-0.81**	-0.35*	0.57**	-0.92**						
SAB	0.80**	0.29**	-0.54**							
SI	-0.67**	-0.30**								
pH	0.52**									

*Significant at $P < 0.05$, **Significant at $P < 0.01$, coefficients without symbol (* or **) are not significant. SV: semen volume, WM: wave motion, PM: progressive motility, TSE: total spermatozoa per ejaculate, Conc: spermatozoa concentration, SL: percentage of live spermatozoa, SAB: percentage of abnormal spermatozoa, SI: semen index. MBRT: Methylene blue reduction time.

MBRT did not have significantly seasonal variations. Moreover, effect of PTP was also observed clearly on semen quantity characteristics (P < 0.01). These seasonal variations in semen quality and quantity were attributive mainly to changes in daylight length throughout the year (Chemineau et al., 1992). No significant difference was found on all traits between the two genetic groups (P > 0.05). Significant differences among the rams within each genetic group (P < 0.05) were found in some of the seminal traits, but, non-significant differences were found between the two genetic groups in any of the other traits which is consistent with the previous reports (Karagiannidis et al., 2000). Our results on mean values of semen characteristics are in agreement with those of other researchers (Zamiri et al., 2010; Gundogan, 2007; Kafi et al., 2004; Ghalban et al., 2004; Karagiannidis et al., 2000). The semen volume of 0.60 to 1.6 ml, SC of 2.6 to 5.5 ×10^9, SAB of 4 to 29% and live or motile spermatozoa of 60 to 90% is on record (Karagiannidis et al., 2000; Kafi et al., 2004; Gundogan, 2007).

Therefore, it could be accepted that there is wide amplitude of semen traits in several breeds of ram. In the current study, mean values for the SAB (9 to 14%) were generally higher than that of other researchers (Karagiannidis et al., 2000) for Chios and Friesian rams and for Akkarman and Awassi rams (Gundogan, 2007). Zamiri et al. (2010) reported in Moghani breed a minimum SAB of 7.9% in September much which is lesser than the value observed in our study, 11.42% in BL × GH rams in September and 8.91% in the AM × MG genetic group in November. LS in the two genetic groups was lower than the values recorded by Kafi et al. (2004) in Iran (29°25'N, 52°46'E). The semen volume in the BL × MG (0.84 ± 0.09) did not coincide with the results of Kafi et al. (2004) and was lower (1.03 ± 0.08) in AM × GH rams than reported value (Kafi et al., 2004) making the comparison of seminal attributes often difficult. Thus, it is not surprising that wide variations have been reported in the seminal attributes of rams (Gundogan, 2007; Zamiri and Khodaei, 2005; Kafi et al., 2004; Karagiannidis et al., 2000). In BL × GH genetic group, the SC remained high (3.952 ± 0.180) during winter and low in spring (3.443 ± 0.195), summer (3.625 ± 0.177) and autumn (3.796 ± 0.176), a trend comparable to that reported by Karagiannidis et al. (2000) and Talebi et al. (2009). These findings confirmed the previous records of seasonal variations of SC in BL × MG rams at 38°N latitude. In both crosses, circumstance of seasonal fluctuations of semen color and SC was similar. In our study, most of the mean values for the semen characteristics of BL × MG and AM × MG rams, were almost similar to those reported by other authors (Barkawi et al., 2006; Zamiri et al., 2010; Gundogan, 2007), in similar temperate regions. The semen quantity and quality attributes in the crossbreed rams differed in breeding (late summer to middle of autumn) and non-breeding seasons. SC did not follow a quite similar trend

with that of the ejaculate volume in this study and was comparable with the results obtained by Talebi et al. (2009). Mean values of MBRT in our study were quite different with the reports of Galal et al. (1978) in Egypt. BL × MG and AM × MG rams had best performance in breeding seasons. Galal et al. (1978) recorded in their study on Merino, Ossimi and their crosses the best metabolic activity in spring (76.8 ± 1.04 s) and autumn (77.2 ± 1.04 s). While summer (102.2 ± 1.04 s) had the greatest mean values in these breeds. On the contrary, Galal et al. (1978) did not observe significant difference in MBRT traits between several seasons of the year.

In the present study, the semen characteristics were generally better towards the end of summer (onset of improvement) and in the two first months of autumn, than during the winter (onset of decrease in quality) and spring (usually with lowest quality and quantity). In both genetic groups, PM was lowest in winter and spring in contrast to the findings of Karagiannidis et al. (2000) at 40°N. The data suggest that summer and autumn with decreasing daylight length and winter and spring with increasing daylight length influenced the seminal traits of BL × MG and AM × MG rams. PTP effects on seasonal breeders have been reported to be determined by the latitude at which they are kept. At latitudes above 40°N, marked variations in seminal traits and increased sperm production with decreasing daylight length have been observed (Zamiri et al., 2010). Seasonal variations, although less marked, were observed between 30 and 40°N latitude, with higher sperm production during the summer and autumn (Corteel, 1977). Although, the crossbreed rams were capable of ejaculating throughout the year, Rosa and Bryant (2003) illustrated that seasonal breeding animals occurred in middle latitudes. High correlation between WM and sperm progressive motility with live sperm (r = 0.74, r = 0.90, P < 0.01, respectively) demonstrated that concurrent with improved PM (as one of the most important semen quality indicators) there is increased LS which resembled the findings of Kafi et al. (2004). The significant correlation found between motility and TSE is in agreement with the results of Kafi et al. (2004). A high negative correlation of MBRT with motility traits (r = -0.76 to r = -0.84, P < 0.01) was similar to the findings of Chandler et al. (2000) but was inconsistent with the results of Kishk (2008). MBRT is an evaluator of the metabolic status of the spermatozoa (Salisbury et al., 1978).

The observed high negative correlation between SC and MBRT (r = -0.67, P < 0.01) and between MBRT and LS (r = -0.81, P < 0.01) were similar to the findings of Kishk (2008). This can be attributed to the rate of release of hydrogen upon fructose utilization by sperm cells. Thus, these samples might become acidic and not reliable for long-term storage. Most relationship among semen traits with semen pH could be well correlated with SC and color (r = -0.60 and –0.56, P < 0.01), and watery sample that is alkaline. Among the quantitative traits, a high correlation between MBRT with color and SC was observed. Karagiannidis et al. (2000) also reported a significant correlation between SC and SAB (r = -0.19, P < 0.05). The results of the present investigations suggest that ewes exhibiting estrus could be artificially inseminated by fresh semen throughout the year and consequently, reproductive performance of herd increased considerably. Seasonal fluctuations of environmental conditions markedly influenced reproduction of animals at higher latitudes and altitudes (Rosa and Bryant, 2003). Robinson (1981) argued that breeds located between 35°N and have the tendency to breed at all times of the year. Evans and Maxell (1987) reported 30 and 40°N latitudes for the breeds to follow this tendency. Latitudes above 35°N (Hafez, 1952; Goot, 1969) or higher than 40°N (Talebi et al., 2009; Zamiri et al., 2010) considerably influenced the seminal traits.

However, in some studies, for example in Jordan (at 31.5°N in Damascus bucks) and Iran (34°18'N, 47°3'E in Markhoz bucks) the PTP was found to have significant effect on breeding behaviour of sheep. In temperate latitudes (40 to 50°N) sperm production of rams is a continuous process, although the total number of sperm produced per testis is usually higher in autumn than in spring (Dacheux et al., 1981). The present study showed that the reproductive activity of the seasonal breeding animals example, rams, may be improved by exploitation of PTP synchronized circannual reproductive rhythm (endogenous mechanisms) and exogenous factors. Reproductive activity is not a direct function of day length, but is affected by the photoperiodic history of the animal, the direction of photoperiodic changes and the stage of the circannual rhythm at which a photoperiodic signal is received (Robinson and Karsch, 1987; Gorman and Zucker, 1995a). Our study clearly showed linkage between PTP and reproduction of BL × MG and AM × MG genetic groups located at 38°02'N, 46°27'E in Iran.

Conclusion

Semen evaluation does have an important role in AI programs or in flocks where single sire joining groups are used. This will be useful for identifying rams with poor performers. Thus, it will provide optimum breeding selection of males in herd. The seminal traits of BL × MG and AM × MG rams in Northwest of Iran showed a significant seasonal variation in semen characteristics. The best semen is produced during late summer to November. Nonetheless, the magnitude of these seasonal effects should not prevent the animals from been used for semen collection for AI throughout the year. But it is necessary to perform semen evaluation on an individual basis for every ram used for AI or breeding.

REFERENCES

Al-Ghalban AM, Tabbaa MJ, Kridli RT (2004). Factors affecting semen characteristics and scrotal circumference in Damascus bucks. Small Rum. Res. 53:141-149.

Asadpour R, Pourseif MM, MoghAddam GH, Jafari SR, Tayefi H, Mahmodi H (2012). Effect of vitamin B12 addition to extenders on some physicochemical parameters of semen in crossbred rams. Afr. J. Biotechnol. 11:11741-11745.

Asadpour R, Pourseif MM, MoghAddam GH, Tayefi H (2012a). Activity of antioxidant enzymes and malondialdehyde in seminal plasma and their relationship with semen characteristics in crossbred rams. Indian J. Anim. Sci. 82:710-712.

Aurich C, Burgmann F, Hoppe H (1996). Opioid regulation of luteinizing hormone and prolactin release in the horse-identical or independent endocrine pathways? Animal Reprod. Sci. 44:127-134.

Barkawi AH, Elsayed EH, Ashour G, Shehata E (2006). Seasonal changes in semen characteristics, hormonal profiles and testicular activity in Zaraibi goats. Small Rum. Res. 66:209-213.

Candler JE, Harrison CM, Canal AM (2000). Spermatozoal methylene blue reduction: an indicator of mitochondrial function and its correlation with motility. Theriogenology 54:261-271.

Chemineau P, Malpaux B, Delgadillo JA, Guerin Y, Ravault JP, Thimonier J, Pelletier J (1992). Control of sheep and goats reproduction: use of light and melatonin. Anim. Reprod. Sci. 30:157-184.

Colas G, Lefebvre J, Guerin J (1988). Recherche d'une prevision precoce de l'amplitude des variations saisonnieres du diameter testiculaire et du pourcentage de spermatozoids anormaux chez le belier Ile-de-France: 1. Animaux nes en fevrier. Reprod. Nutr. Dev. 28:589-601.

Colas G, Lefebvre J, Guerin J (1990). Father–male offspring transmission of seasonal variations in testis diameter and percentage of abnormal sperm in the Ile-de-France rams: male offspring born in February. Reprod. Nutr. Dev. 30:589-603.

Corteel JM (1977). Production, storage and insemination of goat semen. Management of Reproduction in Sheep and Goat symposium. University of Wisconsin, Madison, pp. 41-57.

Dacheux JL, Pisselet C, Blanc MR, Hocherau-de-Reviers MT, Courot M (1981). Seasonal variation in rete testis fluid secretion and sperm production in different breeds of rams. J. Reprod. Fert. 61:363-371.

Deldar Tajangoookeh H, Zare shahneh A, Moradi shahrebabak M, Shakeri M (2007). Monthly variation of plasma concentration of testosterone and thyroid hormones and reproductive characteristics in three breeds of Iranian fat-tailed rams throughout one year. Pak. J. Biotechnol. Sci. 10:3420-3424.

Ernst LK, Dmitriev NG (2007). Sheep: transbalkan finewool. ftp: //ftp.fao.org /docrep/ fao/009/ ah759e/ah 759e16.pdf.

Evans G, Maxwell WMC (1987). Salamon's artificial insemination of sheep and goats. pp. 17-30.

Galal ESE, El-Gamal AA, Aboul-Naga A (1978). Male reproductive characteristics of Merino and Ossimi sheep and their crosses. Anim. Prod. 27:261-267.

Goot H (1969). Effect of light on spring breeding of mutton Merino ewes. J. Agric. Sci. Camb. 73:177-180.

Gorman MR, Zucker I (1995a). Seasonal adaptations of Siberian hamsters: II. Pattern of change in day length controls annual testicular and body weight rhythms. Biol. Reprod. 53:116-125.

Gundogan M (2007). Seasonal variation in serum testosterone, T_3 and andrological parameters of two Turkish sheep breeds. Small Rumin. Res. 67:312-316.

Hafez ESE (1952). Studies on the breeding season and reproduction of the ewe. J. Agric. Sci. Camb. 42:189-265.

Ibrahim SA (1997). Seasonal variations in semen quality of local and crossbred rams raised in the United Arab Emirates. Animal Reprod. Sci. 49:161-167.

Janett F, Thun R, Niederer K, Burger D, Hassig M (2003). Seasonal changes in semen quality and freezability in the Warmblood stallion. Theriogenology 60:453-461.

Kafi M, Safdarian M, Hashemi M (2004). Seasonal variation in semen characteristics, scrotal circumference and libido of Persian Karakul rams. Small Rum. Res. 53:133-139.

Karagiannidis A, Varsakeli S, Alexopoulos C, Amarantidis I (2000). Seasonal variation in semen characteristics of Chios and Friesian rams in Greece. Small Rumin. Res. 37:125-130.

Karagiannidis A, Varsakeli S, Karatzas G (1999). Characteristics and seasonal variations in the semen of Alpine, Saanen and Damascus goat bucks born and raised in Greece. Theriogenology 53:1285-1293.

Kishk WH (2008). Interrelationship between ram plasma testosterone level and some semen characteristics. Slovak J. Anim. Sci. 41:67-71.

Mandiki SNM, Deriscke G, Bister JL, Paquay R (1998). Influence of season and age on sexual maturation parameters of Texel, Suffolk, and Ile-de- France rams. 1. Testicular size, semen quality and reproductive capacity. Small Rumin. Res. 28:67-79.

Moghaddam GH, Pourseif MM, Asadpour R, Rafat SA, Jafari R (2012). Relationship between levels of peripheral blood testosterone, sexual behavior, scrotal circumference and seminal parameters in crossbred rams. Acta Sci. Veterin. 40:1049.

Robinson JE, Karsch FJ (1987). Photoperiodic history and a changing melatonin pattern can determine the neuroendocrine response of the ewe to daylength. J. Reprod. Fert. 80:159-165.

Robinson JJ (1981). Photoperiodic and nutritional influences on the reproductive performance of ewes in accelerated lambing systems. Pages 1-10 in Proc. 32nd Annu. Mtg., of the European Association for Animal Production (EAAP), Zagreb, Croatia.

Rosa HJD, Bryant MJ (2003). Seasonality of reproduction in sheep. Small Rumin. Res. 48:155-171.

Salisbury GW, VanDemark NL, Lodge JR (1978). Physiology of reproductive and artificial insemination of cattle. Second edition. San Francisco, CA: WH freeman and company.

SAS (1996). SAS/STAT software: Changes and Enhances Through Release 6.12. SAS Institute Inc., Cary, NC, USA.

Simplicio AA, Riera GS, Nelson EA, Pant KP (1982). Seasonal variation in seminal and testicular characteristics of Brazilian Somali rams in the hot semi-arid climate of tropical northeast Brazil. J. Reprod. Fert. 66:735-738.

Talebi J, Souria M, Moghaddam A, Karimi I, Mirmahmoodi M (2009). Characteristics and seasonal variation in the semen of Markhoz bucks in western Iran. Small Rumin. Res. 85:18-22.

Zamiri MJ, Khalili B, Jafaroghli M, Farshad A (2010). Seasonal variation in seminal parameters, testicular size, and plasma testosterone concentration in Iranian Moghani rams. Small Rumin. Res. 94:132-136.

Zamiri MJ, Khodaei HR (2005). Seasonal thyroidal activity and reproductive characteristics of Iranian fat-tailed rams. Anim. Reprod.. Sci. 88:245-255.

Zarazaga LA, Guzman JL, Domınguez C, Perez MC, Prieto R (2005). Effect of plane of nutrition on seasonality of reproduction in Spanish Payoya goats. Anim. Reprod. Sci. 87:2.

26

Growth curve estimation in pure goat breeds and crosses of first and second generation in Tunisian oases

Amor Gaddour*, Mabrouk Ouni and Sghaier Najari

Arid Land Institute, Medenine, Tunisia.

Five non-linear statistical models were tested to fit the growth curve parameters of the kids of indigenous, Alpine and Damascus goats and their crosses. Data from 16 years' periodical weight study was used to adjust the growth curve of 1,687 suckling kids before they attained the age of five months. Among the tested models, the iterative procedure made it possible for the Gompertz model to be identified as the best for use to adjust kids' growth evolution. Brody, Richards, Logistic and Polynomial models showed some convergence problems of accuracy. Curve parameters were fitted by Gompertz model after about 16 iterations with a coefficient of determination (CD) value of 71%. Growth parameters were established by genetic groups and the shape of the curve changed with kids' genotypes. Crossbreeding allowed for a better growth kinetic in indigenous kids. After birth, kids' weights increased rapidly to an asymptotic weight at an early age. The best growth performances were obtained in the first generation of crossbreeding due to heterosis. The growth curve adjustment helped in better flock management and in the fattening of kids according to the potentialities of each genotype.

Key words: Goat, kids' growth, curve model, Gompertz, crossbreeding, Tunisian oases.

INTRODUCTION

In many marginal regions, goats often constitute the only source of protein through their meat production. In Tunisia, goat flocks contain about 1,500,000 females and more than 60% of goats are raised in the semi-arid and arid zones (Najari et al., 2007). The indigenous goat is genetically considered as a population that has a wide phenotypic variability; it is essentially raised via pastoral and agro-pastoral modes. The lactated kid's meat is the main product that results from indigenous goat breeding in Oasian conditions and it contributes about 75% to the regional meat production in very low input systems (Najari et al., 2007d). Under Oasian conditions, goat husbandry plays a key role through its various significant contributions to the farmers' incomes. Goats thrive in an

intensified breeding mode with low climatic risks which characterize the arid area (Trangerud et al., 2007).

To increase the production of Oasian goat flocks, some high yielding exotic breeds were introduced in 1980 in the arid regions (Gaddour et al., 2008c). The objective of this program was either to produce meat where goats were not milked or to increase dairy yields where milk contributes to the income of farmers (Gaddour et al., 2008a). This goal was achieved by upgrading local breeds to different levels through crossbreeding so as to produce new goat genotypes that have high performances and are adapted to local environments (Serradia, 2001).

The model assessment of the growth of kids is particularly important in animal production because of its practical implications in genetic evaluation and flock management (Gipson and Wildeus, 1994; Schinckel and de Lange, 1996). Like other animal phenotypes, growth curve parameters change with all factors affecting the weight, especially the genetic potentials of the breed (Alexandre et al., 1997a; De Lange et al., 1998; Oltenacu,

---

*Corresponding author. E-mail: gaddour.omar@yahoo.fr or amor.gaddour@ira.agrinet.tn.

1999).

During an animal's lifetime, essential weight gain is reached before maturity stage, and it is well known that animals achieve the target mature size in a well-defined sigmoid or S-shaped curve (Najari et al., 2007, 2007b). Thus, typical curves are used to describe animal growth due to the general predictable pattern followed by the growth process (De Lange et al., 1998). A typical growth curve can be divided into two phases, an early phase where the weight gain rate increases and a later phase where the weight gain rate decreases (Trangerud et al., 2007). The point of inflexion is the point where the curve turns from concave to convex. Several non-linear functions have been proposed for various domestic livestock species and breeds to model the growth curve per genetic group (Barbato and Vasilatos-Younken, 1991; Bathaei and Leroy, 1996).

The present study aims at adjusting the growth curve of kids from local population, introduced breeds and crosses so as to evaluate the meat production kinetics and potentials for each genotype. Establishing curve parameters leads to an optimised use of the genetic resources of local and introduced animal breeds, and subsequently, to increased incomes for farmers in the southern Tunisian oases.

MATERIALS AND METHODS

Data base

For the last 16 years, crossing scheme has been in use and an individual periodical weighing control has continuously been realized from birth until the weaning of the kids at the beginning of summer. A total of 1,687 data files of kids were registered and used as the data base for this study. The data of each kid included genotype and control dates with respective observed weights (Gaddour et al., 2007a, b, c, d).

Growth curve assessment and curve parameters estimation

Due to the fact that the basic aspects of the physiological growth process are identical, some developed functions are largely used to describe the general growth curves (Wahi and Lal, 2004). The models used in our study are Gompertz, Richards, Logistic, Brody and Polynomial. These mathematical functions are considered as non-linear regression models and are solved by iterative procedures that minimize the residual variance (Yang et al., 2006). The residual values are assumed to be independent with a constant variance (Trangerud et al., 2007).

The evaluation criteria used to compare the accuracy of studied models were computing difficulty and goodness of fit. Computing difficulty is defined as the number of iterations needed to converge (Najari et al., 2007a). Except for the Richards' model, the starting values of parameters are null to allow the same convergence conditions (Wahi and Lal, 2004; Yang et al., 2006). Goodness of fit is defined as the magnitude of the residual mean squares (RMS) at convergence, which provides a measure of the estimation precision. The accuracy is evaluated by the non-linear coefficient of determination (CD). Statistical analysis was done by using SPSS 12.0.

RESULTS AND DISCUSSION

Growth model choice

The tested models' convergence performances and criteria are Gompertz: $A*Exp^{(-Exp^{(-bt-c)})}$, Richards: $A*(1+(b-1)*Exp(-c*((âge)-d)))^{(1/(1-b))}$, Logistic: $A/(1+b*Exp\{-c*(âge)\})$, Brody: $A*(1-b*Exp^{(-c*âge)})$ and Polynomial: $A+b*âge+c*âge^2+d*âge^3$. For each tested model, the iteration number, CD as well as RMS were considered. Note that the convergence criterion value was fixed to 10^{-8}. Among the tested models, only the Brody showed a convergence problem up to 300 iterations; the other three models met the convergence criterion after an iteration number varying between 12 and 26. The starting values were set to "zero" except for the Richards' model; this can be considered as a constraint on the use of this model. The choice of the starting values can inhibit the convergence when the estimation is not adequate (Najari et al., 2007a).

The most rapid convergence was obtained with the Logistic and Polynomial functions which needed only 12 iterations to generate the best possible estimation of the growth curve parameters. However, the Gompertz function seemed to be the most accurate; the CD value, estimating the goodness of fit, was 71%. The RMS values ranged from 5.92 to 6.34; the Logistic model generated the best as well as the worst values. The Polynomial regression model provided a good curve fit, but its parameters had no meaningful biological interpretations (Trangerud et al., 2007).

In view of the foregoing results, the Gompertz equation seems to be the most appropriate to adjust the growth curve of the kids. According to de Lange et al. (1998), this model is suitable for describing growth curve because domestic animal meat generally comes from animals that do not achieve mature or asymptotic weight (Najari et al., 2007a; Trangerud et al., 2007). The model takes into account the exponential decay of the specific growth rate of the animal based on initial body weight and inflexion point parameters. The Gompertz model confirms, in our case, that it can be considered a typical representation of the S-shaped growth curve as proposed by de Lange et al. (De Lange et al., 1998). Indeed, this model has been shown to be valid for a wide range of mammalian species and Aves (Barbato, 1991).

Shape and parameters of the growth curve of kids

The growth parameters of kids, which are derived from the assessment of the curve from the Gompertz functions are asymptotic weight (A), Gompertz curve parameters (b and c), age of inflexion (days) and weight at inflexion (kg). The growth curve of the kids, which is adjusted by the Gompertz function is presented in Figure 1 and includes lower and upper limits of weights' estimation.

Having the curve parameter values A, b and c, the growth

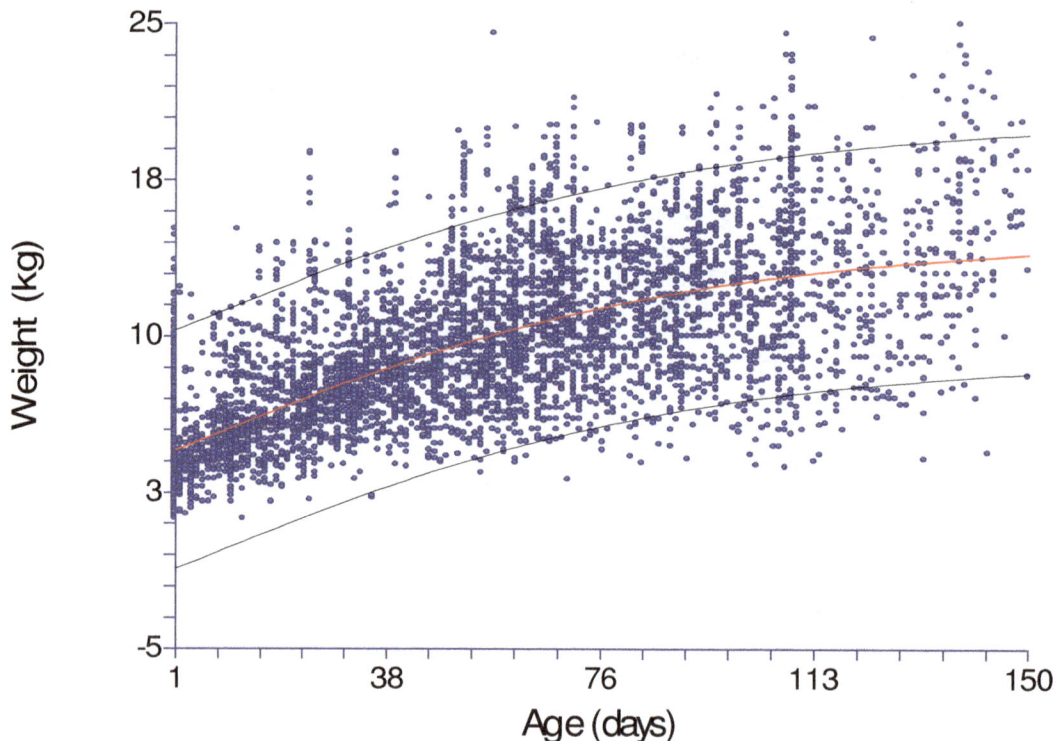

Figure 1. Growth curve of kids adjusted by the Gompertz model, with lower and upper limits.

curve equation is as shown in equation 1:

$$P = 15.74\ e^{(-e^{(-0.03^*t+0.31)})} \qquad (1)$$

Where, P is the kids' weights (kg), and 't' the kids' ages (days).

The model's function allows for the estimation of some crucial growth curve parameters: the asymptote A value represents the adult weights, while the age 't' tends to infinity; the inflexion point corresponds to the point at which the second derivative becomes "zero" and the growth rate is maximum (Wang and Zhang, 2005; Najari et al., 2007a). The weight and the age at inflexion are, consequently, calculated as shown in equations 2 and 3:

Age at inflexion (days) = c/b   (2)

Weight at inflexion (kg) = $A\ e^{(-e^{(-b^*age-c)})}$ (3)

The inflexion point is located at 10 days, at a weight of 5.79 kg; the asymptotic weight is estimated to be 15.74 kg.

The curve asymptote is usually used to estimate the adult weight, while the Gompertz model is used to adjust the growth (Najari et al., 2007a; 2007b). As shown in equation (1), the A value is 15.74 and seems to be less than the adult goat's real weight estimated through other studies (Wang and Zhang, 2005; Gaddour et al., 2007c). It is well known that the Gompertz model can underestimate the A constant, especially when a

relatively belated age is used to estimate the curve parameters (Trangerud et al., 2007).

Kids' growth curve of caprine genotypes

The asymptotic value (A) seems to be the highest for Alpine kids and the crossed Alpine*local. These kids reach more than 16.5 kg of body weight before the age of 5 months whereas the indigenous kids' asymptotic weight is estimated to be 12.47 kg.

Among the pure breeds, the Alpine kids showed the heaviest weight inflexion with 6.08 kg, reaching an average at the age of 15 days (Figure 2). The Damascus and indigenous kids showed the lowest weight inflexion starting from one week of age. Barbato (1996) related the age at which the curve inflexion occurred with the value of the corresponding weight which can affect the maturity age of the animals.

The most important period of growth seems to be the first two months after birth for all genotypes; the kids' weights tended rapidly to the asymptotic value. Consequently, keeping kids that are over four or five months of age in the flock does not provide any additional meat production, but rather induces more production costs per kg of kids' meat.

The weight inflexion was heaviest for the crosses of the first generation (F_1). The asymptotic and inflected weights of the F_1 kids were both higher than those of the paternal and F_2 genotypes (Figures 3 and 4). This illustrates a

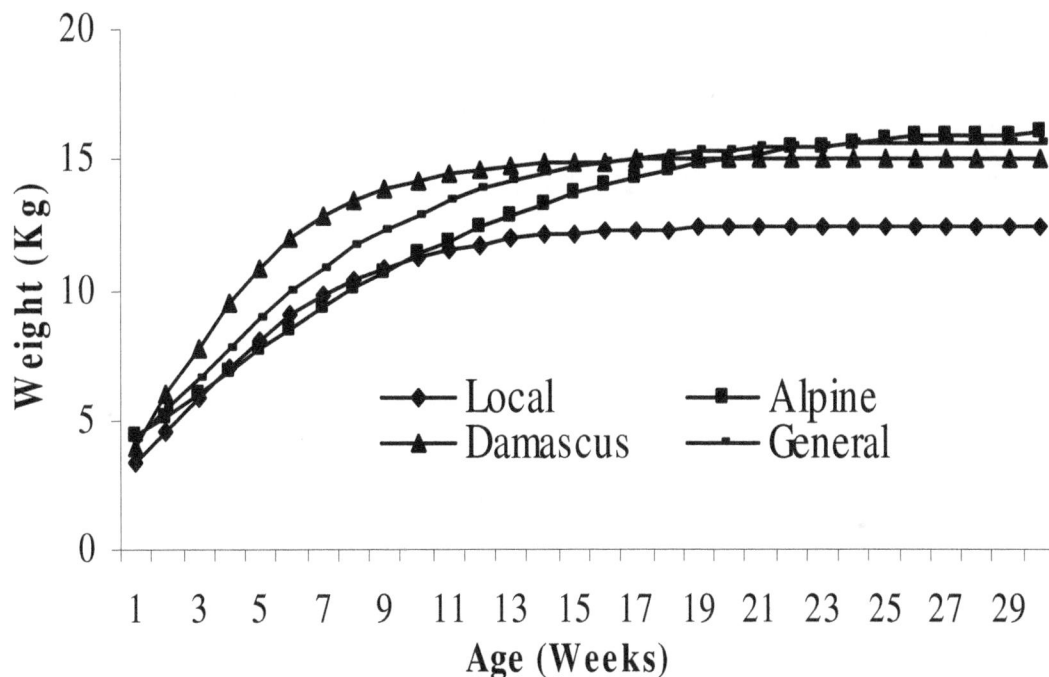

Figure 2. Growth curve of the kids of indigenous goat and pure breeds adjusted by the Gompertz model.

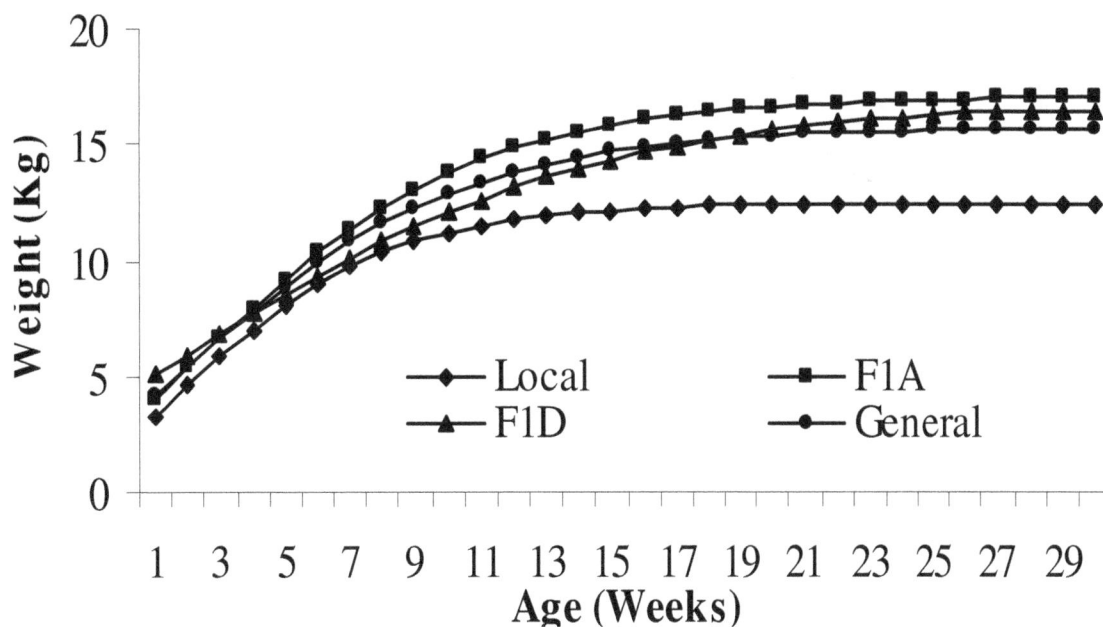

Figure 3. Growth curve of the kids of indigenous goat and first generation crossbreed adjusted by the Gompertz model.

clear effect of heterosis. This result agrees with that of Najari et al. (2007a), underlining the superiority of the performances of the F_1 to the parental breeds. Again, all the crossbred kids performed better than the indigenous genotypes; therefore, crossbreeding can improve caprine meat production (Serradia, 2001; Trangerud et al., 2007).

The parameters and shape of the growth curve illustrate a specific growth behavior for the studied genotypes. Apparently, some groups were able to produce an additional weight with age, while others stopped weight gain at an early age. This aspect has to be considered to optimize the genotypes' management to

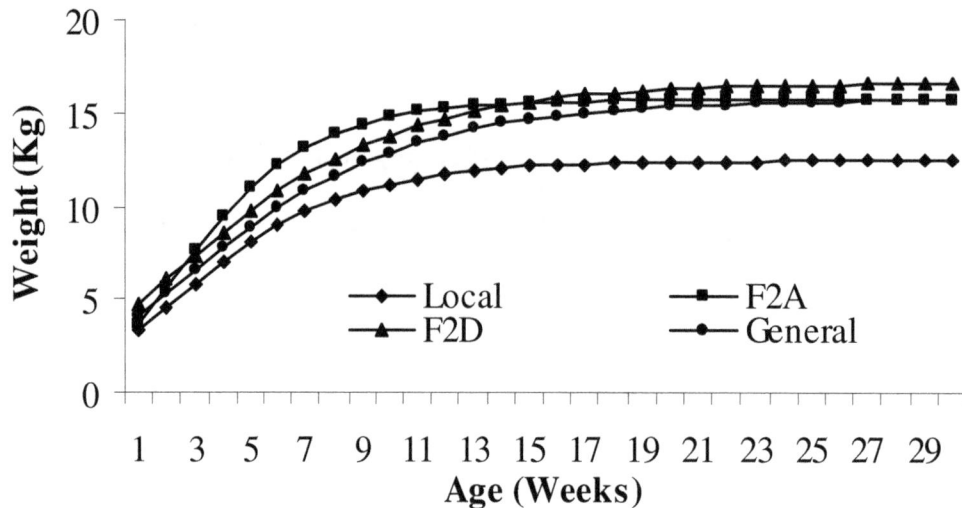

Figure 4. Growth curve of the kids of indigenous goat and second generation crossbreed adjusted by the Gompertz model.

ensure better meat production and more income for farmers.

REFERENCES

Alexandre G, Aumont G, Despois P, Mainaud JC, Coppry O, Xandé A (1997a). Productive performances of Guadeloupe in Creole goats during the suckling period. Small Rum. Res., 34: 157-162.

Barbato GF (1991). Genetic architecture of growth curve parameters in chickens. Theor. Appl. Genet., 83: 24-32.

Barbato GF (1996). Genetics of the growth curve in poultry: Physiological implications. Proc. 2nd European Poultry Breeders Roundtable. Landbrugs, 73: 153-165.

Barbato GF, Vasilatos-Younken R (1991). Sex-linked and maternal effects on growth in chickens. Poult. Sci., 70: 709-718.

Bathaei SS, Leroy PL (1996). Growth and mature weight of Mehraban Iranian fat-tailed sheep. Small Rumin. Res., 22: 155-162.

De Lange CFM, Szkotnicki B, Morphy J, Dewey C (1998). Establishing feed intake and growth curves for individual growing-finishing pig units. Community center (eds). Proceedings of 17th Annual Centralia Swine Research Update, Kirkton-Woodham, p. 341.

Gaddour A, Najari S, Ouni M (2007a). Dairy performances of the goat genetic groups in the southern Tunisian. Agric. J., 2: 248-253.

Gaddour A, Najari S, Ouni M (2007b). Kid's growth of pure breeds and crossed caprine genotypes in the coastal oases of southern Tunisia. Res. J. Agro., 2: 51-58.

Gaddour A, Najari S, Ouni M (2007c). Reproductive performances and kid's mortality of pure breeds and crossed caprine genotypes in the coastal oases of southern Tunisia. Paki. J. Biol. Sci., 14: 2314-2319.

Gaddour A, Najari S, Ouni M (2007d). Kid's growth and dairy performances of pure breeds and crossed caprine genotypes in the coastal oases of southern Tunisia. Pakistan J. Biol. Sci., 17: 2874-2879.

Gaddour A, Najari S, Ouni M (2008a). The genotype-environment interaction effects on dairy performances of goat genetic groups in the Tunisian oases. Res. J. Dairy Sci., 1: 22-26.

Gaddour A, Najari S, Ouni M (2008c). Genotype effect upon crossed goats' performances in the oases of southern Tunisia. "Sustainable Mediterranean grasslands and their multi-functions. Options Mediterrannennes, Series, A: 234-238.

Gipson TA, Wildeus S (1994). Growth curve analysis of body weight and scrotal circumference in dairy and meat-type goats. J. Anim. Sci., 72: 258.

Najari S, Gaddour A, Abdennebi M, Abdennebi M, Ouni M (2007). Specificities of the local kid's genotypes expression towards arid conditions in southern Tunisia. J. App. Sci., 3: 301-306.

Najari S, Gaddour A, Ben Hamouda M, Djemali M, Khaldi G (2007a). Growth model adjustment of local goat population under pastoral conditions in Tunisian arid zone. J. Agron., 1: 61-67.

Najari S, Gaddour A, Ouni M, Abdennebi M, Ben Hammouda M (2007b). Indigenous kid's weight variation with respect to non-genetic factors under pastoral mode in Tunisian arid region. J. Anim. Vet. Adv., 6: 441-450.

Najari S, Gaddour A, Ouni M, Abdennebi M, Ben Hammouda M (2007d). Non-genetic factors affecting local kids' growth curve under pastoral mode in Tunisian arid region. J. Biol. Sci., 6: 1005-1016.

Oltenacu EAB (1999). Using math to see how well your goat is growing. New York state 4H Meat Goat Project Fact Sheet 16. Cornell University, Ithaca, NY,

Schinckel AP, de Lange CFM (1996). Characterization of growth parameters needed as inputs for pig growth models. J. Anim. Sci., 74: 2021-2036.

Serradia JM (2001). Use of high yielding goat breeds for milk production. Livestock Prod. Sci., 71: 59-73.

Trangerud C, Grondalen J, Indrebo A, Tverdal A, Ropstad E, Moe L (2007). A longitudinal study on growth and growth variables in dogs of four large breeds raised in domestic environments. J. Anim. Sci., 1: 76-83.

Wahi SD, Lal Chand Bhatia VK (2004). A growth pattern study in crosses and pure Indian breeds of goats. Indian J. Anim. Sci., 74: 955-958.

Wang CF, Zhang JY (2005). Analysis of body conformation and fitting growth model in Tibetan chicken raised in plain. Sci Agric. Sin., 38: 1065-1068.

Yang Y, Mekki DM, Wang LY, Wang JY (2006). Analysis of fitting growth models in Jinghai Mixed-Sex Yellow Chicken. Int. J. Poult. Sci., 6: 517-521.

27

Genetic and environmental trends in the long-term dairy cattle genetic improvement programmes in the central tropical highlands of Ethiopia

Kefena Effa[1], Zewdie Wondatir[1], Tadelle Dessie[2] and Aynalem Haile[3]

[1]Ethiopian Institute of Agricultural Research, Holetta Agricultural Research Center, P. O. Box 2003, Addis Ababa, Ethiopia.
[2]International Livestock Research Institute (ILRI), P. O. Box, 5689, Addis Ababa, Ethiopia.
[3]International Center for Agricultural Research in Dry Areas, P. O. Box 5466, Aleppo, Syria.

A total of 1979 lactation records from 550 selected crossbred dairy cows that born between 1974 and 2005 were used to estimate annual genetic and environmental trends in milk production and reproduction traits at Holetta Agricultural Research Center, Ethiopia. Annual genetic and environmental trends were estimated by regressing BLUP estimated breeding value on year of birth. Variance components and genetic parameters were estimated using univariate analysis of individual animal model based on restricted maximum likelihood procedures. Annual genetic trends were -3.384 days, -8.00 kg and -5.96 kg, -0.26 months, -0.29 months and -0.88 days, for lactation length (LL), lactation milk yield (LMY), adjusted 305 milk yield (305-days MY), age at puberty (APU), age at first calving (AFC) and calving interval (CI), respectively. Environmental trends for LMY was positive (6.717 kg) and was in the desired direction. Heritability estimates were 0.14, 0.44, 0.39, 0.38, 0.40 and 0.17 for LL, LMY, 305-d MY, APU, AFC and CI, respectively. Negative genetic trends in all milk production traits reflect ineffective selection program and/or lack of using sires that have positive breeding values. The result from the environmental trends shows substantial improvement in the management practices over time. Contrasting directions in genetic and environmental trends reflect ineffective breeding objectives. This warrants reconsideration of the existing breeding program in the country.

Key words: Genetic trends, environmental trends, genetic parameters, breeding objectives, variance components.

INTRODUCTION

From the very outset, in whatever way we implement it, genetic improvement implies change. For a change to be an improvement, however, the overall effects of the change must bring positive benefits to target stake-holders and respond to the broad national development objectives (FAO, 2007). Therefore, planning national genetic improvement program takes into account careful analysis of the short and long-term objectives, socio-economic and environmental context in which it operates

(FAO, 2007). Under the Ethiopian context, livestock, particularly adapted cattle genetic resources are an important element in the livelihood of many resource-poor farmers living in wide arrays of production systems and contribute more than marketable products that are considered in economic statistics. Their special adaptive traits to harsh climates, disease resistance, heat tolerance, ability to utilize poor quality feeds and the multipurpose role they play in ranges production systems are some of their inherent genetic attributes. However, they are poor milk and meat producers. Consequently the demand for milk and milk products remained lagging behind supply for many years in Ethiopia.

As compared to many countries in Africa, Ethiopians

---

*Corresponding author. E-mail: kefenaol@yahoo.com

consume lesser amount of dairy products (Ahmed et al., 2003), which is currently estimated at 19 L per annum (CSA, 2009). Motivation for popularizing crossbreeding between high-yielding European dairy breeds and cattle breeds adapted to local environments was initiated in the national agricultural research system (NARS) of Ethiopia in the early 1970s. As compared to other dairy cattle genetic improvement strategies, this approach was believed to be the only feasible and quick way of increasing milk production in Ethiopia. The outcome of the crossbreeding programs have been amply reported in several literatures with various outcomes (Beyene, 1992; Beyene et al., 1987; Demeke et al., 2004a, b; Kefena et al., 2006; Haile et al., 2009). The effectiveness of any dairy cattle genetic improvement program is measured by the genetic progress obtained (Hallowell et al., 1998). Bakir and Cilek (2009) also stated that the genetic capacity and its progress in dairy cattle breeding are measured by genetic trend. A standard way of measuring progresses in animal breeding is by regressing estimated annual environmental and breeding value on year of birth (FAO, 2007).

However, the dairy cattle genetic improvement program started in Ethiopia in the early 1970s has never been subjected to periodic evaluation for the genetic and environmental trends. Thus, the effectiveness of this program is not clearly known. Moreover, no information is available on the status of the national dairy cattle genetic improvement program that guide policy makers, development planners and breeders to redesign appropriate breeding programs that respond to the current scenarios in Ethiopia. The purpose of this study was, therefore, to investigate genetic and environmental trends for selected crossbred dairy cows produced at Holetta agricultural Research Center in the past two decades.

MATERIALS AND METHODS

Data source and traits recorded

Data for this study were obtained from long-term dairy cattle crossbreeding programme conducted from 1974 to 2005 at Holetta agricultural research center, central Ethiopia. A total of 550 crossbred cows that belong to two genetic groups, derived from crossing two exotic dairy breeds (Friesian and Jersey) with Ethiopian Boran were used. Details of the traits recorded and the number of records are depicted in Table 1.

Description of the study site and animal management

The Holetta agricultural research centre is located at 35 km west of Addis Ababa at 38.5°E longitude and 9.8° N latitude. It is situated at about 2400 m above sea level and is delineated as one of the areas known as "the Addis Ababa milk shed". The average annual rainfall is about 1200 mm and the average monthly relative humidity is 60.6% (Haile et al., 2009). All heifers and cows above six months of age were allowed to graze on natural pasture for about 8 h

during daytime. At night, all animals are housed in an open shade and supplemented with natural pasture hay. Except for the lactating cows, which were supplemented with approximately 1 to 2 kg of concentrate at each milking, no other animal received any regular concentrate supplement.

All animals had free access to clean water. All calves were weighed at birth and allowed to suckle their dams for the first 24 h in order to obtain colostrum, after which they were moved to individual calf pens for bucket feeding until weaning at 94 days of age. Each calf was fed a fixed amount of 260 kg of whole milk during the pre-weaning period. Weaned calves were kept indoors until 6 months, during which they were fed ad lib on natural pasture hay and supplemented with approximately 1 kg per day per animal of concentrate composed of 30% wheat bran, 32% wheat middling, 37% noug seedcake (*Guizoita abysinica*) and 1% salt. There is no partial management option based on genetic groups or level of milk production. All local Boran cows were bred by artificial insemination (AI) while second generation crosses were bred by both natural services and AI as necessary.

Data editing

Data has been collected on both F x Bo and J x Bo crossbred dairy cows that have sufficient information on both milk production and reproduction traits. Animals with limited information, missing pedigree or failed to meet minimum criteria to be parents of next generation were excluded from the final data set. The final data set with complete information consisted of 550 cows (Table 1). Lactation lengths of less than 60 and greater than 1000 days were excluded from the data set following the lactation length truncation points recommended by Kiwuwa et al. (1983) and Sendros (2002), respectively for indigenous and crossbred cows.

Adjusted 305 days milk yield were computed following standard procedures (305 days x total milk yield/actual lactation length). Records of animals with abnormal calving such as abortion and stillbirth were also ignored. Two data sets were prepared and used separately in the analysis. Data set 1 consisted of data structured into six periods (5 years each) from 1974 to 2005 and used to estimate least squares means for milk production and reproduction traits with time. Data set 2 consisted of all data for animals born in each year and was used to compute estimated breeding value (EBV) for yearly genetic and environmental trends.

Statistical analysis

The statistical analysis involved three steps. At the preliminary stage, least squares means analysis was carried out to compare between the genotypes and identify systematic environmental factors that have significant effects on the traits using type III model of the Statistical Analysis System (SAS, 2004). Two genotypes (F x Bo and J x Bo), six periods (1974 to 1980) period 1; (1981 to 1985) period 2; (1986 to 1990) period 3; (1991 to 1995) period 4; (1996 to 2000) period 5 and (2000 to 2005) period 6, and eight parity classes (1, 2, 3, 4, 5, 6, 7 and 8[+]) were identified significant (p<0.05) and retained for the final analysis. For the least squares means analysis in data set 1, all parities greater than 8 were grouped together and analyzed as single parity record denoted as 8[+]. In the second step of statistical analysis, genetic parameters and estimated breeding value for each trait was estimated by running series of univariate analysis using the derivative-free restricted maximum likelihood algorithm fitting individual animal model (Meyer, 1998). Convergence criteria for REML solution were considered to have been reached when the variance function values (-2log-likelihood) in the simplex was less than 10^{-8}. If the likelihood values changed substantially during the analysis, iterations

Table 1. Traits considered and number of Friesian and Jersey derived genotypes.

Traits	Friesian crosses (n)	Jersey crosses(n)	Total
Lactation length (LL)	1258	721	1979
Lactation milk yield (LMY)	1258	721	1979
Adjusted 305 milk yield (305 LMY)	1258	721	1979
Age at first calving (AFC)	399	151	550
Age at puberty (APU)	399	151	550
Calving interval (CI)	847	559	1406

n= number of records in each genetic group.

were restarted using the final parameter estimates from the previous analysis as starting values. The mixed linear model equation in matrix notation for the analysis of each trait was as follows:

$$y_i = X_i \beta_i + Z_i a_i + e_i \qquad (1)$$

where y_i is the vector of observations of the animal for trait i (No. of records x 1); β_i is the vector of unknown fixed effects including, overall mean, genetic classes and year and parity for trait i (total No. of fixed effect levels); X_i is the known design matrix relating fixed effects to y_i (No. of records x total No. of fixed effect levels); a_i is the vector of random animal solutions for trait that is, breeding values (total No. of animals x 1); Z_i is the known design matrix relating animals direct additive genetic effects to y_i (total No. of records x total No. of animals) and e_i is the vector of unknown random residual effects (total No. of records x 1). The variance and covariance structure for the model was assumed to be:

$$V(a) = A\sigma^2_a; \ V \ (e) = I\sigma^2_e \ \text{and} \ \text{Cov}(a,e) = \text{Cov}(e,a) = 0 \qquad (2)$$

where;

I is an identity matrix, A is a numerator relationship matrix, σ^2_a is additive direct and σ^2_e is residual variance.

In the third step, genetic and environmental trends were estimated for each trait. Best Linear Unbiased Prediction (BLUP) estimates of breeding values were estimated fitting a univariate individual animal model as described in Equation (1). The yearly mean estimated breeding values (EBV) that used to predict true breeding value were then calculated. Deviations of yearly mean EBV from the base year (year 1) were taken as estimates of genetic progress in each year and used to plot responses. The base animals with unknown pedigree were assumed to have EBV of zero. Genetic trends (average increase in each year in EBV) were estimated by regressing yearly mean EBV on the birth years. To account for environmental trends, the environmental values were described as the difference between the genetic values from phenotypic values for the year of birth (Table 2).

RESULTS

Estimates of genetic and non-genetic effects

No variation was observed in AFC and APU between the two genetic groups, but considerable differences (p<0.001) were observed in LL, LMY, adjusted 305-day

MY and in CI for genetic groups, birth years and parities (Table 3). Overall LMY showed that F x Bo crossbred cows had longer LL (21 days), produced more milk (376 L), adjusted 305 milk yield (247 L) and had longer CI (20 days) as compared to J x Bo crossbred cows. Though the trend fluctuates, lactation length for aggregated genotype generally showed a declining trend with time. Similarly, LMY showed a declining trend from period 1 to period 4, slightly improved in period 5 and considerably deteriorated in the last period. Though the pattern was irregular, APU and AFC generally showed a declining trend from period to period. Animals born in period 5 (1996 to 2000) had the lowest AFC and APU while the longest AFC and APU were recorded in period 2 (1981 to 1985) (Table 3). Parity was also a significant sources of variation (p<0.001) for LL, LMY, adjusted 305-day MY and CI. Lactation lengths showed an increasing trend from the first parity through to the fourth parity and starts declining from the fourth parity onwards. Other correlated traits such as LMY and adjusted 305-day MY also showed similar trends (Table 3). However, CI showed a declining trend from the first parity to latter parities. The rate of decline was slight from the first to the fourth parity but rapidly decreasing from the fourth parity onwards. Generally, overall evaluation of least square means for milk production traits showed declining trend over the periods considered.

Estimates of variance components and genetic parameters

Table 4 presents variance components and heritability estimates for milk production and reproduction traits. Heritability estimates were higher for LMY (0.44) and 305-days milk yield (0.39), AFC (0.40) and APU (0.38) and relatively lower for LL (0.14) and CI (0.17).

Genetic and environmental trends

Figures 1, 2, 4 and 5 depict genetic trends (average yearly increases in EBV) as estimated by regressing yearly

Table 2. Characteristics of the data and pedigree structure used for genetic trend evaluation.

	$^{\Psi}$LL	LMY	305-day MY	APU	AFC	CI
Number of "base" animals	345	345	345	356	356	283
Number of animals with records	550	550	550	550	550	386
With unknown/pruned sire	140	140	140	241	241	116
With unknown/pruned dam	167	167	167	291	291	126
Number of sires with progeny records	90	90	90	60	60	77
Number of dams with progeny records	269	269	269	180	180	207
Number of grand-sire with progeny records	69	69	69	35	35	66
Number of grand-dam with progeny records	130	130	130	30	30	102
Average	346.4	1958	1720.0	29.5	40.3	488.7
Standard deviation	106.5	883.5	613.9	10.4	11.0	156
CV (%)	30	32	32.5	27.3	21.7	31.8

$^{\Psi}$LL= Lactation length; LMY= Lactation milk yield; 305-days MY= Adjusted 305 milk yield; APU= Age at puberty; AFC=Age at first calving; CI=Calving interval.

Table 3. Least squares means (± S.E) for genetic effects, periods and parity for milk production and reproduction traits.

Effect and level	$^{\Psi}$LL (Days)	LMY (Liter)	305-days MY (Liter)	AFC (Month)	APU (Month)	CI (Days)
Overall mean	333.0±4.1	1900.3±32.9	1734.8±22.0	42.5±0.7	31.7±0.6	462.4±8.7
CV (%)	30.0	42.8	32.7	22.7	28.9	36.2
Genetic groups	***	***	***	NS	NS	*
Frisian crosses	343.8±3.6[a]	2088.7±29.4[a]	1858.3±19.7[a]	43.4±0.6	32.4±0.6	472.8±8.0[a]
Jersey crosses	322.3±4.5[b]	1712.0±36.4[b]	1611.3±24.4[b]	41.7±0.8	31.1±0.7	452.0±9.4[b]
Periods	***	***	***	***	***	***
1974-1980	370.1±5.3[a]	2012.4±43.1[b]	1674.6±28.9[c]	43.9±1.1[b]	32.5±1.1[c]	507.1±10.3[a]
1981-1985	366.4±5.4[a]	1913.2±43.4[bc]	1594.2±29.1[d]	50.9±1.1[a]	39.4±1.0[a]	489.3±10.3[ab]
1986-1990	347.6±6.2[b]	1858.5±50.1[c]	1658.5±33.5[cd]	44.7±1.2[b]	33.5±1.1[c]	465.6±12.2[b]
1991-1995	329.8±8.3[b]	1996.1±66.6[bc]	1845.5±44.6[b]	46.8±1.3[b]	37.1±1.2[a]	425.0±16.8[cd]
1996-2000	309.7±5.6[c]	2140.0±44.8[a]	2033.1±30.0[a]	31.7±0.8[d]	21.4±0.8[e]	415.8±11.4[c]
2000-2005	274.7±7.8[d]	1481.8±63.0[d]	1602.9±42.2[cd]	37.3±0.9[c]	26.6±0.8[d]	471.4±20.5[ab]
Parity	***	***	***			***
1	356.3±4.7[a]	1749.2±37.9[c]	1496.9±25.4[c]	-	-	533.8±9.1[a]
2	352.0±5.4[ab]	1882.7±43.8[b]	1615.7±29.4[b]	-	-	491.7±10.7[b]
3	336.2±6.2[b]	1944.2±49.7[ab]	1760.9±33.3[a]	-	-	482.3±12.2[b]
4	346.5±7.1[b]	2052.4±56.8[a]	1812.4±38.1[a]	-	-	479.1±13.7[bc]
5	334.9±7.9[bc]	2034.6±63.4[a]	1836.2±42.5[a]	-	-	440.7±15.5[cd]
6	332.2±8.9[bc]	1934.7±71.9[ab]	1766.4±48.2[a]	-	-	433.6±18.5[d]
7	310.0±10.7[cd]	1833.0±86.1[bc]	1806.3±57.7[a]	-	-	427.7±24.7[d]
8[+]	296.4±10.5[d]	1771.7±84.7[bc]	1783.5±56.7[a]	-		410.1±25.6[d]

$^{\Psi}$LL= Lactation length; LMY= Lactation milk yield; 305-d MY=Adjusted 305 milk yield; APU=Age at puberty; AFC=Age at first calving; CI=Calving interval; CV=Coefficient of variation.

mean EBV on birth year of animals for LL, LMY and 305-days MY, AFC and APU and CI, respectively. Figure 3 depicts environmental trend taken as a deviation of phenotypic deviations from additive genetic. Regression coefficients for all aggregated genotypes showed negative annual genetic gains of about -3.348 days, -8.0 L, -6.0 L and -0.89 days for LL, LMY, 305-days MY and CI, respectively. On the contrary, a trend in the environmental components was positive (6.72 kg) for LMY and was in the desired direction. Trends in early growth traits such as AFC and APU showed negative annual genetic gains of about -0.295 and -0.263 months,

Table 4. Estimates of variance components and genetic parameters for milk production and reproductive traits.

Estimate	LL	LMY	305-Days MY	APU	AFC	CI
$^{\Psi}\sigma^2_g$	1508	174093.72	122365.01	24.63	31.10	3993.76
σ^2_e	9342.31	225684.84	190914.06	40.07	45.51	20097.96
σ^2_p	10850.31	399778.56	313279.09	64.70	76.61	24091.72
h^2	0.14	0.44	0.39	0.38	0.40	0.17

$^{\Psi}\sigma^2_g$= additive genetic variance; σ^2_e=residual variance; σ^2_p= phenotypic variance; h^2= heritability.

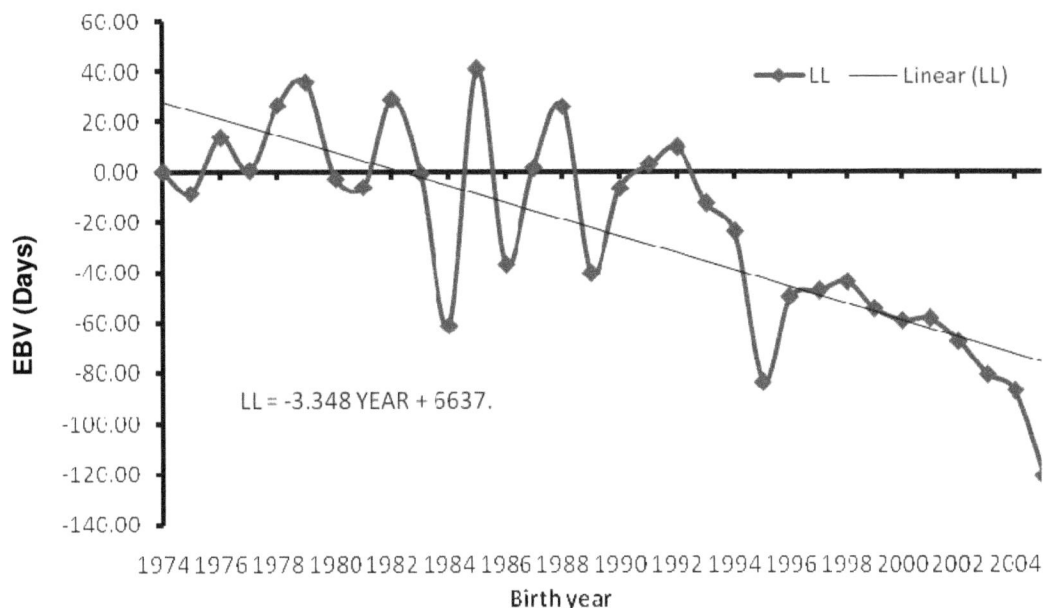

Figure 1. Yearly mean EBV for LL and its trends over birth years.

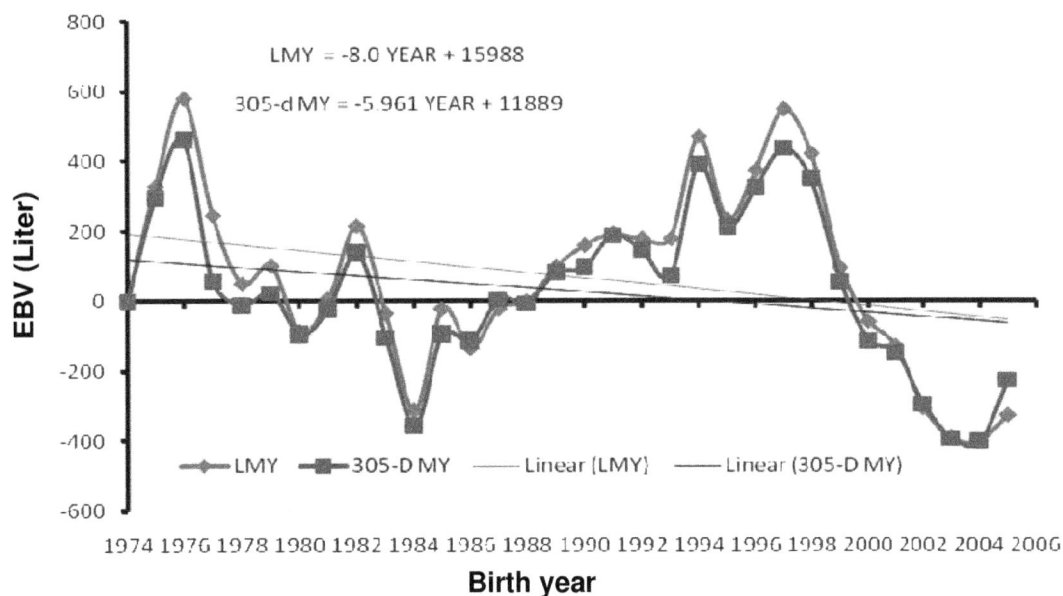

Figure 2. Yearly mean EBV for LMY and 305-MY and their trends over birth years.

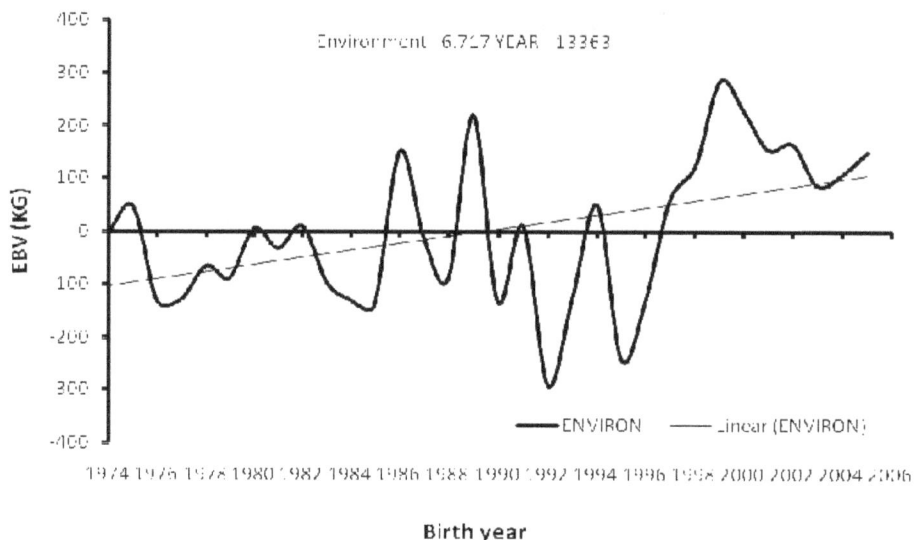

Figure 3. Yearly mean EBV for the environment and its trends over birth years.

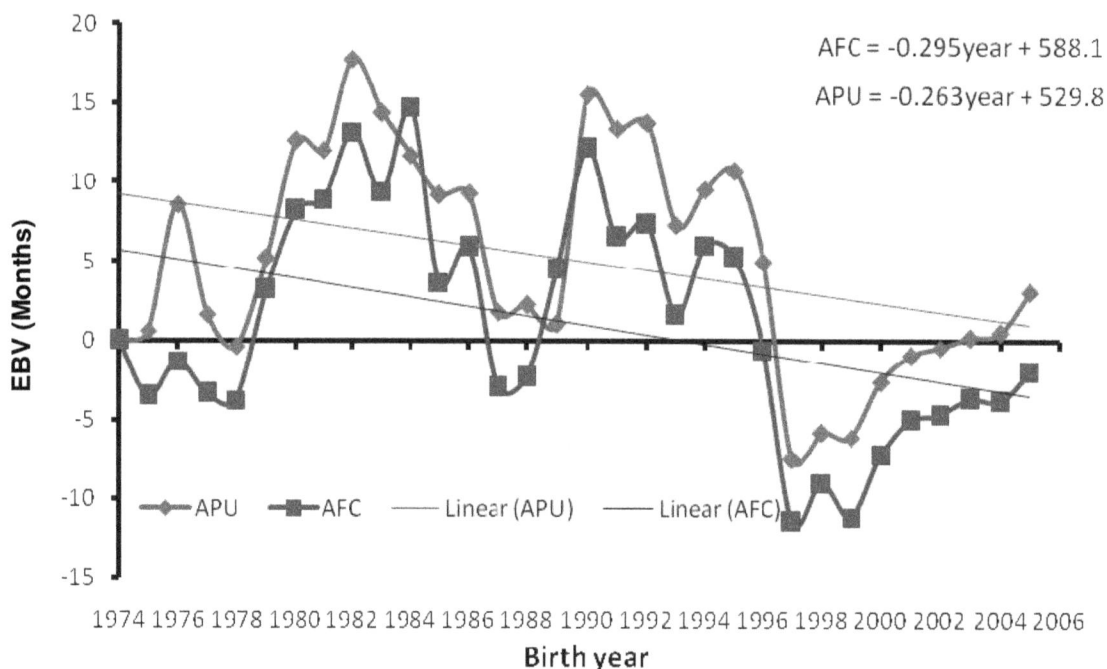

Figure 4. EBV for APU and AFC and their trends over birth years

respectively, and were in the desired direction (Figure 3).

DISCUSSION

Estimates genetic and non-genetic effects

In this study, F x Bo crossbred cows had longer LL (6.8%),

gave more milk (22%), 305-days MY (15.32%) and had longer CI (4.6%) as compared to J x Bo crossbred cows. Superiority of F x Bo crossbred cows in these traits over J x Bo crossbred cows were amply reported in several crossbreeding experiments (Cunningham and Syrstad, 1987; Demeke et al., 2004b; Kefena et al., 2006). Contrary to expectation in dairy cattle genetic improve-ment programs, results of least squares means showed

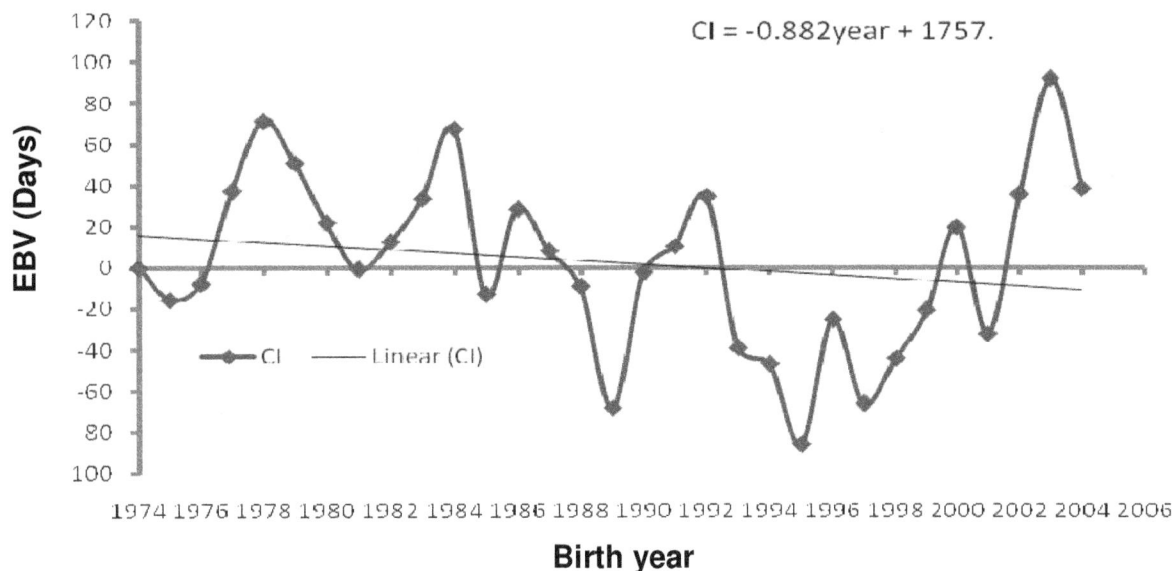

Figure 5. EBV for CI and its trends over birth years

that milk production traits drastically deteriorating in later periods. For instances, the difference in least squares means for LL, LMY and 305-day MY between period 1 and period 6 was to the extent of 95 days (34.7%), 530 L (35.8%) and 71.7 L (4.5%), respectively. On the contrary, reproductive traits substantially improved over the periods. For instances APU, AFC and CI reduced from period 1 to period 6 by about 5.9 months (22.2%), 6.6 months (17.7%) and 36 days (7.6%), respectively.

Improvements in all reproductive traits were in the desired directions and this reflects improvement in herd management practices over the periods. However, declining trends in milk production traits certainly reflect deterioration in the genetic component of the dairy breeding program in the country. Possible explanations for unexpected trends in milk production traits are as follows. Firstly, there are no effective and clearly set selection criteria to identify animals with superior breeding values on the farm. Secondly, most of the semen used in this program has been purchased from the national artificial insemination center (NAIC) of Ethiopia that has virtually no progeny-testing program. Thirdly, there is no clearly established national dairy cattle breeding strategies and breeding goals that are responsive to changing demands and scenarios in Ethiopia. Evidences (Cunningham and Syrstad, 1987; Falconer and Mackay, 1996) indicated that the average performance of a group of animals is determined its inherent genetic makeup, the environment in which it is kept and the interaction between genetic and the environment. Therefore, partial improvement in the environmental components alone may not guarantee improvements in the genetic merits unless all components of the genetic Improvement programs are simultaneously considered.

Estimates of genetic parameters

Effective breeding programmes depend on the accuracy of genetic and phenotypic parameter estimates (Demeke et al., 2004a; Ilatsia et al., 2007). Therefore, accurate estimation of these parameters will help to design an efficient breeding programme for goal traits that are to be improved by selection. Table 4 shows estimates of genetic parameters and variance components. The heritability estimate obtained for LL in the current study is similar with the estimate reported in Demeke et al. (2004b) for pure and crossbred dairy herds in Ethiopia.

However, estimates for LMY and CI were higher than that previously reported by Demeke et al. (2004b). This probably attributed to the differences in data size and structure and models used for estimating these traits in the previous study. Heritability estimate for AFC was also similar with that reported by Demeke et al. (2004b) but higher than that reported by Ilatsia et al. (2007) for Sahiwal cows in Kenya. Generally, selection and breeding programs based on traits with higher heritability estimates improves the genetic components of the breeding program and therefore, their performances in the next generation.

Genetic and environmental trends

Estimated breeding value for genetic and environmental trends in this study indicated that the overall objective of the program was not fully achieved (Figures 1, 2, 4 and 5). Decreasing trends observed in LMY, 305-days MY, LL and CI over the periods for least means squares estimates were consistent with the negative annual genetic trend demonstrated by regressing average annual EBV on years of birth. Negative genetic trend in early growth and fertility traits showed overall improvement in the environmental components the breeding program (Figure 4) implying that reproductive efficiencies were improving with time. The overall trend for positive environmental trends (Figure. 3) shows that emphasis was given to the environmental components of the breeding programme than to the genetic components.. Reports for genetic and environmental trends for crossbred dairy cows are scarce in the literature with variable outcomes. Peixoto et al. (2006) reported a positive annual genetic gain ranging from 6.47 to 7.09 in local Brazilian Guzerat herd under selection program.

An annual genetic trend ranged from 9.13 to 183.14 L was also reported for cows born and raised in the multiple ovulation and embryo transfer (MOET) program in the same Brazilian Guzerat herd showing the potential of using MOET for rapid genetic improvement programs. On the other hand, Freitas et al. (1995) reported negative annual genetic trends in 20 crossbred dairy herds in Brazil. Musani and Mayer (1997) reported positive genetic trend of about 0.8 kg per year for milk production in a large commercial Jersey herd in the central Rift Valley of Kenya. Hallowell et al. (1998) also reported a positive genetic trend of about 19 kg per year for first lactation milk yield in Ayrshire dairy herd in South Africa. Variable results in different dairy cattle genetic improvement programs probably showed differences in selection intensity in different farms, type of herd involved and extent of monitoring genetic progress over time in a particular program.

Conclusion

Comparison of milk production and reproduction performances in this study showed that Frisian crossbred dairy cows were more productive than Jersey crosses. However, Jersey crossbred dairy cows have shorter lactation lengths and calving interval than Frisian crossbred dairy cows that reflects better reproduction efficiencies in Jersey crosses. Continuous decline in the milk production traits over time accompanied by substantial improvements in reproduction traits showed gradual deterioration in the genetic components of the breeding programmes. From this viewpoint, improvement in the environmental components of the breeding program alone would not guarantee improvements in milk productions.

Moreover, non-uniform and generally negative genetic trends in milk production traits followed by positive trend in the environmental components provide addition evidences of gradual deterioration of the genetic components of the dairy cattle genetic improvement programme. It generally reflects the lack of efficient selection program, absence of periodic monitoring of the genetic progresses attained

and use of sires with low breeding value. Therefore, the dairy cattle genetic improvement programmes need to be subjected to national evaluations to redesign appropriate strategies that would be more responsive to the currently changing scenarios in the country.

ACKNOWLEDGMENTS

Our great thanks goes to Dr. Sendros Demeke who developed an excellent data management system at Holetta agricultural research center, Ethiopia. We also thank W/ro Roman Haile Silasie for her careful and tireless data management and recordings at the center for the last twenty years.

REFERENCES

Ahmed MM, Ehui S, Yemesrach A (2003). Dairy development in Ethiopia. Socio-economics and Policy Research Working Paper 58. ILRI Nairobi, Kenya. Int. Livest. Res. Inst., P. 47.

Bakir G, Cilek S (2009). Estimates of genetic trends for 305-days milk yield in Holstein Friesian cattle. J. Anim. Vet. Adv., 8(12): 2553-2556.

Beyene K (1992). Estimation of Additive and Non-Additive Genetic effects for Growth, Milk Yield and Reproduction Traits of Crossbred (Bos taurus x Bos indicus) Cattle in the Wet and Dry Environments in Ethiopia. Ph.D. Dissertation, Cornell University, USA.

Cunningham EP, Syrstad O (1987). Crossbreeding Bos indicus and Bos taurus for milk production in the tropics. FAO Animal Production and Health Paper 68. Food Agric. Org. United Nations, Rome, P. 90.

Demeke S, Neser FWC, Schoeman SJ (2004a). Estimation of genetic parameters for Boran, Friesian and crosses of Friesian and Jersey with Boran cattle in the tropical highlands of Ethiopia: milk production traits and cow weight. J. Anim. Breed. Genet., 121: 163-175.

Demeke S, Neser FWC, Schoeman SJ (2004b). Estimation of genetic parameters for Boran, Friesian and crosses of Friesian and Jersey with Boran cattle in the tropical highlands of Ethiopia: Reproductive traits. J. Anim. Breed. Genet., 121: 57-65.

Falconer DS, Mackay TFC (1996). Introduction to Quantitative Genetics. Longman, Harlow, England, P. 480.

FAO (2007). The State of the World's Animal Genetic Resources for Food and Agriculture, edited by Barbare Rischkowsky and Dafydd Pilling. Rome, pp. 381-427.

Freitas de AF, Wilcox CJ, Costa CN (1995). Genetic trends in the production of Brazilian dairy crossbreds. Brazil J. Genet., 18(1): 55-62.

Haile A, Joshi BK, Ayalew W, Tegegne A, Singh A (2009). Genetic evaluation of Ethiopian Boran cattle and their crosses with Holstein Friesian in central Ethiopia: milk production traits. Animal, 3(4): 486-493.

Hallowell GJ, van der Westhuizen J, van Wyk JB (1998). Genetic and environmental trends for first lactation traits milk traits in the South African Ayrshire breed. S. Afr. J. Anim. Sci., 28(1): 38-45.

Ilatsia ED, Muasya TK, Muhuyi WB, Kahi AK (2007). Genetic and phenotypic parameters and annual trends for milk production and fertility traits of the Sahiwal cattle in semi arid Kenya. Trop. Anim. Health Prod., 39: 37-48.

Kefena E, Hegde BP, Tesfaye K (2006). Lifetime production and reproduction performances of Bos taurus x Bos indicus crossbred cows in the central highlands of Ethiopia. Ethiopian J. Anim. Prod., 6(2): 37-52.

Kiwuwa GH, Trail JCM, Kurtu MY, Getachew W, Anderson MF, Durkin J (1983). Crossbred dairy productivity in Arsi Region, Ethiopia. ILCA Research Report No. 11, ILCA, Addis Ababa, P. 58.

Meyer K (1998). DFREML (Dereivative Free Restricted Maximum Likelihood) programme. Version 3.0β. User Notes. University of New

England, Armidale, NSW 2351, Ausralia, P. 48.

Musani SK, Mayer M (1997). Genetic and environmental trends in a large Jersey herd in the Central Rift Valley of Kenya. Trop. Anim. Health Prod., 29: 108-116.

Peixoto MGCD, Verneque RS, Teodoro RL, Penna VM, Martinez ML (2006). Genetic trend for milk yield in Guzerat herds participating in progeny testing and MOET nucleus schemes. Genet. Mole. Res., 5(3): 454-465.

SAS (Statistical Analysis System Institute) (2004). SAS guide for personal computers, version 6. SAS Institute, Kary, NC.USA.

Sendros D (2002). Genetic Factors Affecting Milk Production, Growth, and Reproductive Traits in *Bos taurus x Bos indicus* Crosses in Ethiopia. PhD Thesis. Faculty Natural and Agricultural Sciences, Department of Animal, Wildlife and Grassland Sciences. University of Free State, Bloemfontein, South Africa.

28

Detection of antibiotic resistance genes of *Escherichia coli* from domestic livestock in south east Nigeria with DNA microarray

Chijioke A. Nsofor[1], Christian U. Iroegbu[2] Douglas R. Call[3] and Margaret A. Davies[3]

[1]Department of Biotechnology, Federal University of Technology, Owerri, Nigeria
[2]Department of Microbiology, University of Nigeria Nsukka, Enugu State, Nigeria
[3]Department of Veterinary Microbiology and Pathology, Washington State University, Pullman, USA.

DNA microarray was developed for detection of up to 90 antibiotic resistance genes in *Escherichia coli* by hybridization. Each antibiotic resistance gene was represented by two specific oligonucleotides chosen from consensus sequences of gene families. A total of 203 oligonucleotides (50-100 base) were spotted onto the microarray. The sequence identity of each gene was compared with GenBank sequences, biotin was used as the positive control and 16s rRNA as orientation. Of the 40 *E. coli* isolates analyzed in this study, 37 were identified as having, at least, one antibiotic resistance gene. Among the different antibiotic resistance genes detected, *bla-CMY-2* and *strA* were the most prevalent occurring in 28 (70%) of the isolates, respectively. Other common genes included were *TEM1* 11(27.5%), *Sul2* 14 (35%) and *TetA* 21(52.5%). The microarray genotyping corresponded with the phenotype of the strains. The disposable microarray presents the advantage of rapidly screening bacteria for the pre-sence of known antibiotic resistance genes. This technology has a large potential for applications in basic research, food safety, and surveillance programs for antimicrobial resistance.

Key words: DNA microarray, antibiotic resistance, *Escherichia coli*.

INTRODUCTION

During the past decades, the worldwide use of antibiotics in animal husbandry for purposes of prophylaxis, chemotherapy and growth promotion has created enormous pressure for the selection of antibiotic resistance among bacteria (Vincent et al., 2005). Today, there is increasing concern about the severity of antibiotics resistance in *Escherichia coli*, which is an important reservoir of antibiotic resistance genes; many other enteric pathogens and commensal bacteria may also play a role as reservoirs for antibiotics genes (Greg et al., 2010; Ma et al., 2007). It is therefore important to follow the evolution of antibiotic resistance in the bacterial population in order to prevent and repress the emergence of multidrug-resistant strains of those bacteria that can still be treated with antibiotics.

The disc diffusion assay technique is commonly used to determine the resistance of pathogenic or commensal bacteria because of its simplicity and because it provides information that is useful in prescribing appropriate antibiotics. Phenotypic testing such as disc diffusion assay technique, however, will not detect "silent" antibiotics resistance genes that might be expressed *in vivo* or disseminated to other bacteria (Frye et al., 2006; Nsofor and Iroegbu 2012, 2013). Molecular testing methods offer similar information more quickly and provides for more discriminatory information.

Because of the large number of recognized antibiotics resistance genes, parallel detection systems such as microarray are well suited to this task (Call et al., 2003).

Presently, PCR and hybridization analysis are common methods used to detect antibiotic resistance genes in bacteria. However, the detection of specific resistance genes remains a tremendous amount of work if every possible resistance gene has to be assessed, and therefore microarray technology is most suitable for resistance gene analysis (Holzman, 2003). A few microarrays have been developed for identifying antibiotics resistance genes (Call et al., 2003; Frye et al., 2006; Moneeke et al., 2003). This study describes a microarray technique for detecting the genes that confer resistance to aminoglycosides, beta-lactam, chloramphenicol, sulfonamide and tetracycline.

MATERIALS AND METHODS

Specimen collection, cultivation and identification of Escherichia coli

Fresh fecal droppings were randomly collected from goats, cattle, pigs and chicken; and care was taken to avoid collecting more than one fecal sample per individual animal. One gram of each animal's feces was homogenized in 9 ml of sterile saline solution, then the volume of the homogenate was made up to 10 ml to get a 10% suspension. The contents were mixed thoroughly and 10-fold serially diluted and 0.2 ml inoculums from each dilution plated out on Eosin Methylene Blue agar (EMB) (Oxoid, England). No antibiotic was included in the EMB agar plates used for the cultivation. The inoculated plates were incubated overnight at 37°C. A single colony on EMB with green metallic sheen taken to be E. coli was selected from an individual fecal sample for further characterization. E. coli was fully identified using conventional microbiological tests-Indole positive, methyl red positive and citrate negative (Cheesbrough, 2000). The cattle and goat specimens came from the herd at Obinze Owerri, Imo State while the Madonna University Poultry Okija, Anambra State was the source of poultry specimens. The specimens from swine came from a farm located at the Ogborhil area of Aba, Abia state.

Antibiotics susceptibility testing

The antibiotics susceptibility pattern of the isolates was determined using the disk diffusion method (Cheesbrough, 2000), on Mueller-Hinton agar (Oxoid, England). Inhibition zone diameter values were interpreted using standard recommendations of the Clinical Laboratory Standard Institute (CLSI, 2006). Susceptibility was tested against ampicillin (10 µg), amoxicillin/clavulanic acid (20/10 µg), tetracycline (30 µg), gentamicin (10 µg), cefpodoxime (10 µg), cefoxitin (30 µg), cefpirome (30 µg), streptomycin (10 µg), chloramphenicol (30 µg), nalidixic acid (30 µg), sulfamethoxazole-trimethoprim (10 µg), cephalothin (30 µg), nitrofurantoin, ceftriaxone (30 µg), and cefotaxine (30 µg) (Oxoid, England). Escherichia coli ATCC 25922 was included as a reference strain.

Preparation of microarray slides

Multiple DNA microarrays were printed on glass slides so that independent arrays were contained within ten individual wells defined by Teflon masking slides (Erie Scientific, Portsmouth, N.H.

USA); the hydrophobic nature of the masking permitted independent samples to be hybridized within each well. Slides were derivatized with epoxysilane (3-glycidoxypropyltrimethoxysilane; (Sigma-Aldrich, Milwaukee, WS, USA) as described by Call et al. (2001). Prior to printing, the slides were soaked in 2.5% Contrad 70 detergent (Fisher Scientific, Pittsburgh, PA, USA.) for 2 min, rinsed three times with distilled water, and dried using compressed air. Slides were then soaked for 1 h in 3 N HCl, rinsed three times with deionized water, and dried with compressed air.

Construction of DNA microarray

Oligonucleotide probes of known antibiotics resistance genes were reconstituted in TE buffer, diluted to 60 µm in print buffer (0.1 M Na_2HPO_4, 0.2 M NaCl, 0.01% sodium dodecyl sulfate) with a pH of 11 and transferred to 384-microwell plates for printing. Arbitrary biotinylated oligonucleotides (70-mer; 5 µM) were included with every array. These biotin pseudoprobes served as positive controls for the detection chemistry and to orient the array for image processing. All probes were deposited as four replicates at a fixed location within each masked well using a Robotic Microgrid II arrayer (Bio-Robotics, Woburn, Mass.USA) with humidity held at 45%. Printing parameters included washing the pins in a recirculating bath (four pins washed twice for 4 s each time), followed by 0.5 s of flushing and 6 s of drying. This washing procedure was repeated twice between probes to minimize possible probe carry-over. Printed slides were baked under vacuum (22 Hg/mm) for 1 h (130°C) and stored away from light at room temperature until used.

Genomic DNA extraction

The bacterial total DNA was extracted using the Qiagen DNeasy silica-gel adsorption method (Qiagen, Valencia, CA USA).

A 1.0-ml volume of overnight broth culture of the test isolate was pelleted in a 1.5 ml microcentrifuge at 10000 rpm for 10 min and resuspended in180 µl of buffer ATL from the Qiagen DNeasy kit. Then 20 µl of Qiagen proteinase K solution was added, mixed by vortexing and the cell was incubated for 3 h in a 55°C shaker water bath for lysis. After the lysis, 20 µl of RNase A (100mg/mL) (Qiagen, Valencia, CA USA) was added to each tube (to degrade RNA) and the tubes were incubated at room temperature for two minutes. This was followed by the addition 200 µl of buffer AL, vortexing, and incubation at 70°C for 10 minutes. Then, the genomic DNA (gDNA) was concentrated by the addition of 200 µl of 100% ethanol. To separate the DNA from other cellular contaminants, the treated DNA lysate was pipetted into a DNeasy column in a collection tube, and centrifuged for 1 min at 10,000 xg. The remaining contaminants were washed out by using 500 µl each of buffer AW1 and AW2 in a new collection tube at each time. The purified gDNA was eluted in a fresh1.5 ml micro-centrifuge tube by using 200 µl AE buffer and centrifugation for 1 min at 10,000 xg. Finally, the nanodrop spectrophotometer was used to quantify the DNA. DNA was quantified to properly scale the subsequent nick translation and any sample that failed to reach the value of A260/A280 ratio of 1.7 to 2 or below 25 ng/µl was re-extracted. All the buffers, enzymes and columns used in this extraction came from the Qiagen DNeasy kit (Qiagen, Valencia, CA USA; Cat. No. 69504).

Nick translation: Biotinylation and fragmentation of DNA

This reaction is designed to generate small (50-100 base) biotin-labeled DNA probes by nick translation which are important for successful in situ hybridization.

Approximately 1.0 µg (up to 40ul) of the quantified gDNA, 5 µl of 10X dNTP mix [(0.2 mM each of dCTP, dGTP, dTTP; 0.1mM of dATP;

0.1mM of biotin-14-dATP; 500mM of Tris-HCl, pH 7.8; 100mM of β-mercaptoethanol and 100 µg/ml of nuclease-free BSA) (Invitrogen, USA)] and 5 µl of 10X enzyme mix [0.5U/µl of DNA polymerase 1, 0.007 U/µl of DNase 1, 50 mM of Tris-HCl pH 7.5, 5 mM of magnesium chloride, 0.1 mM of phenylmethylsulfonyl fluoride, 5% (v/v) of glycerol and 100 µg/ml of nuclease-free BSA) (Invitrogen, USA)] were combined in 0.2 ml PCR tubes on ice. The total volume was brought to 50 µl with PCR water. The mixture was incubated at 16°C in a thermal cycler for 2 h and then held at 4°C for nick translation of DNA. To precipitate the nick translated DNA, the samples were transferred to 1.5 ml micro-centrifuge tubes followed by the addition of 5 µl of 3 M sodium acetate, (pH 5.2), 110 µl of 100% ethanol and incubation at -80°C for 30 min. After the incubation, the DNA was pelleted by centrifugation at 14000 rpm for 30 min at 4°C. Then, the pellets were resuspended with 400 µl of 70% ethanol. For more purification, the above steps were repeated once and the pellets were dried with a vacuum centrifuge for 10 min. Finally, the purified nick-translated DNA was resuspended with 100 µl 1x hybridization buffer.

Microarray slide pre-hybridization preparation

Microarray slides were prepared by immersing them in 50 ml of 1% BSA blocking solution in a Coplin staining jar followed by incubation at room temperature for 10 min, with shaking at 80 rpm to eliminate bubbles on the slide surface. The slides were rinsed 20 times in double de-ionized after which their back and edges were wiped with a Kimwipe and spin dried with slide centrifuge for 15 s.

Sample application/hybridization

The nick translated gDNA was boiled for 3 min, chilled on ice and briefly vortexed for 15 s. Then, the microarray slides were placed on a humidified chamber (200 µl tip box and lid with de-ionized water covering the bottom of the box) and 45 µl of the gDNA sample was placed in each well (2 wells per nick translated gDNA sample) on the microarray slide. The droplets were carefully spread to fully cover the well without touching the slide surface with the pipette. Carefully, the slide was sealed (face-up and frosted end toward the cap) in a hybridization chamber (50 ml conical tube with filter paper moistened with 1x hybridization buffer). The slide was placed on top of the filter paper in the hybridization chamber without allowing the damp filter paper to touch the wells. The hybridization chamber was placed in a rack and lead weight on top of the rack, then the rack was submerged in the 55°C water bath. Finally, the sample DNA was allowed to hybridize with the probes on the array for 16 h.

Post-hybridization stringency washes

After hybridization, the slides were removed from the hybridization chamber with forceps and excess hybridization solution was aspirated off the slides. Then, the slides were completely immersed (frosted end up) in a 55°C pre-warmed low stringency array wash solution (1X SSC, 0.2% SDS) contained in a Coplin jar. The above procedure was repeated in medium stringency (0.1XSSC, 0.2% SDS) and high stringency (0.1XSSC) array wash solutions, respectively. At each time, the slides were washed for 4 min at room temperature on an Orbital shaker at 80 rpm. After the stringency washes, the slides were transferred to a horizontal staining jar that contains enough TNT buffer to cover the slide and were shaken for 1 min at 80 rpm at room temperature to remove the stringency wash buffers. This TNT buffer washing was repeated three times.

Microarray development

For the following applications, 45 µl of each solution was added directly to each well. The slides were gently tapped to distribute the reagent over the full well surface without allowing the reagents to cross over to other wells. The slides were spin-dried for 5 s using a slide centrifuge followed by incubation with 1:100 Streptvadin-Horseredish peroxidase (SA-HRP) in TNB for 30 min. After the incubation, the slides were washed 3 times for 1 min each in horizontal staining jars at 80 rpm shaking. The above procedure was repeated with 10% FES, 2XSSC; 1:50 BioT, 1xAmp Dil; and 1:500 SA-Alexa 555, 1XSSC, 5X Den. This last incubation was done for one hour in the dark. All incubation was done at room temperature in a humidified chamber (made from a covered tip box with ~10 ml PCR water in the bottom). At the end of these development reactions, the slides were spin-dried for 15 s using the slide centrifuge and were stored in the dark prior to scanning.

Scanning/imaging of slides

After hybridization and development, slides were scanned or imaged by standard DNA microarray slide scanners. The florescence marker used in this experiment (Alexa555) has an optimal excitation wavelength of 555 nm and emission wavelength of 565 nm. The scanner/imager we used (Applied Precision arrayWoRx scanner) has a white light source and an emission filter for Cy3 that functions well for Alex555. We used an excitation wavelength of 540 nm (25 nm bandwidth) and an emission wavelength of 595 nm (50 nm bandwidth).

There were five pairs of Teflon-masked wells on each slide, with each well containing a full array and our normal protocol calls for two wells to be hybridized to the same sample. Within each well there were two spots per probe so in effect there are four individual probe-target hybridizations (2 wells total). Each full array has dimensions of 22 horizontal and 20 vertical spots. The distance between spots is approximately 250 µm. Table 1 shows the oligo-nucluotide probes sequences used in constructing the DNA microarray.

RESULTS

Antimicrobial resistance genes for microarray construction

Ninety antimicrobial resistance genes oligonucluotide probes were employed in the microarray, they include 21 aminoglycoside resistance genes, *aac(3)-Id, aac(3)-III, aac(3)-Iva, aac(6')-Ib, aac(6')-IIa, aacC2, aacCA5, aadA1, aadA2, aadA21, aadA5, aadA7, aadB, aadE, aph(3)-Ia, aph(3)-IIa, aphA7, aphD, AphE, strA* and *strB;* 21 beta-lactam resistance genes, *blaACC-01, bla-CMY-2, blaCTX-M-1, blaCTX-M-12, blaCTX-M-15, blaCTX-M-2, blaCTX-M-8, blaDHA-1, blaFOX-2, blaIMP-2, blaKPC-3, blaMIR, blaOXA-1, blaOXA-2, blaOXA-7, blaOXY-K1, blaPSE-1, blaPSE-4, blaROB-1, blaSHV-37,* and *TEM1;* 10 chloramphenicol resistance genes, *cat4, catB2, catB3, catB8, catI, catII, catP, cmlA, cmlB,* and *floR;* 2 integrase genes, *intI1,* and *intI2,* 4 qinolone resistance genes, *qac delta E, qnrA1, qnrB* and *qnrS;* 11 trimethoprim resistance genes, *dfrA1, dfrA14, dfrA16, dfrA21, dhfrII, dhfrV, dhfrVI, dhfrVII, dhfrXII, dhfrXIII,* and *dhfrXV;* 3 sulfonamide resistance genes, *Sul1, Sul2,* and *sul3;* and 18 tetracycline resistance

Table 1. The oligonucluotide probes used in constructing the DNA microarray (Call et al., 2001, 2003).

Gene	Sequence	Description
aac(3)-Ia	CGTAGCCACCTACTCCCAACATCAGCCGGACTCCGATTACCTCGGGAACTTGCTCCGTAG	Aminoglycoside resistance
aac(3)-Ib	AAACAAAGTTAGGTGGCTCAATGAGCATCATTGCAACCGTCAAGATCGGCCCTGACGAAA	Aminoglycoside resistance
aac(3)-Id	TCAAGGCTATAGGCGCAGCGCGTGGAGCTTATGTGATTTACGTCCAAGCTGATAAAGGCG	Aminoglycoside resistance
aac(3)-III	CGACTGGCACTGTGATGGGATACGCGTCGTGGGACCGATCACCCTACGAGGAGACTCTGA	Aminoglycoside resistance
aac(3)-IVa	ACCATTCTTCAGGATGGCAAGTTGGTACGCGTCGATTATCTCGAGAATGACCACTGCTGT	Aminoglycoside resistance
aac(3)-Vb	ACCCTTCGATCTGGCCACATCCGGTACCTATCCCGGCTTCGGCCTGCTCAACCGGTTTCT	Aminoglycoside resistance
aac(6')-I30	TGGCCTGATATGAAAAGTGCCACCAAAGAAGTTGAAGAATGTATTGAGAAGCCAAACATA	Aminoglycoside resistance
aac(6')-Ib	CAATACACAGCATCGTGACCAACAGCAACGATTCCGTCACACTGCGCCTCATGACTGAGC	Aminoglycoside resistance
aac(6')-IIa	TGCTCCATGATTGGCTCAACCGGCCGCACATCGTTGAGTGGTGGGGTGGTGACGAAGAGC	Aminoglycoside resistance
aac(6')-Ia	TGGCCAGATATGACGAGTGCAACAAAAGAAGTAAAAGAATGTATTGAGAGTCCAAACCTT	Aminoglycoside resistance
aacC1	CCTGACCAAGTCAAATCCATGCGGGCTGCTCTTGATCTTTTCGGTCGTGAGTTCGGAGAC	Aminoglycoside resistance
aacC2	CGACTGGCACTGTGATGGGATACGCGTCGTGGGACCGATCACCCTACGAGGAGACTCTGA	Aminoglycoside resistance
aacCA5	TTGCGTTGGCTGCGGTTGACGAGCAAAAAGTCATTGGCGCTATCGCCGCGTATGAGTTGC	Aminoglycoside resistance
aadA1	GGCCTGAAGCCACACAGTGATATTGATTTGCTGGTTACGGTGACCGTAAGGCTTGATGAA	Aminoglycoside resistance
aadA2	GTTCCTGAACAGGATCTATTCGAGGCGCTGAGGGAAACCTTGAAGCTATGGAACTCGCAG	Aminoglycoside resistance
aadA21	GAGCGCCATCTGGAATCAACGTTGCTGGCCGTGCATTTGTACGGCTCCGCAGTGGATGGC	Aminoglycoside resistance
aadA5	CGGTGATCGAGCGCCATCTGGCTGCGACACTGGACACAATCCACCTGTTCGGATCTGCGA	Aminoglycoside resistance
aadA7	GGATCTCTTCAGCTCAGTCCCAGAAAGCGATCTATTCAAGGCACTGGCCGATACTCTGAA	Aminoglycoside resistance
aadB	TACTTTTACTATGCCGATGAAGTACCACCAGTGGACTGGCCTACAAAGCACATAGAGTCC	Aminoglycoside resistance
aadE	GAAGCATTATTTCTATGCCATCAATTGTTCAGGGCGGTATCCGGTGAGGTGGCGGAAAGG	Aminoglycoside resistance
aafA	CGTTGACAGGAGCGCAAATATCGACCTGAGTTTTACTATTAGACAACCGCAACGCTGCGC	E. coli pathotype
aap	GGGACGGGTCCACATTATCTGCGTTCCAACCGCTACCACCCGCAAAGGCATTCAGGCTGA	E. coli pathotype
aatA	ACAGGGAGGTGCATTGGGTAATATGAGTCTCAGAAAAATGGATTATAGTGCTAGTCTGGG	E. coli pathotype
abe (C2-C3)	TGTCCTATTACCAACAAGACTGCTTGAGTTAATGCCAGCGCTTAAAACGAAATTCTTTAT	Serogrouping
aggA	CGACGACAGAGCAATGTGCTAAAAGCGGTGCAAGGGTCTGGTTATGGGGAACAGGTGCCG	E coli pathotype
aidal	GGCCTACAGTATCATATGGAGCCACTCCAGACAGGCCTGGATTGTGGCCTCAGAGTTAGC	Virulence
aph(3)-Ia	ATCGGGCTTCCCATACAATCGATAGATTGTCGCACCTGATTGCCCGACATTATCGCGAGC	Aminoglycoside resistance

Table 1. Contd.

aph(3)-IIa	TAGCCGAATAGCCTCTCCACCCAAGCGGCCGGAGAACCTGCGTGCAATCCATCTTGTTCA	Aminoglycoside resistance
aph4	GGCGTGGATATGTCCTGCGGGTAAATAGCTGCGCCGATGGTTTCTACAAAGATCGTTATG	Aminoglycoside resistance
aphA-3	TCTTTCACTCCATCGACATATCGGATTGTCCCTATACGAATAGCTTAGACAGCCGCTTAG	Aminoglycoside resistance
aphA7	CCTGGAATGCTGTTTTCCCGGGGATCGCAGTGGTGAGTAACCATGCATCATCAGGAGTAC	Aminoglycoside resistance
aphD	CTGCAGAACACCCTGTGGGACATCGAGGACGGGCTGACGGCGATCGCCCCCTCCCAGATC	Aminoglycoside resistance
AphE	GTCGTCTGCCACGGTGATCTCTGCCTGCCCAACATCGTCCTCCATCCGGAGACCCTGGAG	Aminoglycoside resistance
aphIII	CTCCTGCTAAGGTATATAAGCTGGTGGGAGAAAATGAAAACCTATATTTAAAAATGACGG	Aminoglycoside resistance
bfpA	GGTGCTTGCGCTTGCTGCCACCGTTACCGCAGGTGTGATGTTTTACTACCAGTCTGCGTC	E coli pathotype
bla carb-2	GCGTTACGCCGTGGGTCGATGTTTGATGTTATGGAGCAGCAACGATGTTACGCAGCAGGG	Beta-lactam resistance
blaACC-01	CAGCCGCTGATGCAGAAGAATAATATTCCCGGTATGTCGGTCGCAGTGACCGTCAACGGT	Beta-lactam resistance
bla-CMY-2	TTATGCTGCGCTCTGCTGCTGACAGCCTCTTTCTCCACATTTGCTGCCGCAAAAACAGAA	Beta-lactam resistance
blaCTX-M-1	GCGGCACACTTCCTAACAACAGCGTGACGGTTGCCGTCGCCATCAGCGTGAACTGACGCA	Beta-lactam resistance
blaCTX-M-12	GGGTGTGGCATTGATTAACACAGCGGATAATTCGCAAATACTTTATCGTGCTGATGAGCG	Beta-lactam resistance
blaCTX-M-14	CGATCGGCGATGAGACGTTTCGTCTGGATCGCACTGAACCTACGCTGAATACCGCCATTC	Beta-lactamase CTX-M-14
blaCTX-M-2	GCAACGCTGCATGCGCAGGCGAACAGCGTGCAACAGCAGCTGGAAGCCCTGGAGAAAAGT	Beta-lactam resistance
blaCTX-M-8	TTTCGCTGTTGCTGGGGAGTGCGCCGCTGTATGCGCAGGCGAACGACGTTCAGCAAAAGC	Beta-lactam resistance
blaDHA-1	CGGATTCTATGACAGCCATCCGCATATTGATCTGCATATCTCCACCCATAACAATCATGT	Beta-lactam resistance
blaFOX-2	CAAGATGCAAACTTACTATCGGAGCTGGTCACCGGTTTATCCGGCGGGGACCCATCGCCA	Beta-lactam resistance
blaIMP-2	TTTGTGGAGCGCGGCTATAAAATCAAAGGCACTATTTCCTCACATTTCCATAGCGACAGC	Beta-lactam resistance
blaKPC-3	GTTACGGCAAAAATGCGCTGGTTCCGTGGTCACCCATCTCGGAAAAATATCTGACAACAG	Beta-lactam resistance
blaMIR	TCCGAAAAACAGCTGGCTGAGGTGGTGGAACGTACCGTTACGCCGCTGATGAACGCGCAG	Beta-lactam resistance
blaOXA-1	ACCTTCAGTTCCTTCAAATAATGGAGATGCGACAGTAGAGATATCTGTTGATGCACTGGC	Beta-lactam resistance
blaOXA-2	CCACAATCAAGACCAAGATTTGCGATCAGCAATGCGGAATTCTACTGTTTGGGTGTATGA	Beta-lactam resistance
blaOXA-27	GAAAAGGTCATTTACCGCTTGGGAAAAAGACATGACACTAGGAGAAGCCATGAAGCTTTC	Beta-lactam resistance
blaOXA-7	GCAGGCTAATTTACTGCTACTTTTACAAAGCACGAAAACACCATTGACGGCTTCGGCAGA	Beta-lactam resistance
blaOXA-9	GCTCGTCTTTTAAACTTCCATTGGCAATCATGGGGTTTGATAGTGGAATCTTGCAGTCGC	Beta-lactam resistance
blaOXY-2b	TAAAGAGGTGGTAAATAAAAGGCTGGAGATTAACGCAGCCGATTTGGTGGTCTGGAGCCC	Beta-lactam resistance
blaOXA-61	GGAAAAACTTGGGCGAGTAACGACTTTTCAAGGGCTATGGAGACTTTCTCTCCCGCTTCC	Beta-lactam resistance

Table 1. Contd.

blaOXY-K1	ACCAATGATATTGCGGTTATCTGGCCGGAAGATCACGCTCCGCTGATATTAGTCACCTAC	Beta-lactam resistance
blaPER-2	GAAATGGATGGTTGAAACCACCACAGGACCACAGCGGTTAAAAGGCTTGTTACCTGCTGG	Beta-lactam resistance
blaPSE-1	AGTGAGCATCAAGCCCCAATTATTGTGAGCATCTATCTAGCTCAAACACAGGCTTCAATG	Beta-lactam resistance
blaPSE-4	CGTTCAGTATTGCCGGCGGGATGGAACATTGCGGATCGCTCAGGTGCTGGCGGATTTGGT	beta lactam resistance
blaROB-1	TTGCTGACATTAACGGCTTGTTCGCCCAATTCTGTTCATTCGGTAACGTCTAATCCGCAG	Beta-lactam resistance
blaSHV-37	GCAAATTAAACTAAGCGAAAGCCAGCTGTCGGGCCGCGTAGGCATGATAGAAATGGATCT	Beta-lactam resistance
Cat	CGACATGAAGAGTTCAGGACCGCATTAGATGAAAACGGACAGGTAGGCGTTTTTTCAGAA	Phenicol resistance
cat4	CCTTTATTCACATTCTTGCCCGCCTGATGAATGCTCATCCGAAATTCCGTATGGCAATGA	Phenicol resistance
catB2	TCGGCAGCTTCTGCTCCATCGGATCAGGCGCAGCTTTTATTATGGCTGGGAATCAAGGCC	Phenicol resistance
catB3	GGGCGGTACAGCTATTACTCTGGCTACTATCATGGGCACTCATTCGATGACTGCGCACGG	Phenicol resistance
catB8	GCTTTTGTTCTATAGGAAGCGGGGCTTCCTTCATCATGGCTGGCAATCAGGGGCATCGGC	Phenicol resistance
catI	GGTCTTTAAAAAGGCCGTAATATCCAGCTGAACGGTCTGGTTATAGGTACATTGAGCAAC	Phenicol resistance
catII	TAATATCGAGTTTGGTGGTCAGGCTGAATCCGCATTTAATCTGCTGACGATAAAGGGCAA	Phenicol resistance
catII	TTGTTAAGCTAAAACCACATGGTAAACGATGCCGATAAAACTCAAAATGCTCACGGCGAA	Phenicol resistance
catP	TGGCAATTCAAGTTCATCACGCAGTATGTGACGGATTTCACATTTGCCGTTTTGTAAACG	Phenicol resistance
cblA	AAACATATCAATGACTATATCCACCGGTTGAGTATCGACTCCTTCAACCTCTCGGAAACA	Beta-lactam resistance
Cif	TGAAAGACATTACCCTTCCCCCCCCGACGTCCGCGTCCTGTCTGACAGGGGCCATATCTG	Virulence
cmlA	GGCATCACTCGGCATGGACATGTACTTGCCAGCAGTGCCGTTTATGCCAAACGCGCTTGG	Phenicol resistance
cmlB	TCATCTACGGCTTGCTTGGCTCTATGCTTGCTATGGTTCCGGCGATAGGCCCATTGCTGG	Phenicol resistance
Dfr1	AGCCGGAAGGTGATGTTTACTTTCCTGAAATCCCCAGCAATTTTAGGCCAGTTTTTACCC	Trimethoprim resistance
dfrA1	GCGGTCGTAACACGTTCAAGTTTTACATCTGACAATGAGAACGTAGTGATCTTTCCATCA	Trimethoprim resistance
dfrA14	ACCTACAATCAGTGGCTTCTGGTGGGTCGCAAGACGTTTGAATCTATGGGCGCACTCCCC	Trimethoprim resistance
dfrA16	ATGGCTGCCAAGTCGAAGAACGGTATTATCGGTAATGGACCAGATATTCCATGGAGCGCC	Trimethoprim resistance
dfrA19	GCAGTTAGAAAAGGATGGCGCCGAGGAGCGAATCAAGGAGAAAGGAATTCTCCCCGAACG	Trimethoprim resistance
dfrA21	GGTCGTTATGGGCCGCAAGACATTTGAGTCCATAGGCAAGCCCTTACCAAACCGCCACAC	Trimethoprim resistance
dfrA23	TGGCTTGTGCATTACCGTCATGTGGACTTTTGTGGCAGATGCGAGGGCTTGCACGTACAG	Trimethoprim resistance
dhfrl	GGTTAAAGCATCTTTAATTGATGGAAAGATCAATACGTTCTCATTGTCAGATGTAAAACT	Trimethoprim resistance
dhfrll	GCACAAAACTCACTCCTGAAGGCTATGCGGTCGAGTCCGAATCCCACCCAGGCTCAGTGC	Trimethoprim resistance

Table 1. Contd.

dhfrlII	ACTTGATTGGCAAAGATAATCTTATTCCATGGCATCTACCTGCCGATCTGCGTCATTTCA	Trimethoprim resistance
dhfrIX	AAACAAAACTTATTTTCCAAATTTGGATTAACCCTAACCCTATTAGTGAGGAACCCACAT	Trimethoprim resistance
dhfrV	TTCCGAATATTCCCAATACCTTCGAAGTTGTTTTTGAGCAACACTTTAGCTCAAACATTA	Trimethoprim resistance
dhfrVI	TCTTTGTTTCTGGTGGTGGTGAAATATATAAAGCTTTAATCGATCAAGCAGATGTTATCC	Trimethoprim resistance
dhfrVII	GAACACCCATAGAGTCAAATGTTTTCCTTCCAACAAGGAGCCACTGATTATATGTGAGCG	Trimethoprim resistance
dhfrVII	AATGGCATGGAAGAACATGACCTTCACACTTACTTCACTTACCGTAAAAAGGAGCTTACA	Trimethoprim resistance
dhfrX	ATGTGTATGTACCGGTAGAACTAATGAATAAACTCTATAGTGATTTCAAATATCCAGAAA	Trimethoprim resistance
dhfrXII	ATTGGCAATGGTCCTAATATCCCCTGGAAAATTCCGGGTGAGCAGAAGATTTTTCGCAGA	Trimethoprim resistance
dhfrXIII	AGTGCTTAACGCAGCAGAATTCGAGGTTGTCTCATCCGAAACCATTCAAGGCACAATCAC	Trimethoprim resistance
dhfrXV	CGATAAAGTTGATACTTTACATATTTCAACAATCGACATTGAGCCAGAAGGTGATGTCTA	Trimethoprim resistance
DT104	CTAATGCGTTTGGTCTCACAGCCGATGCGGTGCTGGCGGAATATCGTCACTGGCGTAACG	DT104 marker
Eae	TTATGCGGCACAACAGGCGGCGAGTCTCGGTAGCCAGCTTCAGTCGCGATCTCTGAACGG	*E. coli* pathotype
Eaf	CGTGCAGGTCGCCTGTTCGAAACGCTGGCTCAGGGACGGGTGGATGGTAGCTGGCTTAAT	*E. coli* pathotype
ehxA	TACCAGACCTGGGCCCCCTGGGGGATGGGCTGGATGTTGTCTCCGGAATTCTTTCTGCTG	*E. coli* pathotype
Ent	TTAATCGCGCCGCCATGCTGTTCGATGATATTTTGCACCACAGCCAGCCCCAGGCCTGTC	Virulence
espC	TGGCAGCTTTGTCAACAGCAGCCTGACCCTCGAAAAAGGAGCAAAACTAACGGCTCAGGG	E coli pathotype
estA	AGCTAATGTTGGCAATTTTTATTTCTGTATTATCTTTCCCCTCTTTTAGTCAGTCAACTG	Virulence
f165(1)A	CTGGGCCACAAGTAACGGGGCAGGCTGAAGAATTAGCAACTAACGGCGGTACGGGCACAG	*E. coli* pathotype
fliC	ATGAAGTTTCCGTTGATAAGACGAACGGTGAGGTGACTCTTGCTGGCGGTGCGACTTCCC	Virulence
fliCH7	CCCGCGGTAAACCCAATAGTTTTGCTCAGTACACCGGAATTAAAGGTAATTGAAGATGTC	Virulence
floR	GCGTGGGATGGCGTTGCTTGTTTGCGGAGCGGTCCTGTTGGGGATCGGCGAACTTTACGG	Phenicol resistance
fotA	CCTCTGCGCGCATACATTGGTACCTTAAATGGCCAGCCAGGTGTTTTGGGCAATGCGGCC	*E. coli* pathotype
hlyA	TAGTGCTGCTGCAACGACATCTCTGGTTGGTGCACCGGTAAGCGCGCTGGTAGGGGCTGT	*E. coli* pathotype
IncFII / Ori	TAGCGCTAACCGATGGTTTTGCAAAGCGCTAACCGTCAGTCTTTCAGGGTGCGTGGTTCC	Replicon typing
IncN / kikA	CTTCAATATCGTTAAAAAGAACAAGCACGGCTTTTTACCCAACCACACGAAGGATGCTAG	Replicon typing
IncP / trfA2	ACGGATGTTCGACTATTTCAGCTCGCACCGGGAGCCGTACCCGCTCAAGCTGGAAACCTT	Replicon typing
IncW / trwAB	AGCGTATGAAGCCCGTGAAGGGCGAATTGAAGCGCCTTGGCATTGAGGTTTGGACACCGG	Replicon typing
intl1	CTACTTGCATTACAGTTTACGAACCGAACAGGCTTATGTCAACTGGGTTCGTGCCTTCAT	Integrase gene

Table 1. Contd.

intI2	ATGAATGCTTGCGTTTGCGGGTTAAAGATTTTGATTTTGATAATGGCTGCATCACTGTGC	Integrase gene
invA	GTACCAGCCGTCTTATCTTGATTGAAGCCGATGCCGGTGAAATTATCGCCACGTTCGGGC	Virulence
invX	CAGACAGTGACTCAACTTCAAGAGCAGACACTTCCTTTTGGTATAAAGCTTATAGGTGTC	*E. coli* pathotype
ipaB	GGGGGCAATCGCAGGCGCTCTTGTCTTGGTTGCAGCAGTCGTTCTCGTAGCCACTGTTGG	*E. coli* pathotype
Iterons	CGCGAATCGTCCAGTCAAACGACCTCACTGAGGCGGCATATAGTCTCTCCCGGGATCAAA	Replicon typing
Iterons	AACGGGACGACTATGACAACGGTAGTGACTTGCTGGGCTCACTACCATTGTCACCCTGTG	Replicon typing
Iterons	CGGCGTTGTGGATACCTCGCGGAAAACTTGGCCCTCACTGACAGATGAGGGGCGGACGTT	Replicon typing
leoA	TGTCCTGCGTATTGCTCTGTTGGGGGCGTTCTCCGATGGCAAAACCAGCGTTATCGCCGC	*E. coli* pathotype
Lt	TTTTATGTTTTATTTACGGCGTTACTATCCTCTCTATGTGCACACGGAGCTCCTCAGTCT	Virulence
LTIIa	GTGTGCCGAATAATAAAGAATTTAAAGGAGGGGTGTGCATTTCAGCGACAAATGTGCTAT	Virulence
Mpha	ACCCACCGACGTCCATCGTCGACGGTGGCGATCACGATCCTATAGTCGAGCCCAAGCTCA	macrolide 2'-phosphotransferase
ori γ	GCTGATTTATATTAATTTTATTGTTCAAACATGAGAGCTTAGTACGTGAAACATGAGAGC	Replicon typing
OtrB	GATCAACCTTGACGACACGTCCCTGCTGAACGGCATCGACGCCCGGCTGATGCAGCCGGT	Tetracycline resistance
pagC	TGGTTGGGCCAGCCTATCGATTGTCTGACAATTTTTCGTTATACGCGCTGGCGGGTGTCG	Virulence
papGI2	GCTCAGGTCCAGATGTTGCGAGCGGCGTATATTTCCAAGAGTACCTGGCCTGGATGGCAG	*E. coli* pathotype
parA-parB	TGCTGGTAGACCGCCATCACGGATTCTTCGGCAACATCAAGCTGTTTGGGAGAGCAGAGC	Replicon typing
Pet	ATCTATGTCGCCGGTGGCCCGGGCACAGTACAACTCAATGCAGAGAACGCCCTGGGTGAG	*E. coli* pathotype
Pir	AATTCGCCACCGAAACGAGCTAAATCACACCCTGGCTCAACTTCCTTTGCCCGCAAAGCG	Replicon typing
qac delta E	GCAGTCTGGTCGGGACTCGGCGTCGTCATAATTACAGCCATTGCCTGGTTGCTTCATGGG	Disinfectant resistance
qnrA1	CAGCAAGAGGATTTCTCACGCCAGGATTTGAGTGACAGCCGTTTTCGCCGCTGCCGCTTT	Qinolone resistance
qnrB	AACTCCGAATTGGTCAGATCGCAATGTGTGAAGTTTGCTGCTCGCCAGTCGAAAGTCGAA	Qinolone resistance
qnrS	CGTGCTAACTTGCGTGATACGACATTCGTCAACTGCAAGTTCATTGAACAGGGTGATATC	Qinolone resistance
repA FIB	ACACCGTACAACCTGTGGCGCTGATGCGTCTGGGCGTTTTTGTACCGACCCTTAAATCAC	Replicon typing
repA FIC	CATTTGGGACCAAAAGCGTGAGCACGAAGACCTGTCCAACGCCGTAGTGACGCGACAATG	Replicon typing
repA FIIS	CTGATGGCGAAAGCCGAAGGGTTCACGTCCCGTTTTGATTTTTCCGTCCATGTGGCGTTC	Replicon typing
repA L/M	ACCTACAGCTTTCTGACATTGAGTCAGTAGAAGGTCTTTCGCCGGAGTTCATCTCCTGGC	Replicon typing
repA N	AGCCGTTCTGCGGTAATCTTTTACCCGAAAGAAGGGAGTTTTGACTGCGTCGCGCGCCCC	Replicon typing
repA T	AAGCCCTTCCACGTCTAGAAGTTGCACAAGCCCTGTATACCTTCCTTGCAAGCCTTCCAA	Replicon typing

Table 1. Contd.

repA W	AACAAAGCCCCCGGCCATCGTATCAACGAGATCATCAAGACGAGCCTCGCGCTCGAAATG	Replicon typing
repA Y	ACACTGTGCAGCCTGTAGCGTTGATGCGCTTGGGGGTATTCGTGCCGAAGCCATCAAAGA	Replicon typing
repA2 FIC	GATGAGGAAGGTATTACCCAGGCGCAGATGCTTGAAAAACTGATTGAATCAGAGCTGAAA	Replicon typing
repAB L/M	ATGCGTACCCTATTGCAATACAGCCCGGCCAATATGTGCAGGGGCTGGTGAATCAAAAGA	Replicon typing
repC L/M	GTAGTTGAGCGGCAGGTGCATAAGAGTAACCTGGATAAGCAGAAGGATTACAGGAATCGC	Replicon typing
rfbE	ATGTCTGTTAGTGACATAGAACAAAAAATCACTAATAAAACTAAAGCTATTATGTGTGTC	Virulence
rfbE (A_D)	CCTACCCAGCCTTGATCATAAGTAGCAAACTGTCTCCCACCATACATTGATGAATGCCTG	Serogrouping
RNAI	AACGGCAGAATGCGCCATAAGGCATTCAGGACGTATGGCAGAAACGACGGCAGTTTGCCG	Replicon typing
RNAI	CAGGAGAGATGGCATGTACGGGCAGTAAGTCAGAAGACTGAAGATGTTCCGGAAGCCATA	Replicon typing
RNAI	AGAATGCGCCATAAGGCATTCAGGATGTATGGCAGAAACGACGGCAGTTTGCCGGGGCCG	Replicon typing
RNAI/repA	TGGCTGGCCACGCCGTAAGGTGGCAAGGAACTGGTTCTGATGTGGATTTACAGGAGCCAG	Replicon typing
Saa	CTTGGTAGCGGTAAAACGGAGGCAGGGGGAAGAGCATCTGCTACAGGAGTTGATTCGACC	E coli pathotype
sefA	GGGAGCCAATATTAATGACCAAGCAAATACTGGAATTGACGGGCTTGCAGGTTGGCGAGT	Salmonella-specific
sfaA	GCCCTGACCTTGGGTGTTGCGACAAATGCGTCTGCTGTCACCACGGTTAATGGTGGTACA	E. coli pathotype
sfaD	TCCCGCTGCACTGGCCGGAAACCACTGGCATGTCATGCTTCCGGGAGGAAACATGCGCTT	E. coli pathotype
sfaHII	GACCTTCCGTCCTATCCCGGAGGGCCGGTAACAGTCCCTCTTACTGTACGTTGCGACCAG	E. coli pathotype
sipA	CTCAGCCCCCCGTCATAATGCCAGGTATGCAGACCGAGATCAAAACGCAGGCCACGAATC	Virulence
sipB	GTGGCAACGAAAGCGGGCGACCTTAAAGCCGGAACAAAGTCCGGCGAGAGCGCTATTAAT	Virulence
sipC	AGCGCTAAAGATATTCTGAATAGTATTGGTATTAGCAGCAGTAAAGTCAGTGACCTGGGG	Salmonella-specific
sopA	CCCCTCAGGTATGGACCGACCAGAGCTGGCATCCCAATACGCATCTCCGTGATGCTAACG	E. coli pathotype
spvC	GCGGAAGATGCCGGTATCCCACTTTAAAGAGGCGCTGGATGTGCCTGACTATTCAGGGAT	Virulence
spvR	CTGCCAGAAATTATTTTCATCGGGAATCGCTTGTCTGCCGGACATCAGTGGAGGGTGGGG	Virulence
SSpp	CGTCAAAAAGTGAAGGAAATTACGCTGCATTTATTATGGATCAGAATACGCCCCGTTCGG	Salmonella-specific
Stb	AGAATATCGCATTTCTTCTTGCATCTATGTTCGTTTTTTCTATTGCTACAAATGCCTATG	Virulence
stII	CGCATTTCTTCTTGCATCTATGTTCGTTTTTTCTATTGCTACAAATGCCTATGCATCTAC	E. coli pathotype
strA	ACGCGCCGTTGATGTGGTGTCCCGCAATGCCGTCAATCCCGACTTCTTACCGGACGAGGA	Aminoglycoside resistance
strB	GGTGCCTTTCCGCAGCTTGGAACGCGGATGGAGAAGAGGAGCAACGCGATCTAGCTATCG	Aminoglycoside resistance
stx1A	CTGGTGACAGTAGCTATACCACGTTACAGCGTGTTGCAGGGATCAGTCGTACGGGGATGC	E. coli pathotype
stx1B	CGCTTTCATTTTTTTCAGCAAGTGCGCTGGCGACGCCTGATTGTGTAACTGGAAAGGTGG	E. coli pathotype

Table 1. Contd.

Stx2A	CCATGACAACGGACAGCAGTTATACCACTCTGCAACGTGTCGCAGCGCTGGAACGTTCCG	*E. coli* pathotype
Stx2B	GCAATGGCGGCGGATTGTGCTAAAGGTAAAATTGAGTTTTCCAAGTATAATGAGGATGAC	*E. coli* pathotype
Sul1	CCCGCACCGGAAACATCGCTGCACGTGCTGTCGAACCTTCAAAAGCTGAAGTCGGCGTTG	Sulfonamide resistance
Sul2	GCGCTCAAGGCAGATGGCATTCCCGTCTCGCTCGACAGTTATCAACCCGCGACGCAAGCC	Sulfonamide resistance
sul3	GATTGATTTGGGAGCCGCTTCCAGTAATCCTGATACAACTGAAGTGGGCGTTGTGGAAGA	Sulfonamide resistance
TEM1	CCTTGATCGTTGGGAACCGGAGCTGAATGAAGCCATACCAAACGACGAGCGTGACACCAC	Beta-lactam resistance
tet(C)	GACTGGCGATGCTGTCGGAATGGACGATATCCCGCAAGAGGCCCGGCAGTACCGGCATAA	Tetracycline resistance
tet(Y)	GCGTTTATGCAGGTCTTTTGCGCGCCCGTTTTAGGGCGGTTATCTGACCGCTATGGACGG	Tetracycline resistance
Tet30	CGACCGGTTCGGTCGGCGCCCGGTCTTGTTGCTTTCTTTGGCCGGTACCCTGCTTGATTA	Tetracycline resistance
TetA	GCGGCTTCTATAACAACGTGGAACGGGTGGGCATGGATTGCAGGCGCTGCCCTCTACTTG	Tetracycline resistance
tetB	TGGATGCTGTATTTAGGCCGTTTGCTTTCAGGGATCACAGGAGCTACTGGGGCTGTCGCG	Tetracycline resistance
TetD	GGCTATCGGCGGACTGGCGGGGGGATATCTCACCGCATCTGCCGTTTGTCATTGCGGCAAT	Tetracycline resistance
TetE	GTTGAGGCTGCAACAGCTCCAGTCGCACCGGTAATACCAGCAATTAAGCGTCCCAAATAC	Tetracycline resistance
tetG	ACGGGTTCGCGTTCCTGCTTGCCTGCATTTTCCTCAAGGAGACTCATCACAGCCATGGCG	Tetracycline resistance
TetH	GGCGCATCATTGCGGGGATCACAGGCGCAACAGGTGCCGTATGTGCATCAGCGATGAGTG	Tetracycline resistance
TetJ	CCCATGTTAGGGGGATTACTCGGTGAGATCAGCGCCCATACGCCATTTATCTTTGCGGCT	Tetracycline resistance
TetK	TTGGTAGGTTAGTACAAGGAGTAGGATCTGCTGCATTCCCTTCACTGATTATGGTGGTTG	Tetracycline resistance
TetM	GGATATTAAAGAGAAACTTTCTGCCGAAATTGTAATCAAACAGAAGGTAGAACTGTATCC	Tetracycline resistance
TetO	ACGGAACGTTATTTCCCGTTTATCACGGAAGCGCTAAAAACAATCTGGGGACTCGGCAGC	Tetracycline resistance
TetQ	GTGCCGCCCAACCCTTATTGGGCCACAATAGGGCTGACTCTTGAACCCTTACCGTTAGGG	Tetracycline resistance
TetS	CAGAAATGTATACTTCAATAAATGGAGAATTACGCCAGATAGATAAGGCAGAGCCTGGTG	Tetracycline resistance
TetT	GCTACAACGACAACGGATTCGATGGAACTTGAAAGAGATAGGGGAATAACTATACGGGCG	Tetracycline resistance
TetU	GCAGCTAAGACGTGGCAAAGCAACGGATTGGCATGCGATGGTTCAGGAAAGCTTAGATAG	Tetracycline resistance
TetV	CGTCGCGAAGATCACCTCCATCGAGACCACCTTCGACAGCGGACCCACGATCGCGAATGA	Tetracycline resistance
TetW	AACGATGTATTAGGGGACCAAACCCGGCTCCCTCGTAAAAGGTGGCGCGAGGACCCCCTC	Tetracycline resistance
TetX	CGACCGAGAGGCAAGAATTTTTGGTGGAACCCTTGACCTACACAAAGGTTCAGGTCAGGA	Tetracycline resistance

genes, *tet(C), tet(Y), Tet30, TetA, tetB, TetD, TetE, tetG, TetH, TetJ, TetK, TetM, TetQ, TetS, TetT, TetV, TetW* and *TetX*.

The sequence identity of each gene was compared with GenBank sequences, therefore, all the 90 genes were used to construct the DNA microarray; biotin was used as the positive control and 16s rRNA as orientation.

Few virulence and virulence related genes were also

Table 2. The Prevalence of aminoglycosides resistance genes in *E. coli* Isolates.

Genes	SORC Pig N=12	ESOF Goat N=10	SAM Poultry N=8	PLE Cattle N=10	S Total N=40
aac(3)-Id	03(25)	00	05(62.5)	01(10.0)	08(20)
aac(3)-III	04(33.3)	00	03(37.5)	02(20)	09(22.5)
aac(3)-IVa	00	00	00	00	00
aac(6')-Ib	00	00	01(12.5)	00	01(2.5)
aac(6')-IIa	00	00	00	01(10)	01(2.5)
aacC2	00	00	00	00	00
aacCA5	02(16.7)	04(40)	02(25)	02(20)	10(25)
aadA1	01(8.3)	02(20)	01(12.5)	05(50)	09(22.5)
aadA2	01(8.3)	02(20)	01(12.5)	05(50)	09(22.5)
aadA21	01(8.3)	02(20)	01(12.5)	04(40)	08(20)
aadA5	01(8.3)	00	02(25)	01(10)	04(10)
aadA7	00	00	00	00	00
aadB	00	00	00	00	00
aadE	07(58.3)	07(70)	07(87.5)	07(70)	28(70)
aph(3)-Ia	02(16.7)	00	01(12.50	00	03(7.5)
aph(3)-IIa	04(33.3)	00	05(62.5)	02(20)	11(27.5)
aphA7	02(16.7)	01(10)	03(37.5)	00	06(15)
aphD	05(41.7)	03(30)	06(75)	02(20)	16(40)
AphE	06(50)	01(10)	04(50)	03(30)	14(35)
strA	08(66.7)	07(70)	07(87.5)	06(60)	28(70)
strB	03(25)	04(40)	03(37.5)	03(30)	1332.5)

N = Number of isolates hybridized.

included in the array for differentiating the isolates into various pathotypes. To determine the specificity of micro-array hybridization, all of the labeled genes probes were hybridized to the microarray. In most cases there was a one-to-one correspondence for hybridization signal to respective target, orientation gene, and positive control gene spots. There was minor cross-hybridization between some genes and they were marked as abnormal during analysis, thus these genes are not included in the net results shown here.

Detection of antimicrobial resistance gene with microarray

Forty (40) *E. coli* isolates were tested for antimicrobial resistance genes with the microarray. Thirty seven isolates were identified as having at least one antimicrobial resistance gene. Three remaining isolates (CA2, cow; GO3, goat; PL18, poultry) did not hybridize to any of the resistance genes presented on the array. Multiple antimicrobial resistance genes belonging to same category of antimicrobials were detected in most isolates.

Among the aminoglycosides, the most prevalent resistance genes were *aadE* and *strA*, 28 (70%) respectively,

the most prevalent host were the isolates from poultry 07 (87.5%) (Table 2). The most encountered beta-lactam gene in this study was *bla-CMY-2*, 28(70%). However, *blaCTX-M-12* and *blaIMP-2* were detected only in isolates from poultry specimens (Table 3). The most prevalent chloramphenicol resistance genes observed in this study was *floR*, 22 (55.0%), while Integrase gene, int1 had the highest occurrence rate of 37.5% (15 isolates) (Table 4). In the trimethoprim and sulfonamide resistance gene families, the most prevalent was *dhfrV*, which was detected in 9 isolates (22.5%). For sulfonamide resistance genes, 14 isolates (35%) of the animal specimens harbored *Sul2* at highest rate. The *dhfrII* gene was only detected in isolates from pigs and poultry (Table 5). Among the tetracycline resistance genes, *TetA* was most with 21 isolates (52.5%) of animal specimens bearing this gene (Table 6). A sample micrograph of microarrays hybridized with genomic DNAs of the *E. coli* isolates are shown in Figure 1.

DISCUSSION

DNA microarrays have been used previously to detect resistance genes in bacteria (Call et al., 2003; Frye et al., 2006; Moneeke et al., 2003; Van Hoek et al., 2005; Ma et

Table 3. The prevalence of beta-lactam resistance genes in *E. coli* isolates.

Gene	S O Pig N=12	U E S Goat N=10	S A M Cattle N=10	O F Poultry N=8	P L E Total N=40
blaACC-01	01(8.3)	00	01(10)	01(12.5)	03(7.5)
bla-CMY-2	09(75)	06(60)	06(60)	07(87.5)	28(70)
blaCTX-M-1	00	00	01(10)	04(50)	05(12.5)
blaCTX-M-12	00	00	00	01(12.5)	01(2.5)
blaCTX-M-15	02(16.7)	01(10)	00	03(37.5)	06(15)
blaCTX-M-2	00	00	00	00	00
blaCTX-M-8	01(8.3)	02(20)	02(20)	04(50)	09(22.5)
blaDHA-1	00	00	00	02(25)	02(5.0)
blaFOX-2	00	00	00	00	00
blaIMP-2	00	00	00	01	01(2.5)
blaKPC-3	06(50)	03(30)	02(20)	06(75)	17(42.4)
blaMIR	02(16.7)	02(20)	01(10)	02(25)	07(17.5)
blaOXA-1	03(25)	00	00	04(50)	07(17.5)
blaOXA-2	00	00	00	00	00
blaOXA-7	00	00	00	00	00
blaOXY-K1	01(8.3)	00	00	00	01(2.5)
blaPSE-1	02(16.7)	01(10)	00	02(25)	05(12.5)
blaPSE-4	07(58.3)	07(70)	04(40)	07(87.5)	25(62.5)
blaROB-1	01(8.3)	00	00	00	01(2.5)
blaSHV-37	01(8.3)	02(20)	01(10)	02(25)	06(15.0)
TEM1	02(16.7)	01(10)	03(30)	05(62.5)	11(27.5)

N = Number of isolates hybridized.

al., 2007; Greg et al., 2010). Several types of DNA templates can be used to construct microarrays, depending on the intended use. For example, short oligonucleotide probes can be used to detect single nucleotide polymorphism, long oligonucleotide probes can be used to detect sequences that contain a few mismatches, and PCR probes can be used to detect moderately divergent genes. In the present study, oligonucleotide probes were used to construct microarrays that could identify up to ninety genes that confer resistance to variety of antibiotics used in combating Gram-ve bacteria like *E. coli*.

When compared with phenotypic testing, microarrays have the advantage of detecting the presence of antibiotic resistance genes that are not phenotypically expressed (Peterson et al., 2009). In this study, antibiotic resistance genes of 40 *E. coli* isolates from variety of domestic live stock viz cattle, goats, swine and poultry in south eastern states of Nigeria were detected. It was observed that microarray detected genes that were not phenotypically expressed in the following isolates, PG6, PG 11-Swine (*aadE, floR, OtrB, qnrA1, strA, TetD, strA*); CA 12-Cattle (*Aph E*) and PL 7-Poultry (*aadE, aphA7, bla-CMY-2, blaOXA-1, blaPSE-4, floR, IncFII/OriB, IncP / trfA2, qnrA1, strA, TetE, TetJ*). Ma et al. (2007) observed that two isolates of *Salmonella* which did not phenotypically express resistance to aminoglycosides were harboring

aadA1 and aadA2 genes, while Maynard et al. (2003) found that two *E. coli* isolates harboring the aph(3)-la gene, which confer resistance to Kanamycin and Neomycin, were susceptible to Kanamycin and Neomycin. Thus, our results and those of Ma et al. (2007) and Maynard et al. (2003) indicate that some antibiotic resistance genes are silent in bacteria *in vitro*; however, these silent genes can spread to other bacteria or turn on *in vivo*, especially under antibiotic pressure.

Furthermore, there were also discrepancies between the absence of the antibiotic gene test on the microarray and the phenotypic resistance (false negative). This was observed in isolates GO13-Goat (Am-C-Sxt-S-T-Amc); CA 9-Cattle (Am); and PL 18-Poultry (Am-C-Sxt-S). Resistance was phenotypically observed against the antibiotics written against each of the isolates but the genes were not detected by the microarray. This could be attributed to the non inclusion of the oligonucleotide probes encoding theses genes in the construction of the microarray or the genes encoding the resistance are novel. However, more research is needed in this area before conclusion can be established.

In conclusion, the microarray technique employed in this study proved to be an efficient method that allows for rapid detection and identification of resistance genes in *E. coli* isolates.

Table 4. The prevalence of chloramphenicol and qinolone resistance genes in *E. coli* isolates.

Gene	S O	U E S	O F	S A M	P L E
	Pig N=12	Goat N=10	Poultry N=8	Cattle N=10	Total N=40
cat4	00	01(10)	03(37.5)	01(10)	05(12.5)
catB2	00	00	00	00	00
catB3	00	00	00	00	00
catB8	00	00	02(25)	01(10)	03(7.5)
catI	00	01(10)	03(37.5)	01(10)	05(12.5)
catII	00	00	00	00	00
catP	03(25)	00	02	00	05(12.5)
cmlA	01(8.3)	010)	01(12.5)	02(20)	05(12.5)
cmlB	00	01(10)	02(25)	00	03(7.5)
floR	06(50)	05(50)	07(87.5)	04(40)	22(55.0)
intI1	04(33.3	03(30)	04(50)	04(40)	15(37.5)
intI2	01(8.3)	00	02(25)	01(10)	04(10)
qac delta E	07(58.3)	03(30)	06(75)	04(40)	20(50)
qnrA1	08(66.7)	05(50)	07(87.5)	04(40)	24(60)
qnrB	01(8.3)	00	00	00	01(2.5)
qnrS	00	00	00	01(10)	01(2.5)

N = Number of isolates hybridized.

Table 5. The prevalence of trimethoprim and sulfonamide resistance genes in *E. coli* isolates.

Gene	S O	U E S	O F	S A M	P L E
	Pig N=12	Goat N=10	Poultry N=8	Cattle N=10	Total N=40
dfrA1	00	01(10)	00	02(20)	03(7.5)
dfrA14	01(8.3)	03(30)	03(37.5)	00	07(17.5)
dfrA16	00	01(10)	01(12.5)	01(20)	03(7.5)
dfrA21	00	00	00	00	00
dhfrII	03(25)	00	01(12.5)	00	04(10)
dhfrV	03(25)	03(30)	03(37.5)	00	09(22.5)
dhfrVI	00	01(10)	00	00	01(2.5)
dhfrVII	00	00	02(25)	00	02(5)
dhfrXII	01(8.3)	00	01(12.5)	02(20)	04(10)
dhfrXIII	01(8.3)	00	01(12.5)	01(20)	03(7.5)
dhfrXV	00	00	00	00	00
Sul1	01(8.3)	01(10)	03(37.5)	02(20)	07(17.5)
Sul2	02(16.7)	04(40)	05(62.5)	03(30)	14(35)
sul3	01(8.3)	02(20)	00	02(20)	05(12.5)

N = Number of isolates hybridized.

Table 6. The prevalence of tetracycline resistance genes in *E. coli* isolates.

Gene	S O Pig N=12	U E S Goat N=10	O F Poultry N=8	S A M Cattle N=10	P L E Total N=40
tet(C)	02(16.7)	01(10)	03(37.5)	02(20)	08(20)
tet(Y)	00	00	02(25)	00	02(5.0)
Tet30	00	00	01(12.5)	00	01(2.5)
TetA	05(41.7)	05(50)	04(50)	07(70)	21(52.5)
tetB	01(8.3)	05(50)	03(37.5)	00	09(22.5)
TetD	05(41.7)	02(20)	06(75)	02(20)	15(37.5)
TetE	04(33.3)	00	05(62.5)	02(20)	11(27.5)
tetG	01(8.3)	00	05(62.5)	00	06(15)
TetH	01(8.3)	00	01(12.5)	00	02(5.0)
TetJ	07(58.3)	04(40)	06(75)	02(20)	19(47.5)
TetK	01(8.3)	00	01(12.5)	00	02(5.0)
TetM	00	00	00	00	00
TetQ	00	00	00	00	00
TetS	00	00	00	00	00
TetT	00	00	02(25)	00	02(5.0)
TetV	00	00	02(25)	00	02(5.0)
TetW	00	00	00	00	00
TetX	00	00	02	00	02(5.0)

N = Number of isolates hybridized.

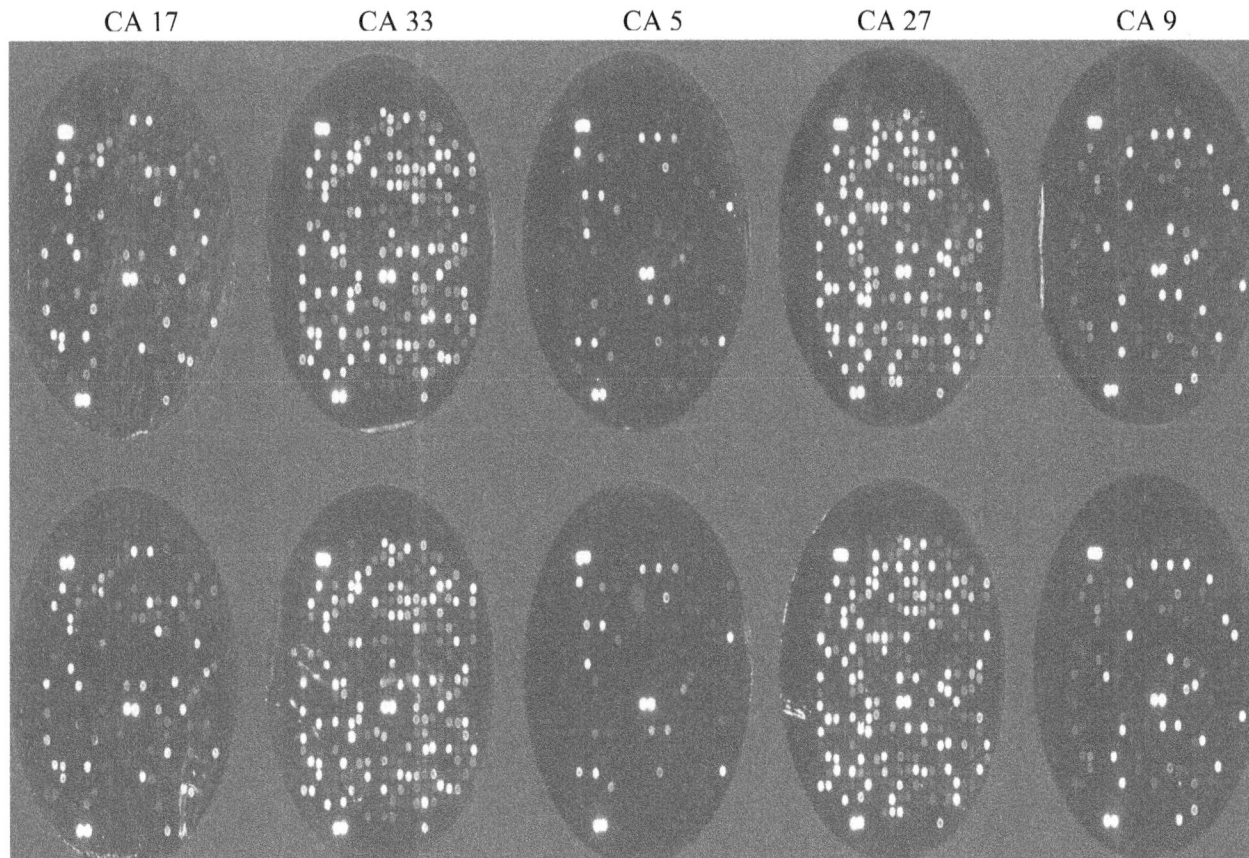

Figure 1. Microphotograph of microarrays hybridized with genomic DNAs of *E. coli* Isolates from cattle.

ACKNOWLEDGEMENTS

This work was supported in part by the Paul G. Allen School for Global Animal Health and the Agricultural Research Center at Washington State University.

REFERENCES

Call DR, Bakko MK, Krug MJ, Roberts MC (2003). Identifying antimicrobial resistance genes with DNA microarrays. Antimicrobial Agents and Chemotherapy. 47:3290-3295

Call DR, Chandler DP, Brockman F (2001). Fabrication of DNA microarrays using unmodified oligonucleotide probes. BioTechniques 30:368-379.

Cheesbrough M (2000). District Laboratory Practice in Tropical Countries, Part 2. Cambridge University Press, Cambridge, UK; 434pp.

Clinical Laboratory Standards Institute (2006): Performance standards for Antimicrobial susceptibility testing. National committee for clinical laboratory standards, Wayne pa.

Frye JG, Jesse T, Long F (2006) DNA Microarray detection of antimicrobial resistance genes in diverse bacteria. Int. J. Antimicrobial. Agents. 27:138-151.

Greg P, Jianfa- Bai, TG, Nagaraja, S N (2010) Diagnostic microarray for human and animal bacterial diseases and their virulence and antimicrobial resistance genes. J. Microbiol Methods. 80:223–230

Holzman, D. 2003. Microarray analyses may speed antibiotic resistance testing. ASM News 69:538–539.

Ma M, Hongning W, Yong Y, Dong Z, Shigui L (2007). Detection of antimicrobial resistance genes of pathogenic Salmonella from Swine with DNA Microarray. J. Vet. Diagn. Invest.

Maynard C, Fairbrother JM, Bakal S (2003) Antimicrobial resistance genes in enterotoxigenic Escherichia coli 0149:K91 isolates obtained over a 23-year period from pigs. Antimicrob. Agents Chemother. 47:3214-3221

Moneeke S, Leube L, Ehricht R (2003). Simple and robust array-based methods for the parallel detection of resistance genes of Staphylococcus aureus. Genome Lett. 2:116-126

Nsofor CA, Iroegbu CU (2012) Antibiotic resistance profile of Escherichia coli isolated from apparently healthy domestic livestock in South-East Nigeria. J. Cell Anim. Biol. 6(8). 129-135

Nsofor CA, Iroegbu CU (2013) Antibiotic Resistance Profile of Escherichia coli Isolated from Five Major Geopolitical Zones of Nigeria. J. Bacteriol. Res. 5(3):29-34.

Peterson G, Bai J, Narayanan S (2009). A co-printed oligomer to enhance reliability of spotted microarrays. J. Microbiol. Methods 77:261-266.

Van Hoek AH, Scholtens IM, Cloeckaer A, Aarts HJ (2005). Detection of antibiotic resistance genes in different Salmonella serovars by oligonucleotide microarrays analysis. J. Microbiol. Methods 62:13-23

Vincent P, Lorianne V, Peter S, Ralf E, Peter K, Joachim F (2005). Microarray-Based Detection of 90 Antibiotic Resistance Genes of Gram-Positive Bacteria. J Clin. Microbiol. 2291-2302.

29

Analysis of longevity traits and lifetime productivity of crossbred dairy cows in the Tropical Highlands of Ethiopia

Kefena Effa[1]*, Diriba Hunde[1], Molla Shumiye[1] and Roman H. Silasie[1]

Ethiopian Institute of Agricultural Research, Holetta Agricultural Research Center, P. O. Box 2003, Addis Ababa, Ethiopia.

Longevity traits, lifetime milk and calf productivities are one of the primary interests of dairy cattle producers. We used lifetime data of 523 crossbred dairy cows that born between 1980 and 2003 in the central tropical highlands of Ethiopia with the purpose to evaluate longevity traits and associated lifetime milk and calf productivity. Based on the type of sires used and level of exotic gene inheritances, the crossbred dairy cows were classified in to six genetic groups. These include F_1 Friesian x Boran (F_1 FxBo), F_1 Jersey x Boran (F_1 JxBo), F_2 Friesian x Boran (F_2 FxBo), F_2 Jersey x Boran (F_2 JxBo) and 75% Friesian and Jersey inheritances. A Generalized Linear Model in the statistical analysis system (SAS, 2004) was used to analyze the data. The overall least squares means ± s. e. for total life (TL), herd life (HL), productive life (PL), lifetime milk yield (LTMY) and lifetime calf crop production were 4036 ± 126.3 days, 2675.74 ± 201.7 days, 1951.00 ± 173.8 days, 10460.6 ± 1117.4 L and 5.70 ± 0.2 calves, respectively. The overall least squares means ± s. e. for lifetime milk yield per day of total life (LTMY/TL), lifetime milk yield per day of herd life (LTMY/HL) and lifetime milk yield per day of productive life (LTMY/PL) were 2.56 ± 0.2, 3.97 ± 0.3 and 5.26 ± 0.3 L, respectively. In conclusion, first generation crosses of all types, particularly those sired by Jersey semen were superior in all the lifetime performance traits considered in this study while second generation (F_2) crosses were inferior in all the lifetime productivity indicators.

Key words: Herd life, lifetime milk yield, longevity, productive life.

INTRODUCTION

Longevity is one of the economically most important functional trait in dairy cattle populations. Nevertheless, the definition given to the term longevity is inconsistent in several literatures. It has been defined as the number of lactations completed (Ibeawuchi, 1984), length of productive life (Arthur et al., 1992; Enyew et al., 1999), the entire lifespan from birth until disposal from the herd (Chaudry and Shafiq, 1995) and survival to certain lactations (Brotherstone et al., 1997; Jairath et al., 1998).

For a close observer, however, the entire lifespan of dairy cows is often partitioned in to two major time periods: (i) the costly period from birth to the first calving and; (ii) the following productive period from first calving to disposal from the herd. In most literatures, the entire lifespan is defined as longevity or total life and part of lifespan from first calving to disposal from the herd is defined as herd life. Productive life is usually defined as the total number of days that dairy cows stay in milking in their entire lifespan. This lifespan classifications patterns seem more informative and therefore, used throughout this report.

In any dairy cattle production enterprise, the lengths of life of a dairy cow have substantial impact on economic performance. Arthur et al. (1992) reported that longer life-span in dairy cows allows producers to be more selective in choosing replacement heifers because only a few have to be chosen each year. Higher longevity also reduces the cost of herd replacements, increases the number of animals available for marketing, and increases the pro-portion of the high-producing, mature animals in the breeding herd (Arthur et al., 1992). Besides, longer average life will lead to a higher proportion of cows in later high-producing lactations and therefore, increase lifetime productivity of dairy cows. Renkema and Stelwagen (1979) showed that an increased length of productive life from about three to four lactations increased milk yield per lactation or profit per year by 11 to 13% and higher percentage of calf crop will born and weaned.

Research evidences (Larroque and Ducrocq, 2001; Zavadilova et al., 2009) showed that type and linearly measured body traits as well as some of the dairy characters in dairy cattle poses negative influence on the length of productive life of a cow. However, in the low-input low-output smallholder dairy production systems of the tropics, the lifespan of dairy cows is often decided not only by cow's milk yield, but also by the owner. For instances, Rufino et al. (2009) reported that evaluation of lifetime productivity is important to target interventions for improving productivity of smallholder dairy systems in the highlands of East Africa, because cows are normally not disposed off based on productive reasons, rather it is determined by feeding strategies and involuntary culling. In such systems, voluntary culling of dairy cows is rarely practiced because farmers that have few animals face difficulties to spread risk. Kebreab et al. (2005) also showed that lifetime productivity needs to be maximized to favor longer stay of cows in the herd. In the high-input dairy systems, however, culling policy is based mainly on unsatisfactory reproduction performance (Bagley, 1993).

In Ethiopia, crossbreeding of indigenous cattle breeds with the commonly known exotic dairy breeds was started in the early 1970s by the National Agricultural Research System (NARS) with various outcomes (Beyene, 1992; Kefena et al., 2006; Sendros, 2002). However, there is paucity of information on longevity traits and lifetime pro-ductivity of crossbred dairy cows. The objective of this study was, therefore, to evaluate longevity traits and provide baseline information on the lifetime milk and calf productivity of crossbred dairy cows with various levels of exotic gene inheritances in Ethiopia.

MATERIALS AND METHODS

Data sources

Data for this study was extracted from the dairy cattle cross-breeding experiment at Holetta Agricultural Research Centre (HARC), central Ethiopia. Holetta is located at 35 km west of Addis Ababa at 38.5°E longitude and 9.8° N latitude and elevation of 2400 m above sea level. The average annual rainfall is about 1200 mm

and the average monthly relative humidity is 60.6%.

Data used for this study were part of the national dairy cattle crossbreeding program that spanned over a period of 23 years from 1980 to 2003. Only cows that exit the herd due to voluntary culling such as reduced milk production, older age and involuntary culling such as accidents or health cases were considered. In either case, only cows that completed at least the first two lactations were included. This is because early culling of cows for other reasons, particularly for economic traits is regularly made at the first com-plete lactation or in the first three months of the second lactation. Lactation lengths of less than 60 milking days were also excluded from the final dataset based on the recommendation made by (Kiwuwa et al., 1983). Records of animals with abnormal calving such as abortions were considered as incomplete lactation and excluded from the final dataset. The data were not censored and we used actual data sets that were collected over the entire lifetime of crossbred dairy cows.

Breeding plan and herd management

In Ethiopia, dairy cattle genetic improvement program was started in the early 1970s. Initially, two imported exotic sire semen sources, namely Friesian (F) and Jersey (J) were used to cross with local Boran (Bo) dam to produce the first generation (F_1) crossbred dairy calves. Secondly, F_1 bulls were selected based on dam milk yield and physical appearances to produce second generations (F_2) crosses (Table 1). Thirdly, semen from pure exotic breeds was used to produce high-grade cows whose level of exotic gene further rose to 75%.

All animals were subjected to almost similar feeding and management practices at the experiment station. They graze from 8:00 to 15:00 h daily on natural pasture of the farm. Up on return to the barn, they were supplemented with conserved hay and green grass harvested from natural pasture or cultivated forages crops as available. In addition, milking cows were supplemented with nearly 2 kg local concentrate feeds per day constituting 30% wheat bran, 31% wheat middling, 35% noug seed cake (Guizota absysinica), 3% bone and blood meal and 1% salt during milking. As the cows were kept for experimental purposes, culling procedures was less stringent and almost identical to the typical smallholder dairy production systems in Ethiopia (Table 1).

Traits studied

Traits studied and the specific definitions used in this particular study were as follows:

(a) Total life or longevity- period from birth to disposal from the herd;
(b) Herd life- period from the first calving to disposal from the herd;
(c) Productive life-number of days in milking during the entire lifetime;
(d) Lifetime milk yield- milk yield during the entire lifespan or longevity;
(e) Lifetime milk yield per day of total life;
(f) Lifetime milk yield per day of herd life;
(g) Lifetime milk yield per day of productive life;
(h) Parity or total number of lactation initiated (corresponds to lifetime calf crop) and;
(i) Age at first calving (AFC).

Statistical model and data analysis

A Generalized Linear Model (GLM) procedure of the statistical analysis system (SAS, 2004) package was used to analyze the

Table 1. Mating design, genotype produced in the breeding program and number of cows considered in this study.

Sire genotype	Dam genotype	Progeny produced	N
F	Bo	F_1 FxBo	112
J	Bo	F_1 JxBo	106
F_1FBo	F_1 FxBo	F_2 FxBo	92
F_1JBo	F_1 JxBo	F_2 JxBo	86
F	F_1 FxBo	3/4F:1/4Bo	69
J	F_1 JxBo	3/4J:1/4Bo	58

Bo = Boran; F = Friesian; J = Jersey; F_1 FxBo = F_1 Friesian x Boran; F_1 JxBo = F_1 Jersey x Boran; F_2 FxBo = F_2 Friesian x Boran; F_2 JxBo = F_2 Jersey x Boran; 3/4F:1/4Bo = 75% Friesian inheritance; 3/4J:1/4Bo = 75% Jersey inheritance.

data. The factor that was used as explanatory variable in the models were the six genetic group of cows (F_1 FxBo, F_1 JxBo, F_2 FxBo, F_2 JxBo, 75% Friesian and 75% Jersey inheritance). The dependent variables that include in the model include longevity, herd life, productive life, and lifetime milk yield, milk production per day of total life, milk yield per day of herd life, milk yield per day of productive life, calf crop and age at first calving (AFC). Generalized linear models used for lifetime traits analysis were:

$$Y_{ij} = \mu + B_j + e_{ij}$$

Where;

Y_{ij} = lifetime traits estimates of ith cow of jth genetic group;

μ = the overall mean;

B_j = the effect of jth genetic group;

e_{ij} = random error associated with ij^{th} cow and assumed to be normally and independently distributed (NID) (0, δ^2_e)

Moreover, ages at first calving (AFC) was also computed for each genetic group to observe its influence on the lifetime milk and calf crop production. AFC was estimated by fitting the data to fixed effects of linear model that consisted of genotype of the cow as follows:

$$Y_{ij} = \mu + B_i + e_{ij}$$

Where;

Y_{ij} = Age at first calving of ith heifer of jth genetic group;

μ = the overall mean;

B_i = the effect of ith genetic group;

e_{ij} = the random error associated with ij^{th} observation and assumed to be normally and independently distributed (NID) (0, δ^2_e).

RESULTS

Longevity traits

Brief results of least square means± s. e. for various longevity traits, lifetime milk yield and calf crop produced by each genetic group as well as age at first calving are summarized in Table 2. The overall least squares means ± s. e. for the entire lifespan of various crossbred dairy cows was 4036 ± 126.3 days, (about 11 years). Statistically significant variation (P<0.05) were observed among different genotype for TL, HL and PL. The result under-

scores the fact that there was a declining trend in total lifespan or longevity of dairy cows as the level of Friesian gene inheritances increased from 50 to 75%. On the contrary, however, high-grade Jersey crossbred dairy cows had longer stayabilty in the herd, which was almost closer to F_1 Friesian crossbred dairy cows.

First generation Jersey (F_1) crosses had significantly longer HL (P<0.05) as compared to any crossbred genotypes considered in this study. Moreover, F_1 F x Bo crosses and crosses with 75% Jersey inheritance had similar HL in the herd indicating their adaption to the low-input low-output dairy production systems in the tropics. Significantly shortest HL (P<0.05) was observed in second generation (F_2) crosses and crosses with 75% Friesian inheritance (Table 2). F_2 JxBo crosses had about 248 days longer herd life than F_2 FxBo demonstrating that Jersey crosses are superior in all the longevity traits and stay longer in the herd as compared to Friesian crosses.

Likewise, F_1 J x Bo crosses had significantly longer (P<0.05) productive life followed by F_1 Frisian crossbred cows and crosses with 75% Jersey inheritances as compared to their contemporary genetic groups. On the other hand comparison between F_2 crosses showed that F_2 JxBo had about 120 days shorter PL than F_2 FxBo (Table 2) indicating the fact that Jersey crosses had shorter lactation length regardless of their relatively longer stayability in the herd. It is also important to note here that genetic groups with extended age at first calving had poorer lifetime performances as compared to early calving genetic groups (Table 2).

Lifetime milk yield and calf crop production

The overall LTMY for the entire crossbred groups and its distribution over different longevity traits is indicated in Table 2. Lifetime milk yield was highly influenced by cow's genotypes (P<0.01). Similar to the longevity traits, the highest LTMY was observed in F_1 crosses with the highest LTMY in F_1 JxBo followed by F_1 FxBo and crosses with 75% of Friesian and Jersey inheritances.

Table 2. Least squares means ± s. e. of longevity traits and lifetime milk and calf productivity of Boran crossbred cows in central Ethiopia.

Lifetime traits	Overall	Genetic groups					
		F₁FxBo	F₁JxBo	F₂FxBo	F₂JxBo	3/4 F:1/4 Bo	3/4J:1/4 Bo
TL, days	4036 ± 126.3	4200.3 ± 135.1[a]	4269.8 ± 135.1[a]	4021.4 ± 179[b]	4036 ± 157.0[b]	3721.9 ± 270.3[c]	3970.3 ± 237.4[b]
HL, days	2675.7 ± 201.7	2876.99 ± 147.5[b]	3107.53 ± 146.6[a]	2334.82 ± 196.3[d]	2582.55 ± 170.7[c]	2435.32 ± 291.4[d]	2717.29 ± 257.8[b]
PL, days	1951.8 ± 173.9	2145.47 ± 127.1[b]	2387.33 ± 126.4[b]	1740.77 ± 169.2[c]	1620.11 ± 147.2[d]	1787.45 ± 251.2[c]	2029.58 ± 222.2[b]
LTMY,lit.	10460.6 ± 1117.4	12816.7 ± 817[b]	13546.50 ± 812.3[a]	8565.8 ± 1087.3[d]	6818.71 ± 945.8[e]	10929.54 ± 1614[c]	10086.34 ± 428[c]
LTMY/TL,lit	2.56 ± 0.2	3.00 ± 0.2[a]	3.04 ± 0.2[a]	2.26 ± 0.2[c]	1.65 ± 0.2[d]	2.84 ± 0.3[ab]	2.54 ± 0.3[abc]
LTMY/HL,lit	3.97 ± 0.3	4.61 ± 0.2[a]	4.38 ± 0.2[a]	3.59 ± 0.3[b]	2.64 ± 0.3[c]	4.68 ± 0.4[a]	3.89 ± 0.4[ab]
LTMY/PL,lit	5.26 ± 0.3	5.77 ± 0.2[ab]	5.57 ± 0.2[ab]	4.86 ± 0.3[c]	4.13 ± 0.2[d]	6.16 ± 0.4[a]	5.07 ± 0.3[bc]
Calf crop	5.70 ± 0.2	6.25 ± 0.1[b]	6.75 ± 0.1[a]	5.32 ± 0.2[c]	5.07 ± 0.2[d]	4.41 ± 0.3[e]	6.37 ± 0.2[ab]
AFC, months	43.21 ± 0.9	44.02 ± 0.7[b]	38.8 ± 0.6[c]	49.8 ± 1.0[a]	48.2 ± 1.0[a]	45.52 ± 1.4[b]	39.72 ± 1[c]

AFC = age at first calving, TL= total life (longevity), HL= herd life, PL= productive life, LTMY= lifetime milk yield, LTMY/TL= lifetime milk yield per day of total life, LTMY/HL= lifetime milk yield per day of herd life, LTMY/PL= lifetime milk yield per day of productive life. Means with the same superscript in each row are not significantly different (p>0.05).

The lowest LTMY was observed in second generation (F₂) crosses. In all the traits investigated, first generation crosses particularly F₁JxBo were superior in most of the desirable dairy traits. For instances, LTMY in F₁JxBo exceeds LTMY in F₁ FxBo, F₂FxBo and F₂JxBo by about 5.7, 58.1 and 98.7%, respectively. Moreover, when LTMY distributed over the entire lifespan of the cows, F₁ crosses were more productive and dominant over all genetic groups whereas second generation crosses remained inferior. When LTMY is distributed over HL and PL, the highest index was observed in 75% Friesian inheritance followed by F₁ crosses (Table 2). Yet, it is important to note that second generation crosses remained inferior when LTMY is distributed over HL and PL. Lifetime milk yield severely declined in F₂ crosses. For instances, reduction in LTMY from F₁JxBo to F₂ JxBo was to the extent of 50%. Similarly, decline in LTMY from F₁ FxBo to F₂ FxBo was up 33%.

The overall calf crop produced for the entire herd was 5.7 ± 0.2 (Table 2). First generation

crosses and crosses with 75% Jersey inheritance completed more number of lactations, i.e. produced more number of calf crops in their lifetime. However, the lowest number of lactation completed was recorded in 75% Friesian inheritance.

DISCUSSION

Evidences show that, for decades, selection in dairy cattle has dealt almost exclusively with production traits. Nowadays, traits like functional longevity are one of the increasingly growing concerns among dairy breeders because of its association with production costs (Arthur et al., 1992). Therefore, evaluation of some of the longevity and lifetime milk and calf crop productivity traits of crossbred dairy cattle with various levels of exotic inheritances would support producers make informed decisions to keep or cull dairy cows.

Literally, there are no sufficient and comprehensive research reports on the longevity and lifetime milk and calf crop productivity of crossbred dairy cows in Ethiopia. Available report shows that the

mean disposal age in the present study is by far longer than the estimate reported in Enyew et al. (1999) for crossbred dairy cows in Ethiopia. They reported that the mean longevity of Arsi crossbred cows with 50, 75 and 87.5% European inheritance was 6.02 ± 0.4 years. Interestingly, they also reported that there were no differences among the crossbred dairy cows with respect to the length of time they stayed in the herd, which is practically unrealistic. Such controversial results may be attributed to the way data was analyzed and sample size considered for each genetic groups. Decline in the lifespan of crossbred dairy cows in tropical climates with an increase in the level of exotic gene beyond 50% was reported in Vaccaro (1990). She noticed that under the Brazilian dairy crossbreeding experiment, there was a tendency of reduction in longevity traits as the level of exotic gene incorporation exceeds 50%. Higher association between longer lifespan of dairy cows and number of lactations completed is useful to obtain more replacement heifers. Lower number of lactation completed in high-grade Frisian crossbred cows might be attributed to their relatively shorter lifespan in the herd. Several other studies (Kiwuwa et al., 1983) reported that high grade cows are characterized by longer calving interval that possibly extend inter-calving interval. Moreover, higher productivity per lactation with inadequacy of meeting the nutritional demand with large body size might have also contributed to the delayed resumption of reproductive activity, which cause delay in estrus and hence resulted in fewer numbers of calves produced or lactations completed in the lifetime of high-grade Friesian dairy cows.

Apparent variability in the length of HL among the crossbred dairy cows might be attributed to the noticeable differences in age at first calving. For instances, both Friesian and Jersey F_2 crosses were characterized extended age at first calving that apparently reduces HL. Other reports (Sendros, 2002; Kefena et al., 2006) showed longer age at first calving, longer days open and longer calving interval in F_2 crosses in Ethiopia. Longer HL in F_1 crosses particularly in F_1 JxBo and crosses with 75% Jersey inheritance indicates that these genetic groups calve earlier, stay for longer period of time in the herd and completed more number of lactations. Moreover, it is an indication of the fact that Jersey crosses are taking the advantage of smaller body size in adapting to the prevailing environmental variables and, therefore, stayed in the herd for a longer period and avoid early culling. Almost similar patterns have been observed in the case of PL.

Lifetime milk and calf productivity

It is indisputable that lifetime milk yield and calf crop production in dairy cows depend on longevity traits, genotypes, management practices and other intrinsic and extrinsic factors. Overall, we noticed that Jersey crossbred dairy cows perform better in most lifetime traits, and in some cases similar to Friesian crosses. This indicates that Jersey crosses are an appropriate genotype for dairying, particularly under low-input low-output dairy production systems.

Higher LTMY observed in F_1 crosses particularly in F_1 JxBo, shows better efficiency of these genotypes for economic traits, longer stay in the herd and adaptation to the prevailing management practices. Moreover, F_1 crosses probably take the advantage of hybrid vigor that arises by crossing two genetically distant populations. Second-generation crosses suffer reduction in hybrid vigor by half than first generation crosses due to segregation and recombination losses (Falconer and Mackay, 1996; Majid et al., 1996; Syrstad, 1996; Cunningham and Syrstad, 1987; Sendros, 2002). Loss of hybrid vigor is the underpinning genetic factor for reduced performances in F_2 crosses. Moreover, poor selection standards to select F_1 bulls for inter se mating may also greatly contributed free fall in the lifetime performances.

Decline in the overall productivity, particularly in LTMY

was noticed as the level of Frisian genes increased beyond 50%. This implies further incorporation Friesian gene might not be resulted in increased lifetime productivity unless the levels of management practices are simultaneously improved. The consistently better performance noticed in F_1 crosses of both Friesian and Jersey is consistent with several other reports. McDowell (1988b) revised crossbreeding results from 25 countries of the tropics involving 57 genetic groups, 15 native breeds and 7 European breeds and reported that F_1 crosses had considerable benefits. They calve earlier, yielded more milk (147%), were milked for more days and had shorter calving interval. Other reports (Kiwuwa et al., 1983; Beyene, 1992; Kefena et al., 2006) agree well with the present results.

Conclusion

Brief overview and estimates of the traits considered in this study (longevity traits, lifetime milk and calf crop productivity) reveals that first generation crossbred dairy cows, particularly those crossed with Jersey semen outperforms their Friesian counterparts. Indeed, Friesian dairy breed is one of most popular dairy breed that outperforms Jersey under ideal production environments. However, higher performances of Jersey crosses underscore their adaptive capability to harsh environment, even with higher level of gene incorporation beyond 50%. On the other hand, this reflects unmet management practices for high yielding Friesian crosses as well as susceptibility to some of the prevailing climatic variables. Drastic decline in lifetime performances of F_2 crosses reiterates continuous reduction in hybrid vigor and recombination losses in inter se mate F_1 populations. F2 crosses are not appropriate genotype of choice for dairy production.

ACKNOWLEDGEMENTS

The authors are grateful to the Ethiopian Institute of Agricultural Research for fully sponsoring this study, and also grateful to W/ro Roman Hailesilasie for her careful data managements and retrieval from the main dairy database platform available at Holetta Agricultural Research Center.

REFERENCES

Arthur P F, Makarechian M, Beng RJ Weingardt R (1992). Longevity and lifetime productivity of cows in a purebred Hereford and two multibred systematic groups under range conditions. J. Dairy Sci. 71:1142-1147.

Bagley CP (1993). Nutritional management of replacement beef heifers: a review. J. Anim. Sci. 72: 3155-3163.

Beyene K (1992). Estimation of Additive and Non-Additive Genetic effects for Growth, Milk Yield and Reproduction Traits of Crossbred (Bos taurus x Bos indicus) Cattle in the Wet and Dry Environments in Ethiopia. Ph.D. Dissertation, Cornell University, USA.

Brotherstone S, Verkamp RW, Hill G (1997). Genetic parameters for a simple predictor of the lifespan of Holstein-Friesian dairy cattle and its relationship to production. Anim. Sci. 65:31.

Chaudry MZ, Shafiq M (1995). Lifetime production performances of Holstein Friesian x Sahiwal crossbreds. Asian J. Anim. Sci. 8 (5): 499-503.

Cunningham EP, Syrstad O (1987). Crossbreeding *Bos indicus* and *Bos taurus* for milk production in the tropics. FAO Animal Production and Health Paper 68. Food and Agricultural Organization of the United Nations, Rome.

Enyew N, Brannang E, Rottmann OJ (1999). Reproductive performances and herd life of crossbred dairy cattle with different level of European inheritance in Ethiopia. *In*: Proceedings of 7th Annual conference of Ethiopian Soc. Anim. Prod. (ESAP) held at Addis Ababa, Ethiopia, pp. 65-74.

Falconer DS, Mackay TFC (1996). Introduction to Quantitative Genetics. Longman, Harlow, England.

Ibeawuchi J A (1984). Longevity and milk production efficiency of Wadara (zebu) cattle in the semi-arid region of Nigeria. East Afr. Agric. J. 59 (1): 11-17.

Jairath L, Dekkers JCM, Schaeffer LR, Liu Z, Burnside EB, Kolstad B (1998). Genetic evaluation for herd life in Canada. J. Dairy Sci. 81: 550-562.

Kebreab E, Smith T, Tanner JC P., Osuji O (2005). Review of under nutrition in smallholder ruminant production systems in the tropics. In Ayantunde et al. (eds) Coping with feed scarcity in smallholder livestock systems in developing countries, Animal Science Group, Wageningen UR, Wageningen, The Netherlands, pp. 3-94.

Kefena E, Hegde BP, Tesfaye K (2006). Lifetime production and reproduction performances of *Bos taurus x Bos indicus* crossbred cows in the central highlands of Ethiopia. Ethiopa J. Anim. Prod. 6 (2): 37-52.

Kiwuwa GH, Trail JCM, Kurtu MY, Getachew W, Anderson MF, Durkin J (1983). Crossbred dairy productivity in Arsi Region, Ethiopia. ILCA Research Report No. 11, ILCA, Addis Ababa.

Larroque H, Ducrocq V (2001). Relationship between type and longevity in the Holstein breed. Genet. Sel. Evo. 33:39-59.

Majid MA, Talukder A, Zahiruddin M (1996). Productive performance of pure breeds, F_1, F_2 and F_3 generations cows raised in Central Cattle Breeding and Dairy Farm of Bangladesh. Asian-Australian J. Anim. Sci. 9:461-464.

McDowell RE (1988b). Strategies for genetic improvement of cattle in the warm climates. In: Second National Livestock Improvement Conference (NLIC) held at Addis Ababa, Ethiopia. pp. 24-26.

Renkema JA, Stelwagen J (1979). Economic evaluation of replacement rates in dairy herds. I. Reduction of replacement rates through improved health. Livest. Prod. Sci. 6: 15-27.

Rufino MC, Herrero M, Van Wijk MT, Hemerik L, De Ridder N, Giller KE (2009). Lifetime productivity of dairy cows in smallholder farming systems of the central highlands of Kenya. Animal 3 (7): 1044-1056.

SAS (Statistical Analysis System Institute) 2004. SAS guide for personal computers, version 6. SAS Institute, Kary, NC.USA.

Sendros D (2002). Genetic Factors Affecting Milk Production, Growth, and Reproductive Traits in *Bos taurus x Bos indicus* Crosses in Ethiopia. PhD Thesis. Faculty of Natural and Agricultural Sciences, Department of Animal, Wildlife and Grassland Sciences. University of Free State, Bloemfontein, South Africa.

Syrstad O (1996). Dairy cattle crossbreeding in the tropics:Choice of crossbreeding strategies. Trop. Anim. Health Prod. 28:223-229.

Vaccaro LPDe (1990). Survival of European dairy breeds and their crosses with zebu in the tropics. Anim. Breed. Abs. 58: 476-494.

Zavadilova L, Nemcova E, Stipkova M, Bouska J (2009). Relationships between longevity and conformation traits in Czech Fleckvieh cows. Czech J. Anim. Sci. 54 (9):387-394.

Permissions

All chapters in this book were first published in JCAB, by Academic Journals; hereby published with permission under the Creative Commons Attribution License or equivalent. Every chapter published in this book has been scrutinized by our experts. Their significance has been extensively debated. The topics covered herein carry significant findings which will fuel the growth of the discipline. They may even be implemented as practical applications or may be referred to as a beginning point for another development.

The contributors of this book come from diverse backgrounds, making this book a truly international effort. This book will bring forth new frontiers with its revolutionizing research information and detailed analysis of the nascent developments around the world.

We would like to thank all the contributing authors for lending their expertise to make the book truly unique. They have played a crucial role in the development of this book. Without their invaluable contributions this book wouldn't have been possible. They have made vital efforts to compile up to date information on the varied aspects of this subject to make this book a valuable addition to the collection of many professionals and students.

This book was conceptualized with the vision of imparting up-to-date information and advanced data in this field. To ensure the same, a matchless editorial board was set up. Every individual on the board went through rigorous rounds of assessment to prove their worth. After which they invested a large part of their time researching and compiling the most relevant data for our readers.

The editorial board has been involved in producing this book since its inception. They have spent rigorous hours researching and exploring the diverse topics which have resulted in the successful publishing of this book. They have passed on their knowledge of decades through this book. To expedite this challenging task, the publisher supported the team at every step. A small team of assistant editors was also appointed to further simplify the editing procedure and attain best results for the readers.

Apart from the editorial board, the designing team has also invested a significant amount of their time in understanding the subject and creating the most relevant covers. They scrutinized every image to scout for the most suitable representation of the subject and create an appropriate cover for the book.

The publishing team has been an ardent support to the editorial, designing and production team. Their endless efforts to recruit the best for this project, has resulted in the accomplishment of this book. They are a veteran in the field of academics and their pool of knowledge is as vast as their experience in printing. Their expertise and guidance has proved useful at every step. Their uncompromising quality standards have made this book an exceptional effort. Their encouragement from time to time has been an inspiration for everyone.

The publisher and the editorial board hope that this book will prove to be a valuable piece of knowledge for researchers, students, practitioners and scholars across the globe.

List of Contributors

Adil M. A. Suliman
Faculty of Veterinary Science, University of Bahr Elgazal, Department of Preventive Medicine and Veterinary Public Health, Sudan

Tawfig El tigani Mohamed
Faculty of Veterinary Medicine, Khartoum University, Khartoum North, Sudan

G. R. Njitchouang
General Biology Laboratory, Department of Animal Biology and Physiology, Faculty of Science, University of Yaoundé 1, P. O. Box 812, Yaoundé, Cameroon
Department of Biochemistry, University of Yaoundé 1, P. O. Box 812, Yaoundé, Cameroon

F. Njiokou
General Biology Laboratory, Department of Animal Biology and Physiology, Faculty of Science, University of Yaoundé 1, P. O. Box 812, Yaoundé, Cameroon

H. C. Nana Djeunga
General Biology Laboratory, Department of Animal Biology and Physiology, Faculty of Science, University of Yaoundé 1, P. O. Box 812, Yaoundé, Cameroon

P. Moundipa Fewou
Department of Biochemistry, University of Yaoundé 1, P. O. Box 812, Yaoundé, Cameroon

T. Asonganyi
Faculty of Medicine and Biomedical Sciences, University of Yaoundé 1, Yaoundé, Cameroon

G. Cuny
Laboratoire de Recherche et de Coordination sur les Trypanosomoses IRD, UMR 177, CIRAD, TA 207/G Campus International de Baillarguet, 34398 Montpellier Cedex 5, France

G. Simo
Department of Biochemistry, Faculty of Science, University of Dschang, P. O. Box 67, Dschang, Cameroon

Karimullah
School of Biological Sciences, University Sains Malaysia, 11800 Penang, Malaysia

Shahrul Anuar
School of Biological Sciences, University Sains Malaysia, 11800 Penang, Malaysia

T. Karuppudurai
Department of Animal Behaviour and Physiology, Centre for Excellence in Genomic Sciences, School of Biological Sciences, Madurai Kamaraj University, Madurai 625 02, India

K. Sripathi
Department of Animal Behaviour and Physiology, Centre for Excellence in Genomic Sciences, School of Biological Sciences, Madurai Kamaraj University, Madurai 625 02, India

Groza I.
Faculty of Veterinary Medicine, University of Agricultural Sciences and Veterinary Medicine Cluj-Napoca Romania

Daria Pop
University of Medicine and Pharmacy Cluj-Napoca Romania

Cenariu M.
Faculty of Veterinary Medicine, University of Agricultural Sciences and Veterinary Medicine Cluj-Napoca Romania

Pall Emoke
Faculty of Veterinary Medicine, University of Agricultural Sciences and Veterinary Medicine Cluj-Napoca Romania

Zewdie Wondatir
Holetta Research Center, P.O. Box, 2003 Addis Ababa, Ethiopia

Yoseph Mekasha
Haramaya University, P.O. Box 38 Dire Dawa, Ethiopia

Bram Wouters
Wageningen UR Livestock Research, P.O. Box 65, 8200 AB Lelystad, The Netherlands

Veeru Kant Singh
Birbal Sahni Institute of Palaeobotany, 53-University Road, Lucknow-226007, India

Rupendra Babu
Birbal Sahni Institute of Palaeobotany, 53-University Road, Lucknow-226007, India

Prabhat Kumar
Department of Zoology, University of Lucknow-226007, India

Manoj Shukla
Birbal Sahni Institute of Palaeobotany, 53-University Road, Lucknow-226007, India

Abbas Doosti
Biotechnology Research Center, Islamic Azad University, Shahrekord Branch, Shahrekord, Iran

Asghar Arshi
Biotechnology Research Center, Islamic Azad University, Shahrekord Branch, Shahrekord, Iran
Islamic Azad University, Shahrekord Branch, Young Researchers Club, Shahrekord, Iran

Mehdi Yaraghi
Biotechnology Research Center, Islamic Azad University, Shahrekord Branch, Shahrekord, Iran

Mehdi Dayani-Nia
Biotechnology Research Center, Islamic Azad University, Shahrekord Branch, Shahrekord, Iran

Barkat Ali Bughio
Department of Zoology, University of Sindh, Jamshoro, Pakistan

Riffat Sultana
Department of Zoology, University of Sindh, Jamshoro, Pakistan

M. Saeed Wagan
Department of Zoology, University of Sindh, Jamshoro, Pakistan

Gamal Mahmoud Bekhet
Department of Zoology, Faculty of Science, Alexandria University, Alexandria 21511, Egypt
Department of Biological Sciences, College of Science, King Faisal University, Al-Hassa 31982, Saudi Arabia

A. B. Saba
Department of Veterinary Physiology, Biochemistry and Pharmacology, University of Ibadan, Ibadan Nigeria

O. A. Oridupa
Department of Veterinary Physiology, Biochemistry and Pharmacology, University of Ibadan, Ibadan Nigeria

Sinan Zhang
Key Laboratory of Southwest China Wildlife Resources Conservation, College of Life Science, China West Normal University, 1# Shida Road, 637009, Nanchong, P. R. China

Yiling Hou
Key Laboratory of Southwest China Wildlife Resources Conservation, College of Life Science, China West Normal University, 1# Shida Road, 637009, Nanchong, P. R. China

Wanru Hou
Key Laboratory of Southwest China Wildlife Resources Conservation, College of Life Science, China West Normal University, 1# Shida Road, 637009, Nanchong, P. R. China

Xiang Ding
Key Laboratory of Southwest China Wildlife Resources Conservation, College of Life Science, China West Normal University, 1# Shida Road, 637009, Nanchong, P. R. China

Chunlian Wu
Key Laboratory of Southwest China Wildlife Resources Conservation, College of Life Science, China West Normal University, 1# Shida Road, 637009, Nanchong, P. R. China

M. S. Sastry
Department of Zoology, Rashtrasant Tukdoji Maharaj Nagpur University, Nagpur University Campus, Nagpur-440033, India

S. B. Pillai
Department of Zoology, Rashtrasant Tukdoji Maharaj Nagpur University, Nagpur University Campus, Nagpur-440033, India

L. S. Yaqub
Department of Physiology, Faculty of Veterinary Medicine, Ahmadu Bello University, Zaria, Kaduna State, Nigeria

M. U. Kawu
Department of Physiology, Faculty of Veterinary Medicine, Ahmadu Bello University, Zaria, Kaduna State, Nigeria

J. O. Ayo
Department of Physiology, Faculty of Veterinary Medicine, Ahmadu Bello University, Zaria, Kaduna State, Nigeria

Lemma Abera Hirpo
Zwai Fishery Research Center, P.O. Box 229, Zwai, Ethiopia

Giannoccaro Alessandra
Institute of Sciences of Food Production (ISPA), National Research Council (CNR), Via G. Amendola 122/O, 70125, Bari, Italy

Lacalandra Giovanni Michele
Department of Animal Production, University of Bari, Strada Provinciale Casamassima km 3, 70010, Valenzano, Bari, Italy

Filannino Angea
Department of Animal Production, University of Bari, Strada Provinciale Casamassima km 3, 70010, Valenzano, Bari, Italy

Pizzi Flavia
Institute of Agricultural Biology and Biotechnology (IBBA), National Research Council (CNR), Via Bassini 15, Milan, Italy

Nicassio Michele
Department of Animal Production, University of Bari, Strada Provinciale Casamassima km 3, 70010, Valenzano, Bari, Italy

Dell'Aquila Maria Elena
Department of Animal Production, University of Bari, Strada Provinciale Casamassima km 3, 70010, Valenzano, Bari, Italy

Minervini Fiorenza
Institute of Sciences of Food Production (ISPA), National Research Council (CNR), Via G. Amendola 122/O, 70125, Bari, Italy

Monica A. Ayieko
Maseno University, P. O. box 333-40105, Maseno, Kenya

Millicent F. O Ndong'a
Maseno University, P. O. box 333-40105, Maseno, Kenya

Andrew Tamale
Busoga University, P. O. box 154 Iganga, Uganda

Mbacké Sembène
Faculty of Science and Technology, University Cheikh Anta Diop, P. O. Box 5005 Dakar, Senegal

Awa Ndiaye
Faculty of Science and Technology, University Cheikh Anta Diop, P. O. Box 5005 Dakar, Senegal

Khadim Kébé
Faculty of Science and Technology, University Cheikh Anta Diop, P. O. Box 5005 Dakar, Senegal

Ali Doumma
Faculty of Science, University Abdou Moumouni Niamey, P. O. Box 10662, Niamey, Niger

Antoine Sanon
Laboratory of Entomology University of Ouagadougou, Burkina Faso

Guillaume K. Kétoh
Laboratory of Applied Entomology, Faculty of Science, University of Lomé, P. B. 1515, Lomé, Togo

Laurent Granjon
Laboratory of Applied Entomology, Faculty of Science, University of Lomé, P. B. 1515, Lomé, Togo

Jean-Yves Rasplus
NRA – UMR 1062 CBGP (INRA / IRD / Cirad / Montpellier SupAgro), Campus international de Baillarguet, CS 30016, 34988 Montferrier-sur-Lez, France

Awad G. Mohammed
Private sector, Khartoum, Sudan

Atif E. Abdelgadir
Department of Preventive Medicine and Veterinary Public Health, Faculty of Veterinary Medicine, University of Khartoum, Sudan

Khitma H. Elmalik
Department of Preventive Medicine and Veterinary Public Health, Faculty of Veterinary Medicine, University of Khartoum, Sudan

O. M. Ogundele
Trinitron Biotech LTD, Science and Technology Complex, Abuja, Nigeria

E. A. Caxton-Martins
Department of Anatomy, University of Ilorin, Kwara State, Nigeria

O. K. Ghazal
Unilorin Stem Cell Research Laboratory, Ilorin, Kwara State, Nigeria

O. R. Jimoh
Department of Anatomy, University of Ilorin, Kwara State, Nigeria

Ghezali Djelloul
Departement Zoologie Agricole et Forestier, Ecole Nationale Supérieure Agronomique El –Harrach, Alger –Algérie

Zaydi Djamel-Eddine
Departement Zoologie Agricole et Forestier, Ecole Nationale Supérieure Agronomique El –Harrach, Alger –Algérie

Weijun Guan
Institute of Animal Science, Chinese Academy of Agricultural Sciences, Beijing, 100193, PR China

Dianjun Wang
Department of Pathology, Chinese PLA General Hospital, Beijing, 100853, PR China

Chunyu Bai
Institute of Animal Science, Chinese Academy of Agricultural Sciences, Beijing, 100193, PR China
College of Wildlife Resources, Northeast Forestry University, Harbin, 150040, PR China

Minghai Zhang
College of Wildlife Resources, Northeast Forestry University, Harbin, 150040, PR China

Changli Li
Institute of Animal Science, Chinese Academy of Agricultural Sciences, Beijing, 100193, PR China

Yuehui Ma
Institute of Animal Science, Chinese Academy of Agricultural Sciences, Beijing, 100193, PR China

Elysée NCHOUTPOUEN
Laboratory of General Biology, Department of Animal Biology and Physiology, Faculty of Science, University of Yaounde I, P. O. Box 812, Yaounde, Cameroon

Guy Benoît LEKEUFACK FOLEFACK
Laboratory of General Biology, Department of Animal Biology and Physiology, Faculty of Science, University of Yaounde I, P. O. Box 812, Yaounde, Cameroon

Abraham FOMENA
Laboratory of General Biology, Department of Animal Biology and Physiology, Faculty of Science, University of Yaounde I, P. O. Box 812, Yaounde, Cameroon

Khitam Jassim Salih
Vertebrate Department, Marine Science Center, Basrah University, Iraq

Majeed Hussein Majeed Al-Sarry
Nursing College, Basrah University, Iraq

M. M. Pourseif
Department of Animal Science, Faculty of Agriculture, University of Tabriz, Tabriz, Iran

G. H. Moghaddam
Department of Animal Science, Faculty of Agriculture, University of Tabriz, Tabriz, Iran

Amor Gaddour
Arid Land Institute, Medenine, Tunisia

Mabrouk Ouni
Arid Land Institute, Medenine, Tunisia

Sghaier Najari
Arid Land Institute, Medenine, Tunisia

Kefena Effa
Ethiopian Institute of Agricultural Research, Holetta Agricultural Research Center, P. O. Box 2003, Addis Ababa, Ethiopia

Zewdie Wondatir
Ethiopian Institute of Agricultural Research, Holetta Agricultural Research Center, P. O. Box 2003, Addis Ababa, Ethiopia

Tadelle Dessie
International Livestock Research Institute (ILRI), P. O. Box, 5689, Addis Ababa, Ethiopia

Aynalem Haile
International Center for Agricultural Research in Dry Areas, P. O. Box 5466, Aleppo, Syria

Chijioke A. Nsofor
Department of Biotechnology, Federal University of Technology, Owerri, Nigeria

Christian U. Iroegbu
Department of Microbiology, University of Nigeria Nsukka, Enugu State, Nigeria

Douglas R. Call
Department of Veterinary Microbiology and Pathology, Washington State University, Pullman, USA

Margaret A. Davies
Department of Veterinary Microbiology and Pathology, Washington State University, Pullman, USA

Kefena Effa
Ethiopian Institute of Agricultural Research, Holetta Agricultural Research Center, P. O. Box 2003, Addis Ababa, Ethiopia

Diriba Hunde
Ethiopian Institute of Agricultural Research, Holetta Agricultural Research Center, P. O. Box 2003, Addis Ababa, Ethiopia

Molla Shumiye
Ethiopian Institute of Agricultural Research, Holetta Agricultural Research Center, P. O. Box 2003, Addis Ababa, Ethiopia

Roman H. Silasie
Ethiopian Institute of Agricultural Research, Holetta Agricultural Research Center, P. O. Box 2003, Addis Ababa, Ethiopia